MODELING AND CONTROL
OF VENTILATION

ADVANCES IN EXPERIMENTAL MEDICINE AND BIOLOGY

MODELING AND CONTROL OF VENTILATION

Edited by

Stephen J. G. Semple and Lewis Adams

Charing Cross and Westminster Medical School
London, England

and

Brian J. Whipp

St. George's Hospital Medical School
London, England

SPRINGER SCIENCE+BUSINESS MEDIA, LLC

Library of Congress Cataloging-in-Publication Data

London Conference on Modeling and Control of Ventilation (1994 :
  Egham, Surrey, England)
    Modeling and control of ventilation / edited by Stephen J.G.
Semple, Lewis Adams, and Brian J. Whipp.
        p.   cm. -- (Advances in experimental medicine and biology ; v.
393)
    "Proceedings of the London Conference on Modeling and Control of
Ventilation, held September 17-20, 1994, in Egham, Surrey, England"-
-CIP t.p. verso.
    Includes bibliographical references and index.
    ISBN 978-1-4613-5792-6   ISBN 978-1-4615-1933-1 (eBook)
    DOI 10.1007/978-1-4615-1933-1
    1. Respiration--Regulation--Congresses. 2. Respiratory organs-
-Pathophysiology--Congresses.   I. Semple, Stephen J. (Stephen John)
II. Adams, Lewis, 1950-   . III. Whipp, Brian J. IV. Title.
V. Series.
QP123.L66   1994
612.2--dc20   95-51225
                                                               CIP

LCMCV

Proceedings of the London Conference on Modeling and Control of Ventilation,
held September 17–20, 1994, in Egham, Surrey, England

© 1995 by Springer Science+Business Media New York
Originally published by Plenum Press New York in 1995
Softcover reprint of the hardcover 1st edition 1995

10 9 8 7 6 5 4 3 2 1

ACKNOWLEDGMENTS

We would like to thank the following sponsers:

- Boeringher Ingelheim - Japan
- The Physiological Society
- The Rayne Foundation
- The Wellcome Trust

PREFACE

The origins of what have come to be known as the "Oxford" Conferences on modelling and the control of breathing can be traced back to a discussion between Dan Cunningham and Richard Hercynski at a conference dinner at the Polish Academy of Sciences in 1971. Each felt that they had benefited from the different perspectives from which the topic of ventilatory control was approached - predominantly physiological in the case of Dr Cunningham and predominantly mathematical in the case of Dr Hercynski. Their judgement at that time was that a conference on the control of breathing which allowed investigators with these different (but related) scientific perspectives to present and discuss their work, might prove fruitful. We would judge that this has amply been borne out, based upon the success of the series of conferences which resulted from that seminal dinner conversation.

The first conference, entitled "Modelling of a Biological Control System: The Regulation of Breathing" was held in Oxford, UK, in 1978. Subsequent conferences were: "Modelling and the Control of Breathing" at Lake Arrowhead, California, in 1982; "Concepts and Formulations in the Control of Breathing" in Solignac, France, in 1985; "Respiratory Control: A Modeling Perspective" at Grand Lakes, Colorado, in 1988; and "Control of Breathing and Its Modelling Persepctive" at the Fuji Institute in Japan in 1991. The conferences, subsequent to the one in Oxford, have all resulted in well-received published proceedings.

The decision to establish a continuing three-year cycle of meetings was made at the Lake Arrowhead meeting and the term "Oxford" Conference to characterise the meeting, regardless of its particular location, is attributed to Dr George Swanson at the Grand Lakes meeting. However, the term was first used for the conference itself at the Fuji meeting. And even though the formal title for this conference was chosen to be the "London Conference on Modelling and Control of Ventilation," the rubric of "Oxford" Conference is now, by general acceptance and usage, applied to the conferences collectively.

We chose the title "London Conference on Modelling and Control of Ventilation" in recognition of the fact that London was established as Londinium by the Romans. The choice was influenced by the fact that "LCMCV" would represent not only a cluster of Roman letters but also of Roman numerals. This was thought to be appropriate for a conference for which numeracy plays such as important role.

It is clear that part of the success of previous meetings is attributable to the location of the meeting: in the terms of the Introduction to the Lake Arrowhead meeting, such a location should "provide a sufficiently placid and contemplative milieu to be conducive to discussions which would extend beyond the formal sessions." We tried to ensure that the conference delegates at this meeting would have the time for informal discussions with colleagues in addition to the more formal presentations. The Organising Committee recognised that, in selecting London for the "Modelling and Control" meeting, the ready access to its cultural (and other) attractions could well prove diversionary to those in attendance. In an attempt to obviate such potential conflicts and to maintain the conference traditions, we chose the Royal Holloway College for the conference site, rather than central London. The College has fine Victorian buildings set in a hundred acres of wooded parkland, and has a highly acclaimed picture gallery. It is considered the "country" campus of the University of London.

We were pleased that the "founding fathers" of these meetings, Dan Cunningham and Richard Hercynski, were able to join us for the Conference Banquet. Their presence and after-dinner comments were greatly appreciated by the delegates.

The Organising Committee, comprised of S.J.G. Semple (Chairman), L. Adams, A. Guz, and B.J. Whipp, wish to express appreciation to the organisations that provided contributions to the Conference (these are cited at the end of the volume), the Department of Medicine at the Charing Cross and Westminster Medical School and the Department of Physiology at St George's Hospital Medical School for providing support during the planning phase of the conference, the Governors of the Royal Holloway College for the use of the College facilities and especially to Mrs. Liz Murray for her excellent administrative efforts on behalf of the conference and her consistently cheerful accomplishment of the myriad tasks that cropped up unexpectedly.

The "Oxford" Conferences continue, and continue to grow. They are successful despite (although some may prefer, because of) no elected committee or formal constitution. We hope that the London "Oxford" Conference successfully met the goals articulated informally at the Warsaw dinner in 1971 and, furthermore, that it sustained the high standards set by the previous meetings.

L. Adams
S. J. G. Semple
B. J. Whipp

CONTENTS

Pathophysiology of Breathing Control and Breathing Awake and Asleep

Exercise and Pulmonary Ventilation

Chemical Control of Breathing

Neurophysiology of Breathing Control

NEUROBIOLOGY OF BREATHING CONTROL

Where to Look and What to Look For

Jack L. Feldman

Department of Physiological Science
1812 Life Science Building, Box 951527
Los Angeles, California 90095-1527

INTRODUCTION

Control of breathing is the domain of the brain with its input sensors and output muscles. Understanding control is therefore a neurobiological problem first and foremost. The challenge is connecting the physiology of control, e.g., CO_2 response curves, to measurable properties of the brain. Movements, such as those that underlie (the control of) breathing, are the product of a finely tuned and coordinated pattern of motoneuronal activity, not the product of readily measurable properties, such as the firing rates of individual neurons. Neurobiologists have delineated many properties of the constituents of the brain: molecules, membranes, synapses and neurons, yet very little is known about how these properties underlie integrative behaviors like control of breathing. The difficulty lies in the unique topology of the brain. Neurons are highly interconnected, and although these connections can be grossly delineated, the rules they follow appear almost entirely *ad hoc*.

What questions can neurobiologists address to illuminate the brain mechanisms controlling breathing?

ARE THERE SPECIFIC SITES IN THE BRAIN UNDERLYING CONTROL OF BREATHING?

Whether or not functions are localized within the brain has been controversial. Recently, however, the preponderance of evidence suggests that many unified behaviors, such as speech, result from the binding of disparate regions seemingly concerned with (highly) limited processes. This may be the case too for breathing. The *sine qua non* of respiratory pattern, the respiratory rhythm, is postulated to be centered in the rostral ventrolateral medulla in a region termed the preBötzinger Complex. Central processing of chemoreceptor information appears to require a small region near the ventral medullary surface, the retrotrapezoid nucleus. Certainly, the motoneuron pools innervating respiratory muscles are important in determining the ultimate breathing pattern. Identification of such

Modeling and Control of Ventilation, Edited by S. J. G. Semple, L. Adams, and B. J. Whipp
Plenum Press, New York, 1995

sites is absolutely essential for real progress. There are simply too many ways any given function can be processed by a generic network of neurons; measurement of the neurons in the proper circuit is the only way the actual properties can possibly be revealed. No one is clever enough to deduce the actual mechanisms from first principles or from a black box reverse engineering approach.

ARE THERE SPECIAL NEURAL PROPERTIES THAT MAY PERMIT THERAPEUTIC MANIPULATIONS?

By definition, neurons with respiratory-modulated patterns of impulse activity that are part of the circuits controlling breathing (respiratory neurons) are phenotypically different from nonrespiratory neurons. If this is solely due to their connectivity, then there is little hope of developing therapies. However, there is some reason to expect that there are differences in key molecules, such as neurotransmitter receptors and ion channels, in respiratory neurons compared to other neurons; manipulation of these molecules could modulate respiratory function. At present, most studies related to receptors and channels have not been designed to reveal such differences, instead they focus on identification of properties similar to those already identified in other regions of the brain. For example, glutamate appears to be the transmitter of inspiratory drive to respiratory neurons, which have a variety of potassium and calcium channels that affect their excitability. Yet, (almost) all neurons have glutamate receptors and calcium and potassium channels, so these features do not distinguish respiratory neurons from neurons involved in other functions, such as speech or defecation. However, the explosive growth in molecular biological techniques will soon allow us to determine if respiratory neurons have identical receptors and channels compared to other neurons, or have (subtle) differences in their molecular structure or their coding mRNAs. Such differences, even if they were functionally neutral, might allow for manipulation by probes designed to bind to the unique sites on the molecule or the associated mRNA.

HOW CAN THE GAP BE BRIDGED BETWEEN THE SINGLE NEURON AND BREATHING?

Even today, some respiratory physiologists believe that they can ignore the brain, inferring its mechanisms by building functional models that reproduce breathing behavior. Part of the attraction of such models is that there is a large body of techniques developed mostly by engineers that can be applied. However, such efforts in top-down modeling have been rarely illuminating, and I do not know of a single instance where it resulted in a meaningful neurobiological experiment related to the control of breathing. Recently, due in large part to cheap computing, a neurons up approach has become possible. Models can now be constructed based on measured properties of synapses, neurons and networks. The principal benefit of such models in the demonstration of plausibility, i.e., one can determine if neurons with specified properties connected in a particular manner can produce the appropriate behavior. Here, phenomenology is important, since models must be able to account for a broad range of behavior. Since manipulation of parameters is easier in a computer than *in vivo*, novel experiments can be performed with models, and if sufficiently clever, can lead to predictions that can be tested experimentally.

REFERENCES

1. Feldman JL, Smith JC (1994). Neural control of respiratory pattern in mammals: an overview. In: *Lung Biology in Health and Disease, Vol. 79: Regulation of Breathing* (Dempsey JA, Pack AI, eds). pp. 39-69. New York: M Dekker.
2. Smith JC, Funk GD, Johnson SM, Feldman JL (1995). Cellular and synaptic mechanisms generating respiratory rhythm: insights from in vitro and computational studies In: *Lung Biology in Health and Disease: Ventral Brainstem Mechanisms and Control of Respiration and Blood Pressure* (Trouth CO, Mills RM, eds.). pp. 463-496. New York: M Dekker.

2

NEW COMPUTATIONAL MODELS OF THE RESPIRATORY OSCILLATOR IN MAMMALS

J. C. Smith

Laboratory of Neural Control
National Institute of Neurological Disorders and Stroke
National Institutes of Health, Bethesda, Maryland 20892

INTRODUCTION

Unraveling how the brainstem respiratory oscillator generates the rhythm of breathing remains one of the fundamental problems in understanding the neural control of breathing. With the recent expansion of information on biophysical and network properties of brainstem respiratory neurons has come the understanding that a synthesis of neural processes at cellular, synaptic, and network levels will be required to solve this problem (4, 7, 9). Modeling is an essential tool for achieving such a synthesis. Computational approaches that allow modeling of biologically realistic neurons and networks, closely based on experimental data of cell biophysical properties and network architecture, will be particularly powerful in this regard (1, 6). There are now well established approaches for modeling networks of neurons incorporating the complex biophysical properties and spatiotemporal interactions that are required to describe the behavior of real neurons and networks (ibid.). Recent technical developments, including the availability of neuron/network simulation software, have made computer simulation of these more realistic types of models practical (e.g., 2, 6). In this paper we briefly describe new models for the respiratory oscillator that incorporate this approach.

THE HYBRID PACEMAKER-NETWORK MODEL

Neural oscillators range from pacemaker-driven networks, where intrinsic properties of pacemaker neurons play a critical role, to network oscillators where the oscillation is an emergent, dynamic property of synaptic interactions. There is growing evidence that both pacemaker and network mechanisms need to be considered to explain the generation and control of respiratory oscillations (3, 4, 7, 9). We propose a new model that incorporates both mechanisms, and thus we call it a hybrid pacemaker-network (9). In addition to explaining oscillatory mechanisms, the model attempts to explain how the oscillation propagates through the respiratory network and is translated into the detailed spatiotemporal pattern of

Modeling and Control of Ventilation, Edited by S. J. G. Semple, L. Adams, and B. J. Whipp
Plenum Press, New York, 1995

respiratory (pre)motoneuronal activity underlying breathing. Hybrid oscillators of this type can have complex behavior, including transformational behavior. That is, the oscillator can dynamically reconfigure, transforming between pacemaker-driven and more network-like states as the milieu of input control signals (e.g., from afferent input systems) change. The behavior of the respiratory oscillator is known to change in different brain states (e.g., sleep vs. wakefulness), during development, or between different experimental conditions (e.g., intact vs. deafferented CNS; anesthetized vs. decerebrate; *in vivo* vs. *in vitro*) (see discussions in 4, 9). Such changes may involve transformation. The hybrid pacemaker-network model provides a framework for analyzing transformations and in this sense can provide a unified model for understanding the dynamic behavior of the respirator oscillator in different functional states.

Fig. 1 summarizes the network architecture and its major components. The network can be viewed conceptually as consisting of two basic components: the rhythm generator, and a pattern formation network. Thus the functions of generating the oscillation and translating it into a spatiotemporal pattern are mechanistically separate. The rhythm generator, which is the kernel of the oscillator, is a population of synaptically coupled,

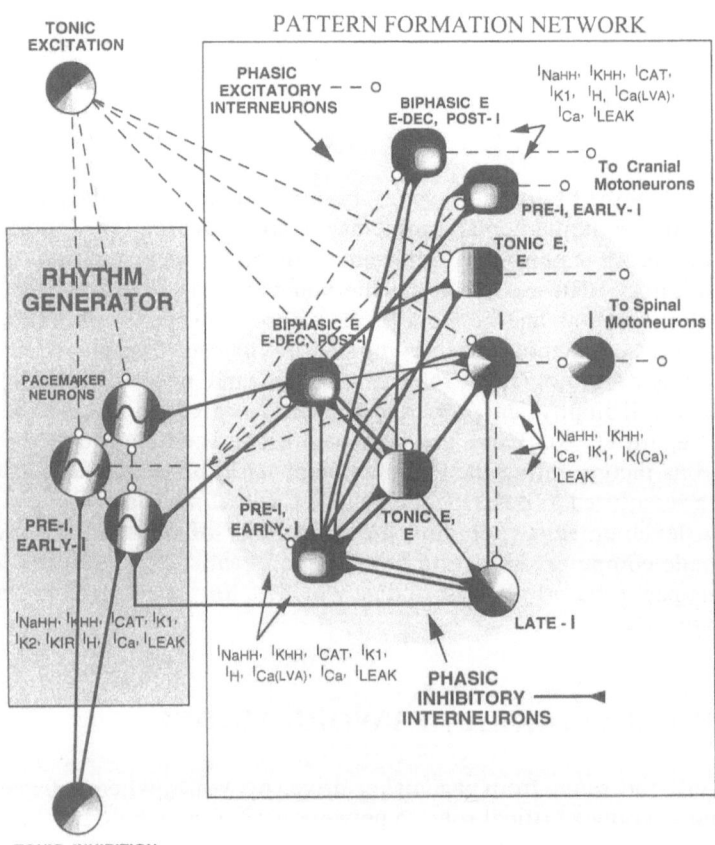

Figure 1. Schematic of hybrid pacemaker-network model showing classes of excitatory and inhibitory respiratory interneurons and their synaptic connections. (Refer to refs. 3, 7, 10 for definition and nomenclature of neuron classes. Multiple naming of a given class reflects different nomenclature used for different experimental preparations and reflects heterogeneity in patterns of cell activity). Intrinsic membrane conductances used in modeling each neuron type are indicated.

bursting pacemaker neurons with complex voltage-dependent behavior (8, 9) that generate the oscillatory drive during the inspiratory phase of network activity. These neurons are bilaterally distributed in the locus for rhythm generation in the ventral medulla (8), and are coupled by fast excitatory connections. The pattern formation network consists of two major, functionally distinct populations of interneurons: (1) excitatory interneurons which form the polysynaptic, inspiratory drive transmission pathways to spinal and cranial motoneurons; and (2) inhibitory interneurons that phasically inhibit a number of neuron populations, including the pacemaker cell population. These inhibitory interactions, in combination with intrinsic membrane properties, play critical roles in pattern formation (i.e., in shaping temporal patterns of discharge and membrane potential trajectories of different neuron classes). Within these two major populations are the known classes of respiratory interneurons (Fig. 1) and their synaptic interactions in the regions of the ventrolateral medulla involved in rhythm and pattern generation (4, 7-10). The oscillatory behavior of the pacemaker cells are regulated by network interactions including tonic inhibitory and excitatory synaptic inputs, as well as the phasic inhibitory inputs. The configuration depicted in Fig. 1 is a simplification which is not meant to include all elements of the respiratory network. The large array of afferent inputs critical for controlling breathing are not represented. The model should be viewed as the kernel that provides the rudimentary mechanisms for rhythm and pattern generation upon which other control elements would operate.

COMPUTATIONAL MODELS OF THE HYBRID PACEMAKER-NETWORK

In developing computational models, our approach has been to use simplified models of the different types of neurons and their synaptic interactions in relatively small networks with connections depicted in Fig. 1. Although highly simplified, these models contain what is currently known or hypothesized to be essential dynamic properties for simulating realistic behavior.

In modeling neuron electrophysiological properties, we have used Hodgkin-Huxley (HH) type of equations to simulate membrane ionic conductances, currents, and voltages. These equations have been extensively and successfully employed to develop biologically realistic models of neurons (e.g. see reviews in 1). Each type of neuron in Fig. 1 is modeled as a one-compartment element represented by standard equivalent electrical circuits incorporating intrinsic ionic and synaptic conductances. Voltage (E)- and time (t)-dependence of the ionic conductances [G(E, t)] and currents [I(E,t)] are described by the HH-like equations of the general form

$$I(E,t) = G(E,t)(E-E_r), \text{ where } G(E,t) = G_{max} \bullet m^x(E,t) \bullet h^Y(E,t) \text{ , and}$$

$$dm/dt = \alpha_m (1-m) - \beta_m m, \quad dh/dt = \alpha_h(1-h) - \beta_h h$$

where the kinetic parameters α_m, β_m, α_h, β_h are functions of E. Concise discussions of these equations and parameter definitions can be found in ref. 5. Synaptic conductances (G_s) and currents (I_s) were modeled by analytical functions (alpha function) that include a presynaptic transmitter release function (T_s) which is initiated by presynaptic activity and activates G_s and I_s:

$$I_s = G_s(E-E_s); \quad G_s = G_{max}T_s t^\alpha exp(-\alpha t/\tau)$$

Parameters (α, τ) and synaptic reversal potentials (E_s) used mimicked conventional fast excitatory or inhibitory conductances.

Simulations done with these equations give membrane potentials and currents that can be directly compared with intracellular recording measurements for the different types of neurons specified in Fig. 1. Sets of conductances, listed in Fig. 1, were used that are known to produce complex spiking and oscillatory bursting behavior in neurons (e.g., see 5 for reviews). These included generic Na^+ and K^+ conductances producing action potentials (INa_{HH}, IK_{HH}), K^+ conductances modulating spiking (IK_1), persistent cationic conductances (I_{CAT}) producing prolonged depolarization (e.g. during the burst phase of pacemaker neurons), and K^+ conductances (IK_2) that interact dynamically with I_{CAT} to regulate the duration of depolarization phases. The most elaborate set was for pacemaker neurons to mimic the complex voltage-dependent oscillatory behavior observed experimentally for the candidate pacemaker cells (8, 9). None of the equation parameters have been directly measured for respiratory neurons. The initial objective of simulations has been to choose plausible parameter values that give time courses of network oscillations, spiking, membrane potential trajectories, and synaptic currents for each neuron type that mimic those measured experimentally. Simulations were performed for the network in both the adult and neonatal systems, and compared to available intracellular measurements from preparations in the neonatal (rat) and adult (cat) mammal for which there exists the largest databases (3, 7, 9, 10). Some of the parameter values were changed between neonatal and adult versions to incorporate postulated developmental differences.

Various aspects of the dynamic behavior of individual neurons and networks, including transformational behavior, have been simulated and explored (see ref. 9 for other examples). Here we simply illustrate that complex patterns of spiking and oscillatory behavior can be simulated to mimic respiratory neurons. Examples of simulations, run with the neural simulation software NODUS (2), are shown in Fig. 2 for the neonatal system and Fig. 3 for the adult.

The simulated time courses of spiking, membrane potential trajectories, and synaptic interactions for each neuron type resemble those measured experimentally. The model neurons exhibit complex behavior including spike frequency adaptation, post-inhibitory rebound, and in the case of pacemaker neurons, voltage-dependent oscillations (9). Although the same sets of membrane conductances and synaptic connections were used for the neonatal and adult simulations, there are differences in the simulated oscillatory frequencies, spiking patterns, and membrane potential trajectories. This reflects, in part, an increased strength of synaptic interactions used in simulations for the adult, particularly inhibitory interactions in the pattern formation network. This increased inhibition has been incorporated to account for changes in neuron and network behavior postulated to occur during development. The model is thus able to mimic possible developmental transformations.

SUMMARY

Understanding how respiratory oscillations are generated and controlled in the mammalian brainstem requires synthesis of neural function at cellular, synaptic, and network levels. Modeling the respiratory network has entered a new era where simulations incorporating complex cellular properties and network interactions exhibited by respiratory neurons are possible. Given the complexity and unknowns of the real system, there are obvious limitations of the modeling in its present form. Nevertheless, if we are to produce neurobiologically satisfying, mechanistic explanations that synthesize cellular and network phenomena, the modeling approach outlined here should have distinct

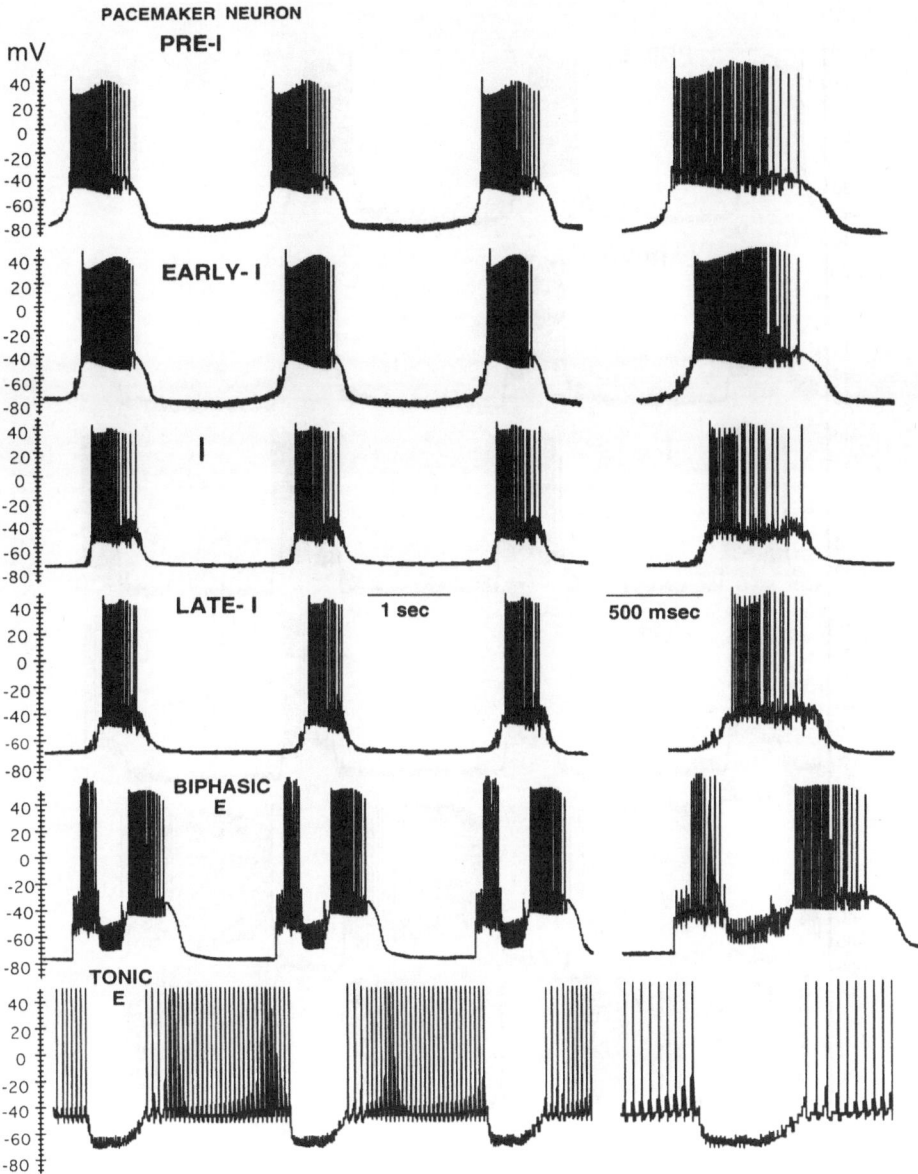

Figure 2. Simulated oscillatory and spiking patterns of several of the neuron types for the neonatal system. Spiking behavior is shown at right in greater detail at expanded time scale.

advantages. The approach allows a close interaction between experimental and computational studies, and will enable simulations with increasing neurobiological realism as additional experimental data becomes available. The modeling summarized here represents the initial stages of establishing a computational framework for storing information about neuron and network properties, and for exploring complex dynamical properties in ways not possible experimentally.

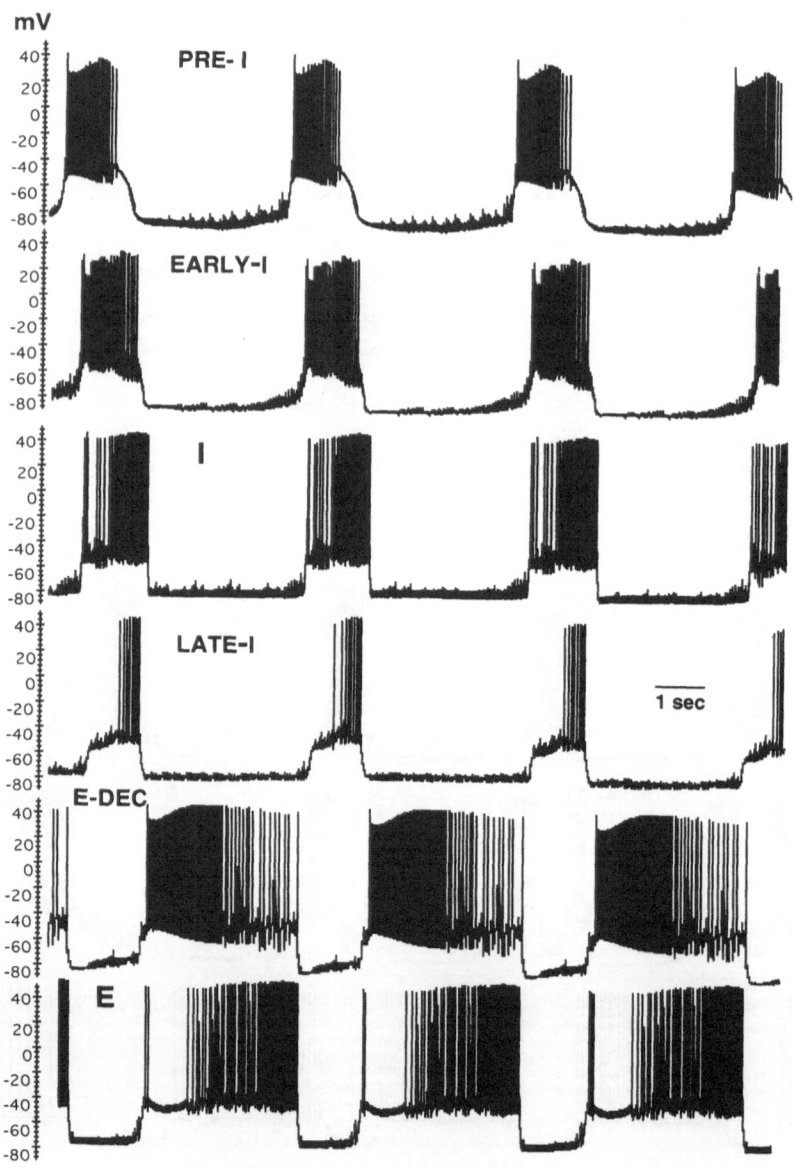

Figure 3. Simulated oscillations and spiking patterns for several neuron types for the adult nervous system. Neuron types shown are homologous to those in Fig. 2 for the neonatal system.

REFERENCES

1. Bower, J.M., ed. Modeling the Nervous System. Trends Neurosci.,1992, vol.15.
2. De Schutter, E. Computer software for development and simulation of compartmental models of neurons. Comput. Biol. Med. 19: 71-78, 1989.
3. Ezure, K. Synaptic connections between medullary respiratory neurons and considerations on the genesis of respiratory rhythm. Prog. Neurobiol. 35: 429-450, 1990.

4. Feldman, J.L. and J.C. Smith. Neural control of respiratory pattern in mammals: An overview. In: Regulation of Breathing, edited by J.A. Dempsey and A.I. Pack. Lung Biology in Health and Disease Series, New York: Dekker, 1995, p. 39-69.
5. Hille, B. Ionic Channels of Excitable Membranes .2nd ed., Sunderland: Sinauer, 1992.
6. Koch, C. and I. Segev, eds. Methods in Neuronal Modeling. From Synapses to Networks, Cambridge, Mass: MIT Press, 1989.
7. Richter, D.W., K. Ballanyi, and S. Schwarzacher. Mechanisms of respiratory rhythm generation. Curr. Opin. Neurobiol. 2: 788-793, 1992.
8. Smith, J.C., H.H. Ellenberger, K. Ballanyi, D.W. Richter, and J.L. Feldman. Pre-Bötzinger Complex: A brainstem region that may generate respiratory rhythm in mammals. Science 254: 726-729, 1991.
9. Smith, J.C., G.D. Funk, S.M. Johnson, and J.L. Feldman. Cellular and synaptic mechanisms generating respiratory rhythm: Insights from in vitro and computational studies. In: Ventral Brainstem Mechanisms and Control of Respiration and Blood Pressure, edited by O. Trouth, R. Millis, H. Kiwull-Schone, M. Schlafke. New York: Dekker, 1995, p. 463-496.
10. Smith, J.C., J. Greer, G. Liu, and J.L. Feldman. Neural mechanisms generating respiratory pattern in mammalian brain stem-spinal cord in vitro. I. Spatiotemporal patterns of motor and medullary neuron activity. J Neurophysiol. 64: 1149-1169, 1990.

3

IS THE PATTERN OF BREATHING AT REST CHAOTIC?

A Test of the Lyapunov Exponent

R. L. Hughson, Y. Yamamoto, and J. -O. Fortrat

Department of Kinesiology
University of Waterloo
Waterloo, Ontario N2L 3G1, Canada

INTRODUCTION

It is possible that the pattern of breathing can provide some information about the characteristics of the controller. Benchetrit and Bertrand (1) observed an autocorrelation that was consistent with the hypothesis of a short-term memory in the respiratory control centre. Recently, the newer mathematical techniques of fractal geometry and chaos theory have been applied to studies of the breathing pattern. We described the pattern of ventilation ($\dot{V}_E$) and gas exchange of humans during exercise to be fractal on the basis of the linearity of the log spectral power versus log frequency relationship (6) during exercise. In this volume, Tuck et al. (15) have described a similar pattern of response for resting ventilatory variability. The linear (so-called 1/f) relationship suggests a long-term correlation. The implication is that the pattern of breathing is influenced over a large number of breaths by past events. As hypothesized for the cardiovascular control mechanisms (2,17), the fractal pattern might be used in effecting control over the homeostasis of the ventilatory events.

A chaotic pattern of breathing, if one exists, would imply that the respiratory control centre is deterministic. That is the random nature of the breathing pattern is not simply stochastic. In an elegent series of experiments with anesthetized rats, Sammon et al. (11-13) identified bifurcation patterns for air flow in the transitions between inspiration and expiration that were consistent with chaos. More recently, Sammon (9,10) has extended these observations with the development of a model of the respiratory central pattern generator based on a set of ordinary differential equations that have the characteristics of chaos.

The Lyapunov exponent is a measure of the degree of divergence of adajcent paths of a variable through multi-dimensional space (16). A chaotic system is one that is not only deterministic as indicated above by the work of Sammon and collegues, but it is also highly dependent on initial conditions. Therefore, if two points within a time series of the ventilatory pattern are observed, there will be exponential divergence of their trajectories as they evolve

Modeling and Control of Ventilation, Edited by S. J. G. Semple, L. Adams, and B. J. Whipp
Plenum Press, New York, 1995

with time. The divergence of trajectories, in spite of their confinement in bounded space, might give rise to the fractal geometry in the ventilatory pattern.

A number of tests are available that could be used to test for the presence of chaos in the time series pattern of ventilatory variability. Calculation of the Lyapunov exponent is one means of quantifying the rate of divergence of adjacent points within multi-dimensional space. Donaldson (3) concluded on the basis of finding positive Lyapunov exponents that human breathing pattern is chaotic. Yet, because researchers have been cautioned about the potential pitfalls of concluding the existence of chaos based on a single test (8,14), we decided that we should re-evaluate these experiments, and test the outcome by the method of surrogate data analysis (14).

METHODS

Resting ventilation and gas exchange measurements were obtained from 7 healthy young men who were seated in a comfortable chair with arm and head rests. All details of the experiments, except that we were studying the pattern of breathing, were provided and subjects signed a consent form approved by the Office of Human Research. Subjects breathed through a mouthpiece connected to a volume turbine (Alpha Technologies, VVM-110) for 60 min. Breath-by-breath values were computed and stored for each of tidal volume (V_T), breathing frequency (f_b), minute ventilation ($\dot{V}_E$), inspiratory and expiratory times (T_I and T_E), end-tidal PO_2 and PCO_2, oxygen uptake ($\dot{V}O_2$), and carbon dioxide output ($\dot{V}CO_2$), as well as beat-by-beat RR-interval. The calibration of this system has been described in detail elsewhere (5).

The data were analyzed in a manner very similar to that described by Donaldson (3). Data were sampled at 2 Hz by holding the value of the current breath. To avoid step changes in the values between breaths, the data were smoothed (Hanning) by a weighted moving function (3). Next, the algorithm described by Wolf et al. (16) for the calculation of the Lyapunov exponent was applied. Rather than use the entire 60 minutes (7,200 points) of data, as Donaldson did, we used only 4096 because we wanted to compare the response with that of surrogate data.

The surrogate data had the same correlation structure as the original data. They were obtained as described by Theiler et al. (14), by first applying a standard fast Fourier transform (FFT) to the original breath-by-breath data (Figure 1). The amplitude components of the FFT were retained, but the phase was scrambled prior to regenerating the data set by the process of inverse FFT (Figure 1). These isospectral surrogate data were then processed in a manner identical to that described for the original data to determine the Lyapunov exponent.

RESULTS

The calculated Lyapunov exponents for the breath-by-breath time series data of $\dot{V}_E$ and its component parts were almost all positive for the individual subject results (Table 1). In all cases, the mean values were positive; a finding that might suggest the presence of chaos in these data. The Lyapunov exponents for the surrogate data were more positive than those for the original data for every subject (Table 1). Similar results were obtained for end-tidal gas concentrations, gas exchange ($\dot{V}O_2$ and $\dot{V}CO_2$), and RR-interval (Table 2).

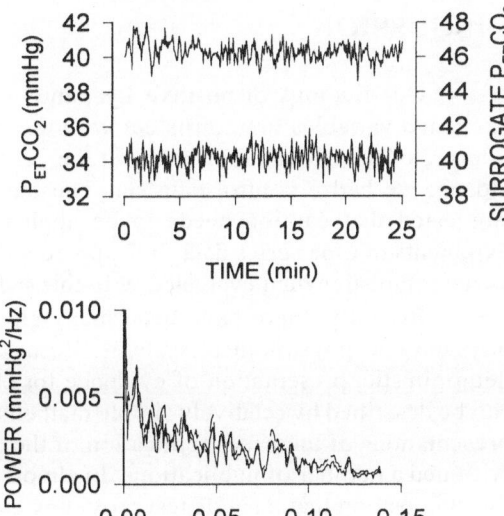

Figure 1. Original time series data for end tidal PCO$_2$ (top series in upper figure, mean $\approx$ 40 mmHg) and corresponding power spectrum (bottom, solid line), and iso-spectral surrogate data constructed from the original time series (lower series in upper figure) and corresponding power spectrum (bottom, dashed line).

Table 1. Lyapunov exponents for original and surrogate data for expired ventilation ($\dot{V}_E$), tidal volume (V_T), breathing frequency (f_b), and inspiratory and expiratory time durations (T_I and T_E)

Subj. #	Orig. $\dot{V}_E$	Surr. $\dot{V}_E$	Orig. V_T	Surr. V_T	Orig. f_b	Surr. f_b	Orig. T_I	Surr. T_I	Orig. T_E	Surr. T_E
1	-0.310	2.485	-0.066	2.386	0.155	2.336	-0.039	2.407	0.150	2.420
2	0.222	2.573	0.126	2.389	0.170	2.423	0.091	2.301	0.048	2.361
3	0.131	2.434	0.114	2.393	0.167	2.269	0.185	2.389	0.135	2.357
4	0.209	2.383	0.096	2.402	0.150	2.388	-0.011	2.398	0.214	2.381
5	0.084	2.353	0.043	2.415	0.148	2.407	0.201	2.519	0.157	2.466
6	0.091	2.430	-0.093	2.438	0.145	2.370	0.106	2.261	0.052	2.374
7	0.153	2.416	0.091	2.358	0.275	2.464	-0.059	2.446	0.295	2.384
mean	0.082	2.439	0.044	2.397	0.173	2.380	0.068	2.389	0.150	2.392
$\pm$ SE	$\pm$.181	$\pm$.072	$\pm$.089	$\pm$.025	$\pm$.046	$\pm$0.63	$\pm$.106	$\pm$0.86	$\pm$0.87	$\pm$.039

Table 2. Lyapunov exponents for original and surrogate data for end-tidal PO$_2$ and PCO$_2$, oxygen uptake ($\dot{V}O_2$), carbon dioxide output ($\dot{V}CO_2$), and RR-interval

Subj. #	Orig. P$_{ET}$O$_2$	Surr. P$_{ET}$O$_2$	Orig. P$_{ET}$CO$_2$	Surr. P$_{ET}$CO$_2$	Orig. $\dot{V}$O$_2$	Surr. $\dot{V}$O$_2$	Orig. $\dot{V}$CO$_2$	Surr. $\dot{V}$CO$_2$	Orig. RR-int.	Surr. RR-int.
1	-0.033	2.431	-0.001	2.370	0.184	3.149	0.119	2.372	0.308	2.463
2	0.153	2.590	0.182	2.413	0.343	2.558	0.220	2.385	0.345	2.463
3	0.130	2.357	0.195	2.330	0.311	2.433	0.282	2.380	0.316	2.410
4	0.075	3.116	0.105	2.399	0.326	2.451	0.327	2.480	0.385	2.412
5	0.016	2.326	0.017	2.405	0.149	2.620	0.045	2.417	0.386	2.368
6	0.122	2.452	-0.106	2.386	-0.103	2.407	0.009	2.488	0.255	2.427
7	0.104	2.863	0.006	2.423	0.144	3.298	0.082	3.233	0.460	2.411
Mean	.081	2.591	.057	2.389	.192	2.702	.154	2.536	.351	2.422
$\pm$ SE	$\pm$0.62	$\pm$.294	$\pm$.109	$\pm$.031	$\pm$.156	$\pm$366	$\pm$122	$\pm$.311	$\pm$.067	$\pm$.033

DISCUSSION

Our findings of positive Lyapunov exponents for each of the ventilatory and gas exchange variables are consistent with the report of Donaldson (3). These data could be interpreted as evidence for chaos in the ventilatory control system. However, unlike Donaldson, we had a control data set. The outcome of the surrogate data analysis strongly suggested that caution needs to be applied in the interpretation of positive Lyapunov exponents in time series data. It is apparent that surrogate, stochastic data, can masquerade as deterministic when evaluated with this specific algorithm (16).

Recently, there have been many claims of evidence of chaos in a wide range of physical and physiological systems. Because of the implication that a chaotic system is deterministic, presentation of evidence for chaos could mean that the system under study can be described by relatively simple mathematical relationships. But, there have been many presentations of incorrect application of the methods of nonlinear mathematics. Ruelle (8) critiqued a number of publications. In almost every case, the weak link in the argument was the supposition that a single test providing evidence of chaos was adequate. The method of surrogate data analysis has provided the opportunity to evaluate a relationship by testing the null hypothesis that chaos is present in the randomly generated data. Theiler et al. (14) evaluated the Lyapunov exponent algorithm of Wolf et al. (16). They found that the algorithm did not yield realistic values with surrogate data. Therefore, it should not be surprizing that the current study found more positive Lyapunov exponents with the surrogate data. The implication of this is that the ventilatory, gas exchange, and RR-interval variables might have been chaotic, but there is no evidence based on the analysis performed.

That there is a pattern in the breath-by-breath variation in breathing pattern has been well established. Lenfant (7) observed fast (2-6 breath), "longer" (25-50 breath), and slow (150-200 breath) oscillations in the pattern of breathing. This implies that there are relationships between breaths such that the current breath influence subsequent breaths. This is consistent with the studies of autocorrelation (1), and fractal (1/f) patterns of breathing (6). Hlastala, Wranne, and Lenfant (4) went further with their analysis and showed that power spectral analysis indicated peaks at some specific frequencies. This type of analysis has set the stage for the fractal and chaotic analyses of today.

Donaldson (3) suggested that "poorly controlled systems will probably let their trajectories diverge at greater rates than tightly controlled systems". This implies that the calculated Lyapunov exponent would become more positive. That is, the system is "more chaotic". But, is this the way in which the ventilatory control system might be expected to behave? Based on the fractal analysis, the less well controlled system will have a greater value of the slope (ß) of the log spectral power vs. log frequency than a well controlled system. That is, spectral power is evidenced as large low frequency oscillations, with less power in the high frequency range. Indeed, a less well regulated cardiovascular system has a greater slope (ß), and syncope is frequently observed (2). For the ventilatory control system, a greater slope is observed with situations in which the pattern of breathing deviates more from the current mean value. Hyperoxia tends to reduce the breath-by-breath coupling of the ventilatory response so that there are greater deviations away from the mean value (15). Whether this is consistent with changes in the dimension of a chaotic attractor has yet to be determined.

In conclusion, carefully controlled studies of the pattern of breathing in anesthetized rats have shown evidence for chaos in the ventilatory controller. However, a previous suggestion that human respiratory pattern was chaotic appears from the outcome of the surrogate data analysis in this study to have been premature. Further testing might find evidence for chaos in human resting ventilation, but this will require more than simply testing for a positive Lyapunov exponent.

ACKNOWLEDGEMENTS

This research was supported by Natural Sciences and Engineering Research Council, Canada.

REFERENCES

1. Benchetrit, G. and F. Bertrand. A short term memory in the respiratory centres: statistical analysis. *Resp. Physiol.* 23: 147-158, 1975.
2. Butler, G. C., Y. Yamamoto, and R. L. Hughson. Fractal nature of short term systolic blood pressure and heart rate variability during lower body negative pressure. *Am. J. Physiol. Regul. Integr. Comp. Physiol.* 267: R26-R33, 1994.
3. Donaldson, G. C. The chaotic behaviour of resting human respiration. *Respir. Physiol.* 88: 313-321, 1992.
4. Hlastala, M. P., B. Wranne, and C. J. Lenfant. Cyclical variations in FRC and other respiratory variables in resting man. *J. Appl. Physiol.* 34: 670-676, 1973.
5. Hughson, R. L., D. R. Northey, H. C. Xing, B. H. Dietrich, and J. E. Cochrane. Alignment of ventilation and gas fraction for breath-by-breath respiratory gas exchange calculations in exercise. *Comput. Biomed. Res.* 24: 118-128, 1991.
6. Hughson, R. L. and Y. Yamamoto. On the fractal nature of breath-by-breath variation in ventilation during dynamic exercise. In: *Control of Breathing and Its Modeling Perspectives,* edited by Y. Honda, Y. Miyamoto, K. Konno, and J. Widdicombe. New York: Plenum Press, 1992, pp. 255-262.
7. Lenfant, C. Time-dependent variations of pulmonary gas exchange in normal man at rest. *J. Appl. Physiol.* 22: 675-684, 1967.
8. Ruelle, D. Deterministic chaos: the science and the fiction. *Proc. R. Soc. Lond. A.* 427: 241-248, 1990.
9. Sammon, M. Geometry of respiratory phase switching. *J. Appl. Physiol.* 77: 2468-2480, 1994.
10. Sammon, M. Symmetry, bifurcations, and chaos in a distributed respiratory control system. *J. Appl. Physiol.* 77: 2481-2495, 1994.
11. Sammon, M., J. R. Romaniuk, and E. N. Bruce. Bifurcations of the respiratory pattern associated with reduced lung volume in the rat. *J. Appl. Physiol.* 75: 887-901, 1993.
12. Sammon, M., J. R. Romaniuk, and E. N. Bruce. Role of deflation-sensitive feedback in control of end-expiratory volume in rats. *J. Appl. Physiol.* 75: 902-911, 1993.
13. Sammon, M., J. R. Romaniuk, and E. N. Bruce. Bifurcations of the respiratory pattern produced with phasic vagal stimulation in the rat. *J. Appl. Physiol.* 75: 912-926, 1993.
14. Theiler, J., S. Eubank, A. Longtin, B. Galdrikian, and J. D. Farmer. Testing for nonlinearity in time series: the method of surrogate data. *Physica D* 58: 77-94, 1992.
15. Tuck, S. A., Y. Yamamoto, and R. L. Hughson. The effects of hypoxia and hyperoxia on the 1/f nature of breath-by-breath ventilatory variability. In: *Modelling and Control of Ventilation,* edited by S. J. G. Semple and L. Adams. New York: Plenum Press, 1995.
16. Wolf, A., J. B. Swift, H. L. Swinney, and J. A. Vastano. Determining Lyapunov exponents from a time series. *Physica D* 16: 285-317, 1985.
17. Yamamoto, Y. and R. L. Hughson. On the fractal nature of heart rate variability in humans: effects of data length and β-adrenergic blockade. *Am. J. Physiol.* 266: R40-R49, 1994.

4

CONTROL OF INTERMITTENT VENTILATION IN LOWER VERTEBRATES

A Computer Dynamic Model

Z. L. Topor[1] and N. H. West[2]

[1] University of Calgary, Faculty of Medicine
   3330 Hospital Drive N.W., Calgary, Alberta, T2N 4N1, Canada
[2] University of Saskatchewan, Department of Physiology
   Saskatoon, Saskatchewan, S7N 0W0, Canada

GENERAL CONCEPT

In an attempt to understand the quantitative aspects of gas exchange for the control of intermittent ventilatory activity in lover vertebrates we developed a conceptual model of this control system. It was assumed that in most species the onset and duration of ventilatory episodes are governed by an oxygen sensitive hypercapnic drive. As a consequence, the controller governing ventilatory activity in the model was designed as an on-off switch coupled with a sensor monitoring a blood compartment. The internal signal used by the controller to begin and terminate a burst of ventilation is partial pressure of carbon dioxide in blood ($PbCO_2$). The controller initiates a breathing episode when $PbCO_2$ rises above a critical value which is influenced by partial pressure of oxygen in blood (PbO_2). Ventilation is terminated when $PbCO_2$ falls below a constant threshold value. The regulated system consists of the stored quantities of oxygen and carbon dioxide in the lungs and the blood compartments. These are separated by a diffusive barrier. The model treats the blood as a uniform pool. The quantities of both respiratory gases, reflected by their partial pressures, are altered by continuous metabolic activity and intermittent bursts of ventilation.

IMPLEMENTATION

This general concept was applied to describe intermittent ventilation in the snapping turtle, *Chelydra serpentina,* using a personal microcomputer (Compaq Deskpro 386s) and an interactive simulation support system - MODSIM (2). The controller relationship was based on the experimental results of West *et al.* (4), and literature values were used for several essential parameters. Careful consideration was given to the issue of the functional representation of the O_2-dissociation curve (ODC). The model successfuly simulated the experi-

Modeling and Control of Ventilation, Edited by S. J. G. Semple, L. Adams, and B. J. Whipp
Plenum Press, New York, 1995

mentally determined ventilatory patterns of *Chelydra serpentina* under normoxic, hypoxic, and hypercapnic conditions. We also used it to simulate differential influences of the hypoxic and hypercapnic drives on respiratory pattern generation. It was observed by others (1,3) that hypoxia generates an increase in minute ventilation by decreasing the interval between bouts of intense breathing with elevated amplitude. It also shortens the length of breathing bouts. A similar type of response to hypoxia was demonstrated in our computer simulations. Hypercapnia, on the other hand, converts intermittent to more continuous breathing by increasing the duration of breathing bouts and decreasing the interval between them. By simulating progressive ventilatory hypercapnia and increased rates of endogenous CO_2 production we also demonstrated that both continuous and intermittent ventilatory patterns can be generated by the model, depending on the simulated level of CO_2. This fact suggests that the evolutionary transition from intermittent ventilation to the continuous pattern associated with endothermy may have depended critically on an increased level of CO_2-related chemoreceptor drive, independent of evolutionary changes in the organization of the respiratory center.

REFERENCES

1. Milsom, W.K., and D.R. Jones. The role of vagal afferent information and hypercapnia in control of the breathing pattern in *Chelonia*. *J. Exp. Biol.* 87:53-63, 1980.
2. Montani, J-P., T.H. Adair, R.L. Summers, T.G. Coleman, and A.C. Guyton. A simulation support system for solving large physiological models on microcomputers. *Int. J. Biomed. Comput.* 24:41-54, 1989.
3. Smatresk, N.J., and A.W. Smits. Effects of central and peripheral chemoreceptor stimulation on ventilation in the marine toad, *Bufo marinus*. *Respir. Physiol.* 83:223-238, 1991.
4. West, N.H., A.W. Smits, and W.W. Burggren. Factors terminating nonventilatory periods in the turtle, *Chelydra serpentina*. *Respir. Physiol.* 77:337-349, 1989.

5

THE INFLUENCE OF CHEMICAL AND MECHANICAL FEEDBACK ON VENTILATORY PATTERN IN A MODEL OF THE CENTRAL RESPIRATORY PATTERN GENERATOR

K. A. E. Geitz,[1,3] D. W. Richter,[4] and A. Gottschalk[2,3]

[1] Department of Bioengineering
[2] Department of Anesthesia
[3] Center for Sleep and Respiratory Neurobiology
   University of Pennsylvania, Philadelphia, Pennsylvania 19104
[4] Department of Physiology
   University of Göttingen
   D-3400, Göttingen, Germany

INTRODUCTION

Much of what is known about respiratory pattern generation has come from preparations that largely isolate the pattern generator from chemical and mechanical feedback. When physiologic feedback loops are preserved, the ventilatory pattern can be quite variable despite a constant metabolic demand. The source of this variability and its role in ventilatory regulation has not been completely determined. However, evidence suggests that it could arise from the interaction of nonlinear system components with the pattern generator rather than from a stochastic source.[9]

To study pattern generation by an integrated respiratory control system, we employ a combined theoretical and computational approach. A theoretical analysis offers a means of identifying the key components of the physiology by limiting the number of assumed system properties to those essential for each specific result.[4] Such models generally do not depend on specific parameter choices or, in the present case, whether the respiratory rhythm is network or pacemaker driven. Unfortunately, analytically tractable formulations of complex problems may not be possible, and solutions are more accessible by computational methods. However, the behaviour of these models depends on the values chosen for many parameters. Testing a model over a full range of parameter choices is impossible because of the number of simulations required. Therefore, complementary theoretical and computational studies may offer a greater opportunity to obtain physiologically meaningful results.

Modeling and Control of Ventilation, Edited by S. J. G. Semple, L. Adams, and B. J. Whipp
Plenum Press, New York, 1995

We use this combined approach to investigate how feedback breaks the symmetry of a constantly repeating respiratory pattern. Initially, in a purely theoretical study, we investigate the characteristic relationship between respiratory amplitude and frequency as the chemical drive to breathe is varied.[2] It is known that for low tidal volumes inspiratory duration remains constant while, as the respiratory drive increases, the inspiratory duration decreases hyperbolically as tidal volume increases. At extremes of drive, a third region of behaviour is sometimes observed, where inspiratory duration increases in tandem with tidal volume. When the vagi, through which much of the respiratory mechanical feedback is relayed, are cut, inspiratory duration remains constant regardless of tidal volume. We incorporate the results of the prior study in an integrated respiratory system model, and investigate the variability in breath to breath amplitude and frequency when chemical and mechanical feedback are preserved and metabolic demand remains constant. This variation is decreased by challenging the system with increased airway resistance or reduced lung compliance.[3]

METHODS

The theoretical model requires only four physiologically reasonable assumptions: 1) The isolated central respiratory pattern generator (CRPG) has a limit cycle output. 2) The amplitude of the limit cycle in the isolated CRPG changes with respiratory drive while the frequency of the limit cycle remains fixed. 3) The structure around the limit cycle grows linearly with the increase in size of the limit cycle. 4) Finally, the lungs are dissipatively attracted to a fixed volume for each fixed input from the CRPG. With these assumptions, the effect of mechanical feedback derived from the three different functions of lung volume summarized in Fig. 1 is determined using averaging theory.

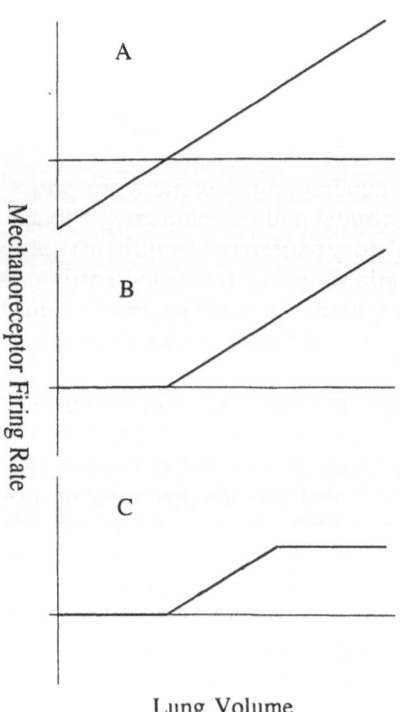

Figure 1. Feedback from pulmonary stretch receptors: A) Linear function. B) Thresholshold. C) Function with a threshold and saturation.

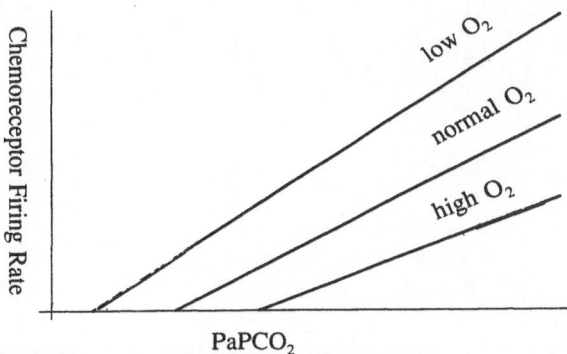

Figure 2. Approximation of the steady state carotid sinus nerve firing rate used in the present work, and shown for low, normal and high O_2 saturations.

The computational model combines a modification of a previously described model of the CRPG[8] with lungs, circulation and feedback from chemical[7] and mechanical sensors. The CRPG is modified by the addition of a sixth neural pool simulating the activity of phase spanning expiratory to inspiratory neurons[10]. The drive to breathe is the sum of a peripheral and central drive. The central drive is a linear function of the central partial pressure of CO_2. The peripheral drive is the product of a linear function of the peripheral oxygen saturation and a non-negative function of the peripheral arterial partial pressure of CO_2 (P_aCO_2). This function is depicted in Fig. 2. A second order response to step changes in P_aCO_2 is also included.[1] Mechanical feedback is determined from a threshold function applied to the instantaneous lung volume. Additional details about both the analytical and computational methods can be found in the referenced articles.[5, 6]

RESULTS

Mathematical analysis of vagal feedback yields a prediction that system behaviour can be well approximated by the system of equations:

$$d\theta_1/dt = \omega + \mu^{-1}H_1(\phi, g, \mu)$$

$$d\theta_2/dt = H_2(-\phi)$$

where H_1 is 2π periodic in ϕ, and depends upon both the shape of the vagal feedback waveform and the pulse response of the CRPG. μ represents the amplitude of the CRPG limit cycle, and $g(\cdot)$ is the vagal output as a function of lung volume. The results from the analytical model are shown in Fig. 3. Purely linear feedback (3A) prevents appropriate modulation of the respiratory pattern with mechanical feedback. However, when linear feedback also undergoes a thresholding operation, the appropriate vertical and hyperbolic ranges of behaviour are produced (3B). If saturation of mechanical feedback is present, the appropriate third range of behaviour is observed (3C).

The above results were used to configure a computational model with both mechanical and chemical feedback. This model produces a ventilatory pattern that is chaotic, and the parameters which characterize the dynamics of its behaviour are similar to published studies. The appearance of such a pattern requires nonlinearities in *both* the mechanical and chemical feedback. Here, we employ a threshold type nonlinearity in mechanical feedback as deduced from the computational study, and a peripheral drive which is non-negative and exhibits adaptation to step changes in P_aCO_2. As shown in Fig. 4, the range of tidal volumes obtainable decreases as pulmonary resistance increases.

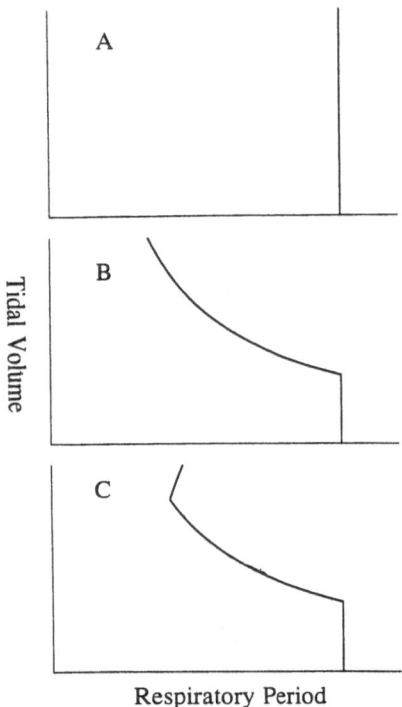

Tidal Volume

Respiratory Period

Figure 3. Analytical predictions of tidal volume as a function of respiratory period for three different stretch receptor functions $g(\bullet)$. A) Linear $g(\bullet)$. B) Linear $g(\bullet)$ with a threshold. C) Linear $g(\bullet)$ with a threshold and a saturation.

DISCUSSION

Nonlinear mechanical *and* chemical feedback to the CRPG are essential for the integrated CRPG to exhibit a full range of physiologic behaviour. These nonlinearities appear

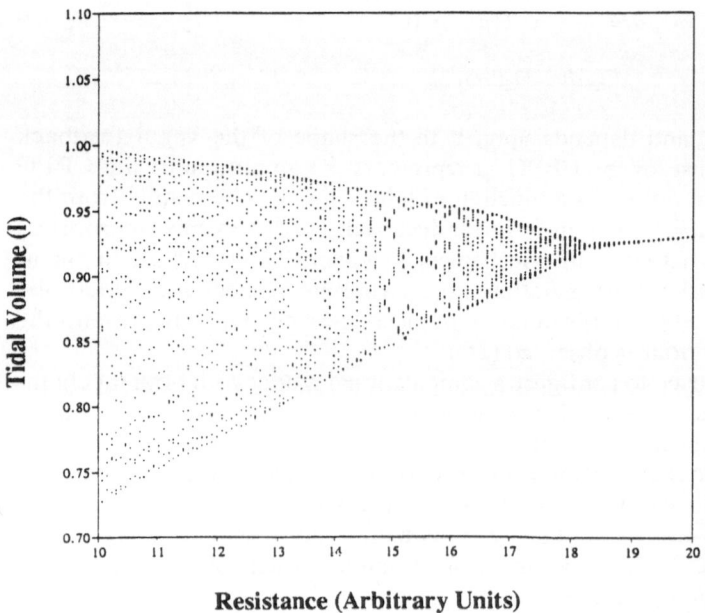

Figure 4. Tidal volumes for each computationally modeled breath are plotted for 20 successive breaths at each of 250 values of resistance. A comparable plot results from reducing the compliance of the lungs.

to have the strongest effect when mechanical and chemical measures of the system are physiologic. When pathologically induced displacements of the system operating point reduce the influence of the nonlinearities, a decreased variability in breathing results (see Fig. 4).

The results concerning mechanical feedback come from an analytical approach that is universally applicable regardless of the source of the CRPG. Consequently, these results remain valid whether respiratory rhythm originates from a network of neurons, an intrinsically oscillating pacemaker neuron, or some combination of the two. The separation of respiratory behaviour into distinct ranges depends on a nonlinear monotonically increasing feedback from mechanoreceptors to the pattern generator. Linear feedback is insufficient to result in any modulation of respiratory timing with amplitude.

The results concerning chemical feedback come from a computational model combining a simple network of neurons that has previously been shown to exhibit many features of respiratory related neurons, and a model of the lungs, gas exchange, metabolism and circulation to the peripheral and central chemoreceptors. When the respiratory drive depends functionally upon the output of the chemoreceptors, specific choices for the output of the peripheral chemoreceptor as a function of P_aCO_2 lead to a physiologic variability in the breathing pattern. An important contributor to this variability is the physiologically observed second order response of the peripheral chemoreceptor to changes in P_aCO_2.

However, this type of peripheral chemoreceptor alone is insufficient to generate the physiologic breathing pattern. Simulations in which either the model vagus or carotid sinus nerve are cut produce a breathing pattern with a fixed amplitude and period of respiration regardless of the source of the respiratory oscillation. Hence, in the current model, vagal feedback works in conjunction with the peripherally derived respiratory drive to produce a breathing pattern with physiologic variability.

The CRPG and lungs with mechanical feedback form one part of a system that generates the frequency of breathing for a given level of respiratory drive. Thus, there is a functional relationship between the drive, and the frequency and amplitude of respiration. Drive depends on the chemical state which is a time delayed function of the product of the frequency and amplitude of respiration. The small oscillatory component from breathing present in the chemical state is sufficient to produce a variable pattern in a manner similar to that of the van der Pol oscillator. When the system is altered by removing either feedback loop, the variability is removed as well.

REFERENCES

1. Black A.M.S., McCloskey D.I. and Torrance R.W., The responses of carotid body receptors chemoreceptors in the cat to sudden changes of hypercapnic and hypoxic stimuli. *Resp. Physiol.* 13: 36-49, 1971.
2. Clark F.J. and von Euler C., On the regulation of the depth and rate of breathing, *J. Physiol.*, 222: 267-295, 1972.
3. Daubenspeck J.A., Influence of small mechanical loads on variability of breathing pattern. *J. Appl. Physiol.*, 50: 299-306, 1981.
4. Ermentrout G.B. and Kopell N., Multiple pulse interactions and averaging in systems of coupled neural oscillators, *J. Math. Biol.*, 29: 195-217, 1991.
5. Geitz, K.A.E. and Gottschalk, A., The role of mechanical feedback in respiratory amplitude and frequency control, *in preparation*.
6. Geitz K.A.E., Richter D.W. and Gottschalk A., Chemical and mechanical feedback in a model of the central respiratory pattern generator, *in preparation*.
7. Khoo M.C.K., Gottschalk A. and Pack A.I., Sleep-induced periodic breathing and apnea: a theoretical study, *J. Appl. Physiol.*, 70: 2014-2024, 1991.
8. Ogilvie M.D., Gottschalk A., Anders K., Richter D.W., and Pack A.I., A network model of respiratory rhythmogenesis, *Am. J. Physiol.*, 263 (*Regulatory Integrative Comp. Physiol.* 32): R962-R975, 1992.

9. Sammon M.P., and Bruce E.N., Vagal afferent activity increases dynamical dimension of respiration in rats, *J. Appl. Physiol.*, 70: 1748-1762, 1991.
10. Schwarzacher S.W., Wilhelm Z., Anders K., and Richter D.W., The medullary respiratory network in the rat, *J. Physiol.*, 435: 631-644, 1991.

6

ROLE OF ACETYLCHOLINE AS AN ESSENTIAL NEUROTRANSMITTER IN CENTRAL RESPIRATORY DRIVE

Homayoun Kazemi and Melvin D. Burton

Pulmonary and Critical Care Unit, Department of Medicine
Massachusetts General Hospital and Harvard Medical School
Boston, Massachusetts 02114

INTRODUCTION

John Scott Haldane in the 1930's suggested that the profound effect of CO_2 on the level of ventilation was related to its ability to change brain hydrogen ion concentration, and that the increase in ventilation seen with a rise in PCO_2 was because of a fall in brain pH. Pappenheimer and associates (12) in the 1960's showed that in the unanesthetized goat brain interstitial fluid pH determined the level of ventilation in chronic acid-base disorders. Site of action of H+ in the brain has been shown to be in at least three superficial areas on the ventral surface of the medulla. How hydrogen ions act at these "chemosensitive" sites is not clear(10). The possibilities are: 1- hydrogen ion gradient across the neuronal membrane; 2- H+ changes affect ions or ionic channels, and 3- H+ causes the release or activation of neurotransmsitters. Our studies have concentrated on neurotransmitter release, particularly acetylcholine.

Earlier experiments in the anesthetized dog with ventriculocisternal perfusion have shown that the effect of acidic CSF on ventilation, whether of respiratory or metabolic origin, can be blocked by the muscarinic receptor antagonist, atropine (2).

Furthermore, other agents that increase ventilation centrally such as salicylates and beta-adrenergic agents can also be blocked with atropine (3,6). Based on these observations, in the recent past we have embarked on studying the role of acetylcholine in central ventilatory drive in a spinal cord brainstem preparation from the neonatal rat.

METHODS

Briefly, the model that we and others have used and reported consists of the brain stem and spinal cord isolated from neonatal rats who are from 0 to 5 days old (4,8). The specimen is superperfused with oxygenated modified Kreb's solution at the rate of 6 ml. per minute. The C4 (phrenic spinal rootlet) is attached to a suction electrode and the signal from

Modeling and Control of Ventilation, Edited by S. J. G. Semple, L. Adams, and B. J. Whipp
Plenum Press, New York, 1995

it is filtered, amplified and monitored on an oscilloscope. This preparation allows for recordings of phrenic nerve output for up to four hours after preparation and its viability has been verified by using NMR spectroscopy where we have demonstrated that its pH stays stable for about four hours and its energy substrates are unchanged(5). Using this preparation, we have looked at the effect of acetylcholine on phrenic nerve output. We have also used blockers of acetylcholine generation and release, as well as M1 and M3 receptor antagonists.

RESULTS

Acetylcholine synthesis is directly linked to carbohydrate metabolism in the brain. Glucose metabolized through the Kreb's cycle leads to formation of citrate which then forms acetyl-CoA and in combination with choline forms acetylcholine. Acetylcholine is stored in vesicles which fuse with cell membrane and ultimately acetylcholine is released for neurotransmission and binds with muscarinic receptors. In generation of respiratory neuronal discharge M1 and M3 receptors are of particular importance. In our studies we have used different blockers at the different steps in formation, release and receptor binding of acetylcholine.

Figure 1 shows the effect of atropine on total phrenic output, burst amplitude and frequency of the preparation, where all parameters fall in a dose dependent manner. When atropine is washed out, neural activity is restored. The same pattern of response was seen when M1 receptors were blocked with pirenzapine or when M3 receptors were blocked with 4, Diphenylacetoxy-N-Methlypiperidine Methiodide (DAMP), the effect with DAMP being more profound than that of pirenzapine (4).

When acetylcholine-release inhibitor, vesamicol, was used, there was again a significant fall in phrenic nerve output, burst magnitude and frequency (Figure 2). The depression of phrenic output was reversed with the facilitator of acetylcholine release tetraethylammonium chloride (TEA). The enhanced phrenic activity with TEA was blocked with atropine.

In order to better identify the site of action of the acetylcholine, a number of microinjection studies with an inhibitor of acetylcholine synthesis, hydroxycitrate, were performed in the neonatal rat brain stem preparation. These injections showed that the site of action was in an angiocentric distribution on the ventral surface of the medulla concentrated along the anterior inferior cerebellar artery and the posterior inferior cerebellar artery (5).

DISCUSSION

The neurochemical basis of spontaneous respiratory neuronal output is not totally clear. A number of amino acid neurotransmitters, neuropeptide or other related substances have been suggested as possible candidates (9). The present study suggests that the cholinergic system is an essential ingredient of the central respiratory neuronal output.

The brain stem spinal cord preparation offers a unique system for study of central output without any feedback loops that would be present in the intact animal. Furthermore it is a biologically viable preparation and results obtained with it are applicable to the intact system (5,8).

The data in this study suggest that cholinergic transmission is critical in the normal ventilatory output. The implication of this observation is that in clinical states where there is central hypoventilation, dampened output of the cholinergic system should also be apparent. Three clinical states that are associated with varying degrees of central hypoventilation are: congenital central hypoventilation, Leigh's disease and Hirschprung's disease.

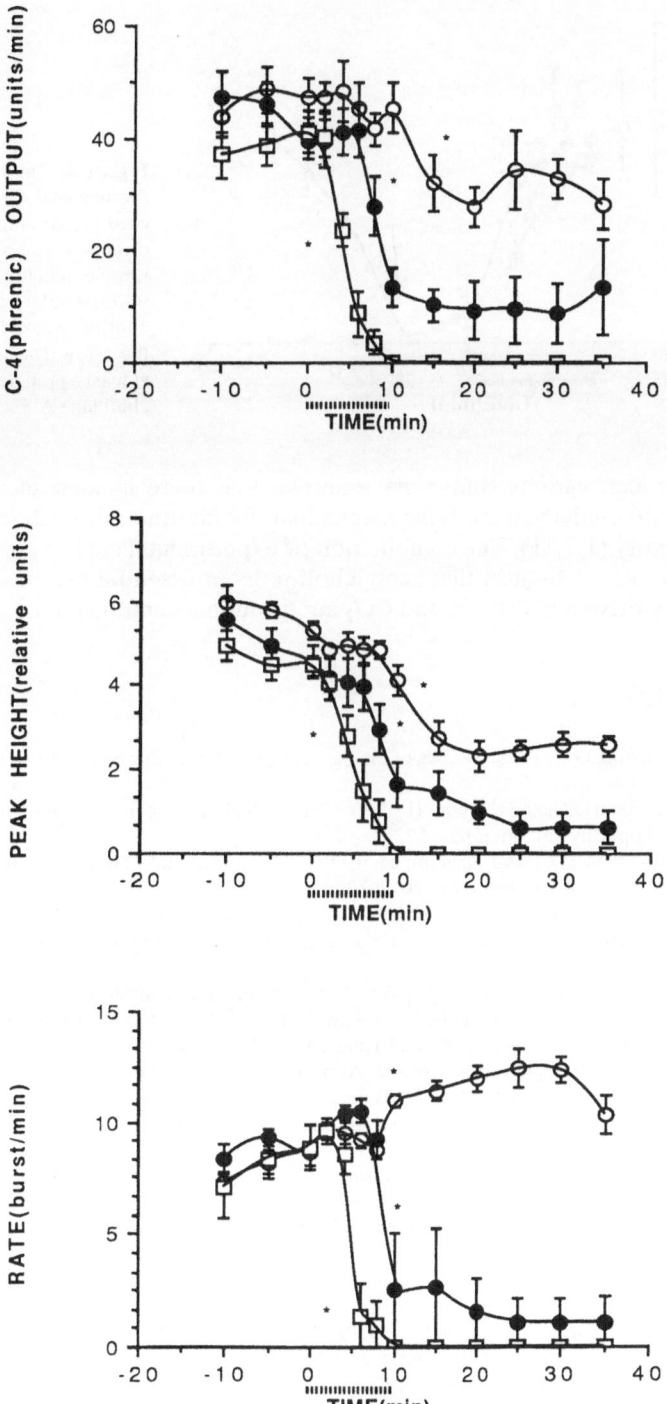

Figure 1. Total phrenic output, burst height and frequency during and after expossure to various concentrations of atropine sulfate (1.0 mM(0), 2.5 mM (●) and 5.0 mM(□). Dashed line on time axis is time of exposure. Mean values ± SEM in 5 preparations. Reprinted with permission from J. Appl. Physiol. (Ref. #4).

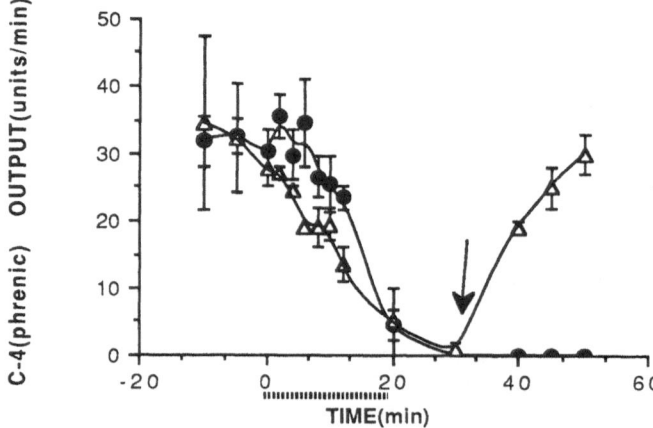

Figure 2. Total phrenic output in the neonatal rat brain stem-spinal cord preparation after exposure to direct acetylcholine release inhibitor, vesamicol, (dashed line) and when challenged with acetylcholine release facilatator, TEA, at the arrow. Mean values ± SEM in at least 5 preparations. ● = no TEA challenge Δ = TEA challenge.

In all these disorders, various studies have shown that there is some element of central hypoventilation, although the underlying mechanism for cholinergic dysfunction is different in each clinical entity (1,7,11). The combination of experimental studies presented here and the clinical observations suggest that acetylcholine is an essential neurotransmitter in the central respiratory drive and that H+ and CO_2 act through a central cholinergic mechanism.

REFERENCES

1. Blass, J.P. and Gibson, G.E. 1978. Studies of the pathophysiology of pyruvate dehydrogenase deficiency. Adv. Neurol. 21:181-194.
2. Burton, M.D., Johnson D.C. and Kazemi, H. 1989. CSF acidosis augments ventilation through cholinergic mechanisms. J. Appl Physiol. 66:2562-2572.
3. Burton, M.D., Johnson D.C. and Kazemi, H. 1990. Adrenergic and cholinergic interaction in central ventilatory control. J. Appl. Physiol. 68: 2092-2099.
4. Burton, M.D., Nouri, K., Baichoo, S., Samuels-Toyloy, S. and Kazemi, H. 1994. Ventilatory output and acetylcholine: perturbations in release and muscarinic receptor activation. J. Appl. Physiol. 77: 2275-2284.
5. Burton, M.D., Nouri, M. and Kazemi, H. 1995. Acetylcholine and central respiratory control: perturbations of acetylcholine synthesis in the isolated brainstem of the neonatal rat. Brain Res. 670:39-47.
6. Fleming, H., Burton M., Johnson, D.C. and Kazemi, H. 1991. Sodium salicylatae centrally augments ventilation through cholinergic mechanisms. J. Appl. Physiol. 71:2299-2303.
7. Haddad, G.G., Mazza, N.M., Defendini, R., Blanc, W.A., Driscoll, J.M., Epstein, M.A.F., Epstein, R.A. and Melons, R.B. 1978. Congenital failure of automatic control of ventilation, gastrointestinal motility and heart rate. Medicine 57: 517-526.
8. Issa, F.G. and Remmers, J.E. 1992. Identification of a subsurface area in the ventral medulla sensitive to local changes in PCO_2. J. Appl. Physiol. 72: 439-446.
9. Kazemi, H. and Hoop, B. 1991. Amino acid neurotransmitters in the central control of breathing. J. Appl. Physiol. 70:1-7.
10. Kazemi, H. and Johnson, D.C. 1986. Regulation of cerebrospinal fluid acid-base balance. Physiol. Rev. 66:953-10377.
11. Minutillo, C., Pemberton, O.J. and Goldblatt, J.1989. Hirschprung's disease and Ondine's curse: further evidence for a distinct syndrome. Clin. Genet. 36:200-203.
12. Pappenheimer, J.R., Fencl, V and Heisey, S.R. 1965. Role of cerebral fluids in control of respiration as studied in unanesthetized goats. Am. J. Physiol. 208:436-450.

7

EFFECTS OF GABA RECEPTOR ANTAGONISTS ON THE RAPHE MAGNUS-INDUCED INHIBITION OF BULBAR AND SPINAL RESPIRATORY NEURAL ACTIVITIES IN THE CAT

Mamoru Aoki, Yasumitsu Sato, Yoshimi Nakazono, and Ikuhide Kohama

Department of Physiology
Sapporo Medical University School of Medicine
South-1, West-17, Chuo-ku
Sapporo 060, Japan

INTRODUCTION

Other investigators (14, 15) and ourselves (1, 2, 4) have previously demonstrated that electrical as well as chemical stimulation of the nucleus raphe magnus (NRM) in the medulla induces marked depressant effects on respiratory activities in cats and other animals. Our previous studies have also provided evidence for the possible involvement of γ-aminobutyric acid (GABA) as a putative transmitter in NRM-induced respiratory inhibition, by intravenous administration of specific receptor antagonists on the NRM-induced respiratory inhibition.

In the present study, we first attempted to clarify whether GABA mediated inhibition is caused by direct actions on the ventral respiratory group (VRG) neurons in the medulla and/or upper cervical inspiratory neurons (UCINs) in C_1-C_2 segments (3, 7, 8, 16, 17). Secondly, in order to determine GABA receptor subtypes, we tested the actions of specific GABA receptor antagonists on the responses to stimulation of the NRM. This was achieved by the use of extracellular recordings from respiratory neurons and iontophoretic application of several GABA receptor antagonists. A brief report on some of the results has been published (4).

METHODS

Experiments were performed on 20 adult cats of either sex, anesthetized with pentobarbital sodium (20-30 mg/kg, i.v. initially, and additional i.v. as required). The animals

Modeling and Control of Ventilation, Edited by S. J. G. Semple, L. Adams, and B. J. Whipp
Plenum Press, New York, 1995

were tracheotomized, paralyzed with gallamine triethiodide, and artifically ventilated. A bilateral pneumothorax was made. Arterial blood pressure (mean pressure < 140 mmHg) was routinely monitored by a catheter in the left femoral artery. End-tidal fractional CO_2 concentrations were continuously monitored and maintained between 4.0-4.8 %. Rectal temperature was maintained at 37°C. The animals were placed in a stereotaxic head holder and fixed in a spinal frame. The spinal cord was exposed by a laminectomy from C_1 to C_2 segments. The medulla oblongata was exposed by an occipital craniotomy and by suction of part of the cerebellum. The left phrenic nerve branch of the C_5 ventral root was exposed by a dorsal approach and the C_5 phrenic nerve discharges were recorded as an indicator of central respiratory rhythm. For electrical stimulation of the NRM, tungsten microelectrodes insulated except for the tip were stereotaxically inserted in the region of the NRM at P_6-P_8 levels. In this series of experiments, the recordings of extracellular unit activities of respiratory neurons and iontophoretic application of GABA receptor antagonists; picrotoxin (5 mM), bicuculline (5 mM), and 2-hydroxy-saclofen (2 OH-Saclofen, 10 mM) were made using a triple-barrel glass microelectrode. One recording pipette contained a tungsten wire, with its tip exposed while other pipettes were filled with drug solutions which were administered iontophoretically with injection currents of 50-300 nA. Extracellular action potentials of single, well isolated neurons were detected from the one recording electrode and displayed on an oscilloscope. Amplified spike discharge was passed through a level discriminator and fed to a pulse counter. Phrenic nerve discharge, its integrated discharge and spike histograms were, together with arterial blood pressure, recorded on a polygraph. At the end of each experiment, recording and / or stimulating sites were marked with electrolytic lesions and later identified histologically.

RESULTS

Effects of GABA$_A$ Receptor Antagonists

In our previous study, a GABA receptor antagonist picrotoxin injection (0.8-1.25 mg/kg, i.v.) produced significant reduction in the magnitude of the raphe magnus (NRM)-induced respiratory inhibition. In the present experiments, we first examined whether the NRM-induced inhibition of the respiratory neuron discharges recorded from the medulla at the obex level and the upper cervical cord at C_1-C_2 segments could be antagonized by iontophoretically applied picrotoxin. As shown in Fig. 1A, in a control before picrotoxin application, both phrenic nerve activity and inspiratory activity of a ventral respiratory group (VRG) neuron were strongly suppressed by electrical stimulation (50 μA) of the NRM.

When picrotoxin application (150 nA) was begun approximately 10 s before the NRM stimulation, the NRM-induced inhibition of the VRG neuron discharge was significantly reduced. This blocking effect of picrotoxin was consistently observed at ejection currents of 50-300 nA and most marked at 150-300 nA Fig. 1B). During the NRM stimulation and picrotoxin application, arterial blood pressure was not changed or only slightly decreased. This blocking effect of picrotoxin was observed in the majority of the respiratory neurons tested, i.e., in 78 % (25/32) of inspiratory neurons and in 83 % (5/6) of expiratory neurons recorded from the VRG. Significant blocking effect by picrotoxin was also observed in 72 % (13/18) of upper cervical inspiratory neurons (UCINs) recorded at C_1-C_2 segments. Another specific GABA$_A$ receptor antagonist bicuculline was also tested. The blocking effect by iontophoretic application of bicuculline (100-300 nA) was similar to picrotoxin, i. e., in 58 % (7/12) of inspiratory neurons, 100% (2/2) of expiratory neurons in the VRG and in 67% (2/3) of UCINs (Fig. 2) .

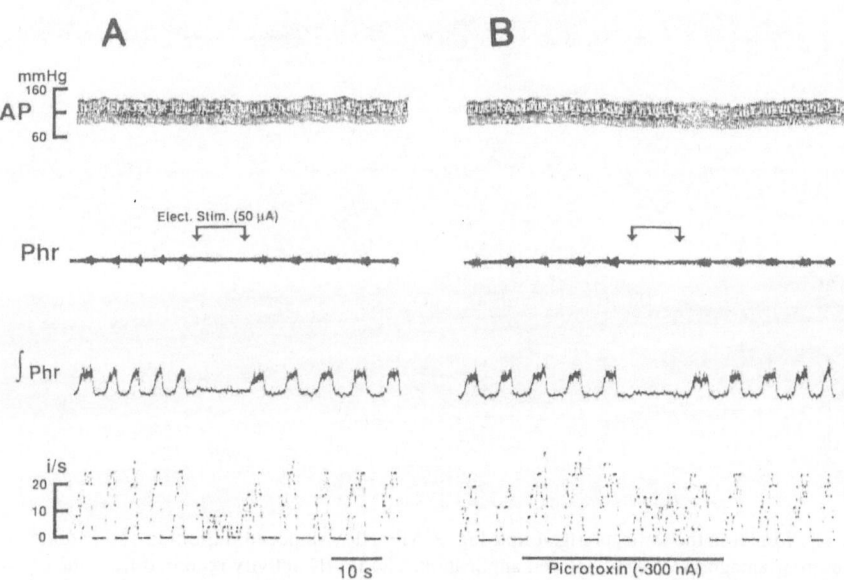

Figure 1. A representative example of the NRM-induced inhibition of respiratory activity and its blocking effect by GABA$_A$ receptor antagonist picrotoxin application. Repetitive stimuli (50 μA, 200 Hz) were delivered to the midline region (P 7.5 level, 4.5 mm deep from the dorsal surface) of the medulla. In A and B, traces from top to down; arterial blood pressure, C$_5$ phrenic nerve discharge (phr), the integrated phrenic nerve discharge (∫ phr) and the frequency histogram of discharge recorded from a VRG inspiratory neuron. Note that picrotoxin application caused slight increases in peak firing frequencies.

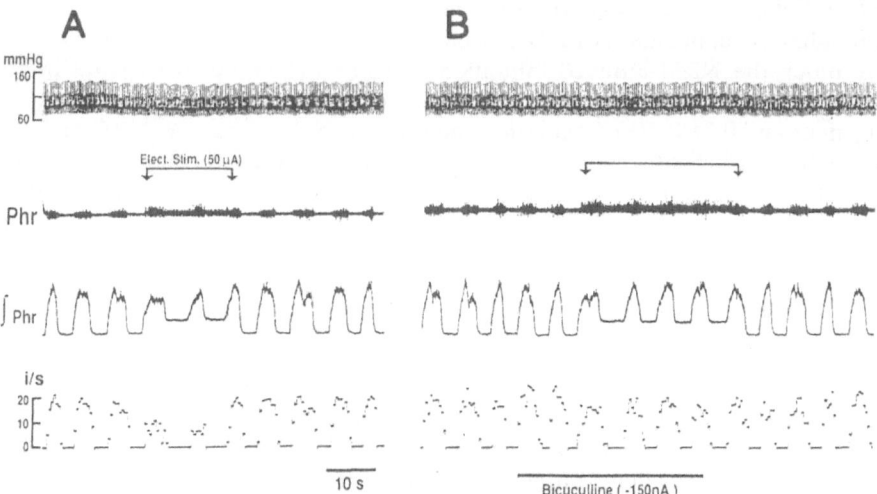

Figure 2. The format is the same as Fig. 1. The NRM-induced respiratory inhibition and its blocking effect by bicuculline application. Respiratory neuron discharge was obtained from a UCIN in C$_1$ segment. Note that the base line of integrated phrenic nerve activity is shifted upward during stimulation due to stimulus artifacts.

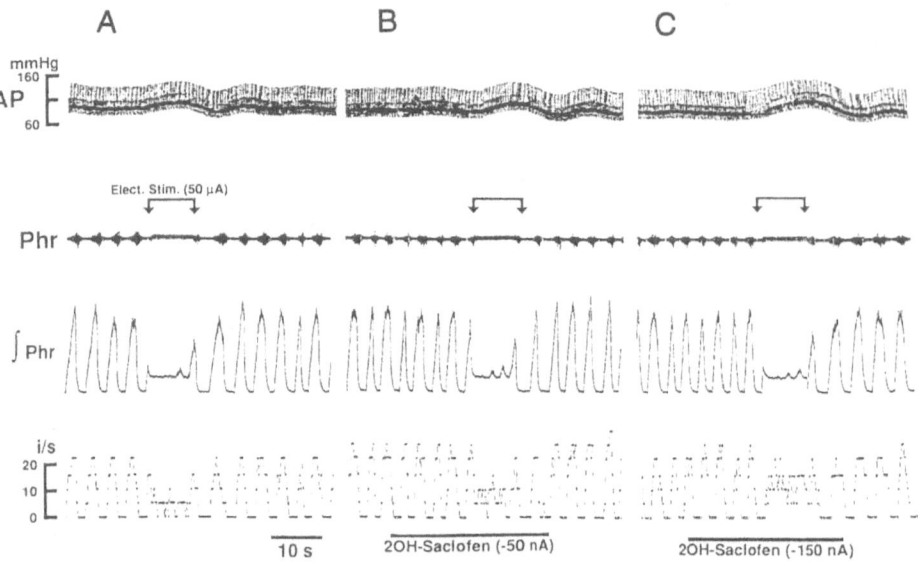

Figure 3. The format is the same as Fig. 1 and Fig. 2. The NRM-induced inhibition and its blocking effect by GABA$_B$ receptor antagonist 2 OH-Saclofen application. The UCIN activity recorded from the C$_1$ segment was strongly suppressed by the NRM stimulation (A). This effect was significantly antagonized by 2 OH-Saclofen in a dose-dependent manner at currents of 50 and 150 nA (B,C).

Effects of GABA$_B$ Receptor Antagonist

The action of a specific GABA$_B$ receptor antagonist, 2 OH-Saclofen, reported to be more effective than phaclofen (18), was examined in VRG neurons and UCINs. A representative result obtained from a UCIN is shown in Fig. 3.

Iontophoretic application of 2 OH-Saclofen with currents above 50 nA was found to effectively block the NRM-induced inhibition of spike discharge of a UCIN in a dose-dependent manner. The blocking effect of 2 OH-Saclofen was observed in 70 % (7/10) of inspiratory neurons, 0 % (0/2) of expiratory neurons recorded from the VRG and in 50 % (2/4) of UCINs recorded at C$_1$ segment. In some VRG neurons (n=5) and UCINs (n=4), the strength of blocking action of picrotoxin and 2 OH-Saclofen was compared on the same neuron. In all neurons tested, 2 OH-Saclofen was much less effective in blocking the NRM-induced inhibition.

DISCUSSION

It has been shown that the medullary raphe nuclei, consisting of the raphe obscurus, raphe pallidus, and raphe magnus, are involved in respiratory control (12, 14, 15). Our previous experiments (1, 2) demonstrated that electrical as well as chemical stimulation of the nucleus raphe magnus (NRM) produces short-latency inhibitory effects on phrenic nerve activity, as well as medullary and upper cervical respiratory neuron discharges. The present experiments have further demonstrated that the NRM-induced inhibition of VRG neuron and upper cervical inspiratory neuron (UCIN) activities were significantly reduced by iontophoretic application of GABA antagonists. These results, together with those of our previous experiments, provide evidence for the involvement of a GABAergic system, probably arising from the NRM, in respiratory inhibition (2, 10, 14, 15).

Various neurotransmitters are involved in the synaptic interrelations in the respiratory neuronal networks. Among putative inhibitory neurotransmitters, GABA is regarded as a most likely transmitter mediating periodic inhibition in medullary respiratory neurons, in which bicuculline-sensitive $GABA_A$ receptor plays a primary role (6, 9, 10, 11, 13, 20). Present experiments revealed that picrotoxin and bicuculline, $GABA_A$ receptor antagonists, were effective in blocking both the NRM-induced inhibition of VRG and UCIN neuron activities. These results suggest that GABAergic neurons in the NRM send their axonal projections to respiratory neurons in the medulla as well as to the spinal cord and the NRM-induced inhibition is the $GABA_A$ receptor-mediated response. In addition, there is now increasing evidence that $GABA_B$ receptors are also involved in respiratory inhibiton (5, 18, 19). Previous studies have demonstrated that a selective $GABA_B$ receptor agonist, baclofen, produced exclusively inhibitory effects on bulbar respiratory neurons and those effects were partly antagonized by $GABA_B$ receptor antagonists such as phaclofen and 2 OH-Saclofen (18, 19). In the present experiments, iontophoretic application of a specific $GABA_B$ receptor antagonist 2 OH-Saclofen was found to antagonize the NRM-induced inhibiton of some VRG neurons and UCINs. When the effects of $GABA_A$ and $GABA_B$ receptor antagonists are compared, $GABA_A$ receptor antagonists seem to be more effective. Thus, the present study supports the view that the NRM-induced respiratory inhibition is mainly $GABA_A$ receptor mediated response and $GABA_B$ receptors are only partly involved.

REFERENCES

1. Aoki, M., Y. Fujito, I. Kosaka, and N. Kobayashi, 1989, Supraspinal descending control of propriospinal respiratory neurons in the cat, In: *Respiratory Control: A Modeling Perspective,* edited by G.D. Swanson, F.S. Grodins, and R.L. Hughson. New York: Plenum, p. 451-459.
2. Aoki, M., Y. Fujito, Y. Kurosawa, H. Kawasaki, and I. Kosaka, 1987, Descending inputs to the upper cervical inspiratory neurons from the medullary respiratory neurons and the raphe nuclei in the cat, In: *Respiratory Muscles and Their Neuromotor Control,* edited by G.C. Sieck, S.C. Gandevia, and W.E. Cameron. New York: Liss, p. 75-82
3. Aoki, M., S. Mori, K. Kawahara, H. Watanabe, and N. Ebata, 1980, Generation of spontaneous respiratory rhythm in high spinal cats, *Brain Res.* 202: 51-63.
4. Aoki, M., and Y. Nakazono, 1992, Raphe magnus-induced inhibition of medullary and spinal respiratory activities in the cat, In: *Control of Breathing and Its Modeling Perspective,* edited by Y. Honda, Y. Miyamoto, K. Konno, and J.G. Widdicombe. New York: Plenum, p. 15-23.
5. Bowery, N.G., A.L. Hudson, and G.W. Price, 1987, $GABA_A$ and $GABA_B$ receptor site distribution in the rat central nervous system, *Neurosci.* 20: 365-383.
6. Champagnat, J., M. Denavit-Saubié, S. Moyanova, and G. Rondouin, 1982, Involvement of amino acids in periodic inhibitions of bulbar respiratory neurones, *Brain Res.* 237: 351-365.
7. Duffin, J., and R.W. Hoskin, 1987, Excitation of upper cervical inspiratory neurons by medullary inspiratory neurons in the cat, In: *Respiratory Muscles and Their Neuromotor Control,* edited by G.C. Sieck, S.C. Gandevia, and W.E. Cameron. New York: Liss, p. 83-87.
8. Duffin J., and R.W. Hoskin, 1987, Intracellular recordings from upper cervical inspiratory neurons in the cat, *Brain Res.* 435: 351-354,.
9. Grelot, L., S. Iscoe, and A.L. Bianchi, 1988, Effects of amino acids on the excitability of respiratory bulbospinal neurons in solitary and para-ambigual regions of medulla in cat, *Brain Res.* 443: 27-36.
10. Haji, A., R. Takeda, and G.E. Remmers, 1991, GABA-mediated inhibitory mechanisms in control of respiratory rhythm, In: *Control of Breathing and Dyspnea,* edited by T. Takishima and N.S. Cherniack. Oxford-New York: Pergamon, p. 61-63.
11. Haji, A. and R. Takeda, 1993, Microiontophoresis of baclofen on membrane potential and input resistance in bulbar respiratory neurons in the cat, *Brain Res.* 622: 294-298.
12. Holtman, J.R., Jr., W.P. Norman, and R.A. Gillis, 1984, Projection from the raphe nuclei to the phrenic motor nucleus in the cat, *Neurosci. Lett.* 44: 105-111.

13. Kirsten, E.B., J. Satayavivad, W.M. St. John, and S.C. Wang, 1978, Alteration of medullary respiratory unit discharge by iontophoretic application of putative neurotransmitters, *Br. J. Pharmac.* 63: 275-281.

14. Lalley, P.M., 1986, Responses of phrenic motoneurones of the cat to stimulation of medullary raphe nuclei, *J. Physiol.* 380: 349-371.

15. Lalley, P.M., 1986, Serotonergic and non-serotonergic responses of phrenic motoneurones to raphe stimulation in the cat, *J. Physiol.* 308: 373-385.

16. Lipski, J., and F. Duffin, 1986, An electrophysiological investigation of propriospinal inspiratory neurons in the upper cervical cord of the cat, *Exp. Brain Res.* 61: 625-637.

17. Nakazono, Y., and M. Aoki, 1994, Excitatory connections between upper cervical inspiratory neurons and phrenic motoneurons in cats, *J. Appl. Physiol.* 77: 679-683.

18. Pierrefiche, O., A.S. Foutz, and M. Denavit-Saubié, 1993, Effects of $GABA_B$ receptor agonists and antagonists on the bulbar respiratory network in cat, *Brain Res.* 605: 77-84.

19. Schmid, K., G. Böhmer, and K. Gebauer, 1989, $GABA_B$ receptor mediated effects on central respiratory system and their antagonism by phaclofen, *Neurosci. Lett.*, 305-310.

20. Yamada, K.A., P. Hamosh, and R.A. Gillis, 1981, Respiratory depression produced by activation of GABA receptors in hindbrain of cat, *J. Appl. Physiol.* 51: 1278-1286.

8

RETROTRAPEZOID NUCLEUS (RTN) METABOTROPIC GLUTAMATE RECEPTORS AND LONG-TERM STIMULATION OF VENTILATORY OUTPUT

RTN Glutamate Receptors and Breathing

Eugene E. Nattie and Aihua Li

Department of Physiology; 706E Borwell Bldg
Dartmouth Medical School
Lebanon, New Hampshire 03756-0001

THE VENTROLATERAL MEDULLA

The Intermediate Area

The importance to breathing of neurons in the ventrolateral medulla (VLM) was first shown by studies that manipulated the environment of the surface of the VLM. For example, cooling probes applied to a specific area on the surface of the VLM can profoundly depress ventilatory output and CO_2 sensitivity in anesthetized (4, 9, 10, 24) and conscious animals (7). The investigators named this surface site the intermediate area because it lay between two adjacent surface areas associated with central chemosensitivity. They proposed it to be integrative in function (4, 9, 10, 24).

Anatomy

Recent work has begun to identify possible anatomical substrates for the intermediate area. Pharmacological manipulation of the rostral aspect of the ventral respiratory group (VRG) can affect respiratory output (see 14, 18, 20 for references). However, these VRG neurons are located 1-2 mm distant to the surface of the VLM and surface cooling is unlikely to alter their function. Lesions of the retrofacial nucleus (RFN), which also is located 1-2 mm deep to the surface of the VLM, depress respiratory output and central chemosensitivity (12). Some RFN neurons, labeled by intracellular injection, have dendrites that reach to the ventral surface (23). Surface cooling could well affect such neurons via these dendrites. There are neurons, possibly involved in the control of breathing, with cell soma lying easily within reach of the low temperatures produced by surface cooling probes placed at the

Modeling and Control of Ventilation, Edited by S. J. G. Semple, L. Adams, and B. J. Whipp
Plenum Press, New York, 1995

39

intermediate area. They are in a region with boundaries medially at the edge of the pyramidal tract extending laterally for 1-2 mm and rostrally at the rostral end of the facial nucleus extending caudally to the level of the rostral aspect of the inferior olive. The predominant anatomical structure within this region is the retrotrapezoid nucleus (6, 13-20, 22, 26). Portions of other structures are also present. These include the nucleus paragigantocellularis lateralis, the parapyramidal aspect of the raphé system, and dendrites from more dorsally located respiratory neurons of the retrofacial nucleus (8, 12, 23, 27). We shall refer to this region, most of which lies ventral and ventromedial to the facial nucleus, as the retrotrapezoid nucleus (RTN) in the following discussion. Some neurons affected by the studies described herein could be included within these other anatomically defined structures.

THE RETROTRAPEZOID NUCLEUS (RTN)

Anatomic Location

The RTN was identified simultaneously by two separate and quite different studies. The first was a retrograde tracing experiment with the initial tracer injections made into the dorsal and ventral respiratory groups. Among the sites identified was a novel location for respiratory control neurons ventral and ventromedial to the facial nucleus (26). In the second study, lesions of the region ventral and ventromedial to the facial nucleus decreased phrenic nerve activity, often to apnea, and virtually abolished the response to increased CO_2. These electrolytic and chemical lesions were unilateral and circumscribed and they were effective in anesthetized and decerebrate cats (15, 17) and in anesthetized rats (19).

Single Units; Physiology

RTN neurons have physiological connections to other traditional groups of respiratory neurons (2) and some RTN single units have respiratory related activity patterns (6, 13, 22). These units, along with some tonic units, increased their firing rates with systemic CO_2 stimulation (13). The production of a focal acidosis within the RTN by a small (one nl) injection of acetazolamide increased phrenic nerve activity while systemic PCO_2 was held constant. This indicated the presence of central chemoreception in this region (5).

Cholinergic Mechanisms

Microinjection studies using muscarinic cholinergic antagonists have shown that RTN neurons containing muscarinic receptor subtypes M1 and M3 appear to be involved in the control of breathing (21); autoradiography has shown the presence of the receptors within the region (16).

GLUTAMATE IN THE RTN

Many investigators have used the microinjection technique with glutamate to examine the role of the rostral ventrolateral medulla in the control of breathing. These studies, which included the RTN, the subretrofacial region, the retrofacial region, and rostral aspects of the VRG, have reported variable effects on respiratory output. This variability probably reflects small but important differences in injection location, volume, and dose (see 14, 18, 20 for discussion and references). The studies reported below focus on the role of glutamate in the RTN.

Ionotropic Antagonists

Here we discuss only results for unilateral injections made into the RTN region. This was determined by postmortem identification of the location of the fluorescent beads that accompanied each injection. For the most part the effects were on the amplitude of the integrated phrenic nerve signal; effects on the duration of the phrenic cycle were more variable and less impressive as were effects on blood pressure. In anesthetized cats (14, 18) and rats (20), unilateral injection (10 nl) of glutamate ionotropic receptor antagonists decreased the amplitude of the integrated phrenic nerve signal and diminished the response to increased PCO_2. In the cat, the effects were dose dependent. The efficacy was non-N-methyl-D-aspartate (NMDA) antagonist, and at high doses NMDA antagonist, 6-cyano-7-nitroquinoxaline-2,3-dione (CNQX)> NMDA antagonist 2-amino-5-phos-phonopentanoic acid (AP5)> nonspecific antagonist kynurenic acid. The doses of CNQX and AP5 were 1 and 10 mM, those for kynurenic acid, 100 and 250 mM. In the rat (20), a significant decrease in the amplitude of the integrated phrenic signal resulted from 10 nl injection of kynurenic acid (100 mM), AP5 (10 mM), and CNQX (10 mM). There was no significant difference between CNQX and AP5 treatments or between either CNQX or AP5 and kynurenate.

Ionotropic Agonists

Unilateral 10 nl injection of glutamate into the cat RTN increased phrenic amplitude in a dose dependent manner; 10 mM had no effect, 100 mM increased phrenic activity, 1 M decreased it. The effect of the 100 mM dose was attenuated by prior injection at the same site of kynurenic acid (18). We concluded that the 1 M dose resulted in significant regions of depolarization block. The observation of decreased firing rates for a small number of RTN units recorded at the center of the injection site supported this interpretation. The 100 mM dose effects were interpreted to reflect a predominant stimulation of RTN units that excite respiratory output, presumably the same units that are inhibited by glutamate receptor antagonists. Unexpectedly, the duration of the glutamate effect varied with the injection duration. Brief injections (3 msec) produced effects lasting a few minutes; longer injections (60 sec) resulted in effects lasting 46+/-9 (SEM) min. A small number of RTN single units showed a prolonged firing rate increase following the 60 sec glutamate injections. In the rat (20), 10 mM glutamate injections (made over 3 secs) increased integrated phrenic activity, as did 50 mM and 100 mM injections. Following the 50 and 100 mM injections, the effect was significant within the first 15 to 30 secs; 18/21 responses were back to baseline within 5 mins and 3 lasted for 15-20 mins. The effect of the 100 mM concentration was significantly greater than the 50 mM and the 10 mM dose (P>0.001). The increase in integrated phrenic activity following RTN glutamate injection was significantly diminished following injection of a mixture of AP5 (10 mM) and CNQX (10 mM). This injection was made at the exact same site using a double barrelled pipette. For 60 sec duration glutamate injections (10 nl), the 50 and 100 mM doses increased the amplitude of the integrated phrenic activity for 49 +/- 11 (SEM) min, as in the cat.

In five experiments (Fig. 1), we made two 100 mM glutamate injections of the same volume (10 nl) at the same exact site via a double barrelled pipette; one barrel connected to a pressure injection system for 3 sec injections, the other connected to a Nanopump for 60 sec injections. For the 3 sec injections, all 5 responses were back to within 11% of the initial baseline value by 5 mins. For the 60 sec injections, two were back to baseline levels by 5 and 10 mins, 1 by 30 mins, 1 by 60 mins, and 1 by 90 mins.

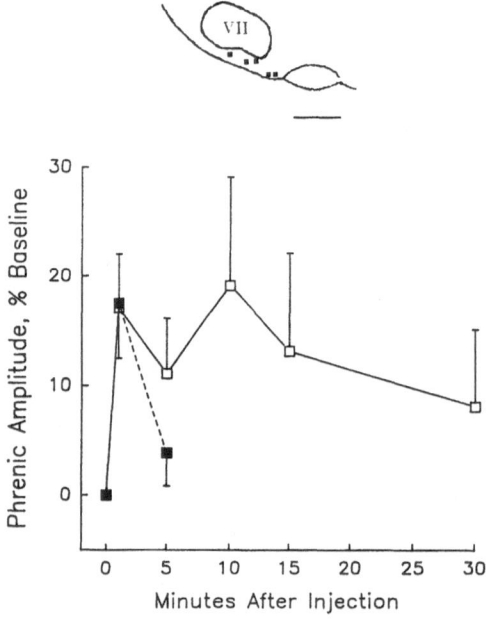

Figure 1. The phrenic amplitude responses (mean +/- SEM) to 5 RTN glutamate injections made either over 3 sec (solid squares) or 60 sec (open squares). The injections were made at the same exact site via double barrelled pipette. At the top, the locations of the center of these injections are shown. They were at 0.22, 0.78, 0.80, 0.86, 0.90 of the rostral to caudal length of the facial nucleus. (From Nattie and Li, 1995, with permission of the American Physiological Society).

Metabotropic Agonists

Glutamate receptors are now divided into two general classes: 1) ionotropic, which are ligand gated ion channels, and 2) metabotropic, which produce effects via G proteins and other cellular messengers (1, 25). It has been suggested that the ionotropic glutamate receptors produce rapid responses while the metabotropic receptors are involved in slower effects (1, 25). We hypothesized that the prolonged (60 sec) injection duration might be stimulating metabotropic glutamate receptors thus accounting for the prolonged phrenic response. As an initial test of this hypothesis, we evaluated the response to RTN injection of a metabotropic receptor agonist, ACPD (1S, 3R-aminocyclopentane-dicarboxylic acid)

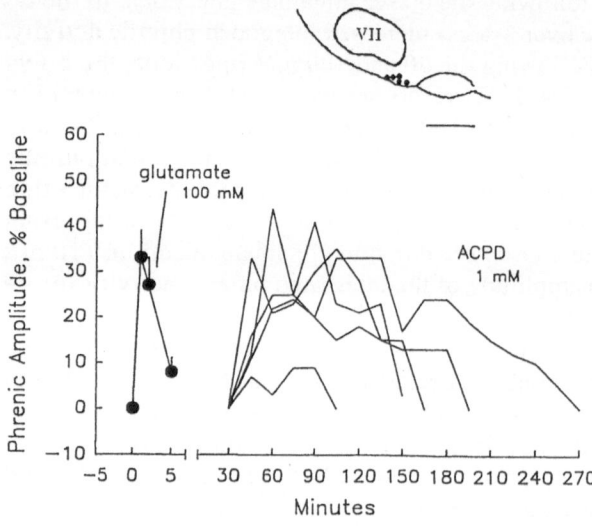

Figure 2. The integrated phrenic amplitude responses, expressed as % baseline, to unilateral 10 nl injection at the same RTN site of 100 mM glutamate (beginning at time = 0, n = 5, solid lines and circles, mean +/- SEM) and 1 mM APCD (beginning at time = 30 min, n = 5, solid lines). All injections were made over 3 secs. At the top, the locations of the center of each injection are shown as a fraction of the rostral to caudal length of the facial nucleus; at 0.22, 0.64, 0.64, 0.73, 0.83. (From Nattie and Li, 1995, with permission of the American Physiological Society).

(1, 25). Using double barrelled pipettes, we first identified an RTN injection site at which glutamate (100mM) injected over 3 secs produced a short-lived increase in the amplitude of the integrated phrenic signal. We then injected 10 nl of ACPD (1 mM) over 3 secs resulting in the responses shown in Fig. 2.

We identified RTN sites physiologically by brief responses to glutamate injection (100 mM) and anatomically by subsequent histological localization. Injection at these sites of the metabotropic glutamate receptor agonist ACPD at a concentration two orders of magnitude lower than the glutamate injection produced stimulation of phrenic activity lasting for 75 to 240 mins.

FUNCTIONAL SIGNIFICANCE OF THE RTN

Endogenous Glutamate; Unknown Afferents

Our results from injection of ionotropic glutamate receptor antagonists indicate that 1) glutamate is released within the RTN region from an endogenous source, 2) this glutamate maintains a tonic stimulation of the respiratory control system, and 3) NMDA receptors are involved in the function of these neurons. The issue of non-NMDA receptor involvement remains an open one. The source of the endogenous glutamate in the RTN region is unknown; the afferents which synapse within the RTN region have yet to be described. Afferents to the nearby juxtafacial portion of the nucleus paragigantocellularis lateralis, which lies contiguous to the RTN, arise from many sources (27) suggesting an integrative role for these neurons. It seems likely that a similar broad based source of afferents will be shown to exist within the RTN region involved with the control of breathing.

Long Term Stimulation

Injection into the RTN of the metabotropic agonist, ACPD, at a dose 100 times lower than necessary for reproducible glutamate effects on phrenic activity, did result in a long lasting stimulation of phrenic activity. This response was similar to that observed following the injection of glutamate (100 mM) provided the glutamate injection was made over a period of 60 secs. The mechanism of the long lasting response to glutamate remains uncertain. We hypothesize that a long duration glutamate injection results in an enhanced availability of glutamate that stimulates metabotropic receptors accounting for the prolonged effect. The details of this process are uncertain. RTN neurons that provide a tonic drive to the respiratory control system could be stimulated to fire for these long durations. Or, the effectiveness of afferent information, presumed to be processed in the RTN, could be enhanced after metabotropic receptor stimulation as is known to occur in the hippocampus (3). This suggests a type of long term potentiation of respiratory afferent activity might be occurring within the RTN. Such a phenomenon has been described for respiration following repeated short duration stimulation of carotid sinus nerve afferents (11).

Role of the RTN: Summary

Respiratory related neurons, present within the RTN, have connections to the dorsal and ventral respiratory groups. RTN neurons provide a tonic drive to the respiratory system and are involved in the response to central chemoreceptor stimulation. These effects involve M3 and/or M1 muscarinic cholinergic receptors and NMDA and, perhaps, non-NMDA ionotropic glutamate receptors. The sources of the afferents that release these neurotransmitters are unknown. Respiratory related and tonically firing RTN neurons are stimulated by systemic hypercapnia and, as determined by the response to focal acidosis, the RTN is one of many central chemoreceptor locations. The potent effects on chemosensitivity of RTN glutamate and acetylcholine receptor blockade and of RTN destruction suggest that: 1) quantitatively important central chemoreceptors are present within the RTN and/or 2) the activity of RTN neurons with glutamate and cholinergic afferent input is needed to allow full expression of the CO_2 response. It appears, based on the response to ACPD injection, that metabotropic glutamate receptors are present on RTN neurons involved in the control of breathing. The significance of this is uncertain at present although the long lasting phrenic response to RTN ACPD injection is reminiscent of the long term potentiation like response to repeated bouts of carotid sinus nerve stimulation. It seems likely that the effects on respiration cause by VLM surface cooling at the intermediate area are due, at least in part, by depressed function of RTN neurons.

ACKNOWLEDGMENTS

The work was supported by NIH grant HL 28066. The assistance of Brian Sites in parts of the data analysis is gratefully acknowledged.

REFERENCES

1. Baskys, A. Metabotropic receptors and 'slow' excitatory actions of glutamate agonists in the hippocampus. Trends Neurosci. 15: 92-95, 1992.
2. Bongianni, F., M. Corda, G.A. Fontana, and T. Pantaleo. Reciprocal connections between rostral ventrolateral medulla and inspiration-related medullary areas in the cat. Brain Res. 565: 171-174, 1991.
3. Bortolotto, Z. A., and G. L. Collingridge. Characterization of LTP induced by the activation of glutamate metabotropic receptors in area C1 of the hippocampus. Neuropharmacology 32: 109, 1993.
4. Bruce, E. N., and N. S. Cherniack. Central chemoreceptors (Brief Review). J. Appl. Physiol. 62: 389-402, 1987.
5. Coates, E. L., A. Li, and E. E. Nattie. Widespread sites of brainstem ventilatory chemoreceptors. J. Appl. Physiol. 75: 5-14, 1993.
6. Connelly, C. A., H. H. Ellenberger, and J. L. Feldman. Respiratory activity in retrotrapezoid nucleus in cat. Am. J. Physiol. (Lung Cell. Mol. Physiol. 2) 258: L33-L44, 1990.
7. Forster, H. V., P. J. Ohtake, L. G. Pan, T. F. Lowry, M. J. Korducki, E. A., Aaron, and A. L. Forster. Effects on breathing of ventrolateral medullary cooling in awake goats. J. Appl. Physiol. 78: (238-265), 1995.
8. Holtman, J. R., Jr., L. J. Marion, and D. F. Speck. Origin of serotonincontaining projections to the ventral respiratory group in the rat. Neurosci. 37: 541-532, 1990.
9. Loeschcke, H. H. Central chemosensitivity and the reaction theory. J. Physiol. 332: 1-24, 1982.
10. Millhorn, D. E., and F. L. Eldridge. Role of ventrolateral medulla in regulation of respiratory and cardiovascular systems. J. Appl. Physiol. 61:1249-1263, 1986.
11. Millhorn, D. E., F. L. Eldridge, and T. G. Waldrop. Prolonged stimulation of respiration by a new central neural mechanism. Respir. Physiol. 41: 87-103, 1980.
12. Nattie, E. E., C. Blanchford, and A. Li. Retrofacial lesions: Effects on CO_2 sensitive phrenic and sympathetic activity. J. Appl. Physiol. 73: 1317-1325, 1992.
13. Nattie, E.E., M. L. Fung, A. Li, and W. M. St. John. Responses of respiratory modulated and tonic units in the retrotrapezoid nucleus to CO_2. Respir. Physiol. 94: 35-50, 1994.
14. Nattie, E. E., M. Gdovin, and A. Li. Retrotrapezoid nucleus glutamate receptors: control of CO_2 sensitive phrenic and sympathetic output. J. Appl. Physiol. 74: 2958-2968, 1993.
15. Nattie, E. E., and A. Li. Fluorescent location of RVLM kainate microinjections that alter the control of breathing. J. App. Physiol. 68: 1157-1166, 1990a.
16. Nattie, E. E., and A. Li. Ventral medulla sites of muscarinic receptor subtypes involved in cardiorespiratory control. J. Appl. Physiol. 69: 33-41, 1990b.
17. Nattie, E. E., and A. Li. Lesions in the retrotrapezoid nucleus decrease ventilatory output in anesthetized or decerebrate cats. J. Appl. Physiol. 71: 1364-1375, 1991.

18. Nattie, E. E., and A. Li. Retrotrapezoid nucleus glutamate injections: long-term stimulation of phrenic activity. J. Appl. Physiol. 76: 760-772, 1994a.
19. Nattie, E. E., and A. Li. Retrotrapezoid nucleus lesions decrease phrenic activity and CO_2 sensitivity in rats. Respir. Physiol. 97: 63-77, 1994b.
20. Nattie, E.E., and A. Li. Rat retrotrapezoid nucleus iono- and metabotropic glutamate receptors and the control of breathing. J. Appl. Physiol. 78: (153-163), 1995.
21. Nattie, E.E., A. Li, J. Mills, and Q. Hwang. Retrotrapezoid nucleus muscarinic receptor subtypes localized by autoradiography. Respir. Physiol. 96: 189-197, 1994.
22. Pearce, R., R. Stornetta, and P. Guyenet. Retrotrapezoid nucleus in the rat. Neurosci. Letters 101: 138-142, 1989.
23. Pilowsky, P. M., C. Jiang, and J. Lipski. An intracellular study of respiratory neurons in the rostral ventrolateral medulla of the rat and their relationship to catecholamine-containing neurons. J. Comp. Neurol. 259: 1388-1395, 1990.
24. Schläfke, M. E., W. R. See, and H. H. Loeschcke. Ventilatory response to alterations of H^+ - ion concentration in small areas of the ventral medullary surface. Respir. Physiol. 10: 198-212, 1970.
25. Schoepp, D., and P. J. Conn. Metabotropic glutamate receptors in brain function and pathology. Trends in Pharm. Sci. 14: 13-20, 1993.
26. Smith, J. C., D. E. Morrison, H. H. Ellenberger, M. R. Otto, and J. L. Feldman. Brainstem projections to the major respiratory neuron populations in the medulla of the cat. J. Comp. Neurol. 281: 69-96, 1989.
27. Van Bockstaele, E.J., H. Akaoka, and G. Aston-Jones. Brainstem afferents to the rostral (juxtafacial) nucleus paragigantocellularis: integration of exteroceptive and interoceptive sensory inputs in the ventral tegmentum. Brain Res. 603: 1-18, 1993.

9

EXPRESSION OF C-FOS IN THE BRAIN STEM OF RATS DURING HYPERCAPNIA

L. J. Teppema,[1] J. G. Veening,[2] and A. Berkenbosch[1]

[1] Department of Physiology
University of Leiden
P.O. Box 9604, 2300 RC Leiden, The Netherlands
[2] Department of Anatomy and Embryology
University of Nijmegen
The Netherlands

INTRODUCTION

Expression of the proto-oncogene c-fos is widely used as a marker of metabolic activation of individual neurones (1,2,12). Expression of this immediate early gene results in the production of the nuclear protein FOS, which forms a heterodimer complex with JUN, the protein product of the immediate early gene c-jun. By acting as a third messenger, the FOS-JUN complex regulates the expression of specific target genes in various types of cells. It is thought that in this way FOS and the products of other early response genes contribute to the genetically determined functional adaptation of the cell to altered stimulus environments (7,15). Various hormones, growth factors and neurotransmitters, as well as depolarisation and voltage gated Ca^{2+} entry into the cell have been shown to induce a rapid but transitional transcription of the c-fos gene in various types of neurones (6,8). Once FOS is produced in sufficient amounts depending on the duration and intensity of the activating stimulus, a FOS-like immunoreactivity (FOS-LI) can be demonstrated in the nuclei of these cells.

The precise neuro-anatomical pathways involved in the processing of afferent input from both the peripheral and central chemoreceptors is largely unknown. Mapping of individual neuronal activity in the brain using FOS immunohistochemistry could be a suitable tool to gain more information about the central respiratory neuronal network. Recently, Erickson and Millhorn (4) and Sato et al. (13) have studied the pattern of FOS-LI in the brain stem of rats after exposure to hypoxia and hypercapnia respectively. Particularly after exposure to hypercapnia application of FOS immunohystochemistry could be useful in identifying the neuro-anatomical sites belonging to the central chemoreflex loop. Sato et al (13) found that exposing rats to 13 and 15% CO_2 resulted in FOS-LI in superficial ventrolateral medullary regions analogous to the classical chemosensitive regions as described for the cat (14). However, we have found that exposure of cats to 10% CO_2 resulted

Modeling and Control of Ventilation, Edited by S. J. G. Semple, L. Adams, and B. J. Whipp
Plenum Press, New York, 1995

in a large concentration of cells immunoreactive to FOS in the retrotrapezoid nucleus, while the remaining parts of the chemosensitive areas were nearly devoid of labelled cells (see 19). In the present study we have re-investigated the effect of hypercapnia (15%) on the pattern of FOS-LI in the brain stem of rats. In the ventrolateral medulla, the largest amount of labelled cells was found in the lateral paragigantocellular nucleus, at a level starting just caudally from the facial nucleus, and extending caudorostrally to the level of the nucleus of the trapezoid bodies.

METHODS

The experiments were performed in ten adult Wistar rats (220-300 g). Five animals were subjected to inhalation of 15% CO_2 in air for two hours. Five control animals were placed in the same flow-through chamber but breathed air, also for two hours. Oxygen and carbon dioxide tensions in the chamber were measured continuously with an O_2 analyser and an infrared CO_2 analyser respectively. Immediately after the exposures, the animals were anaesthetized with pentobarbital and with a mixture of fluanison and fentanyl, and perfused transcardially with heparinized saline at 4°C, followed by 4% paraformaldehyde in 0.1 M Phosphate buffered saline PBS (pH 7.3). The brain stem was removed , postfixed for several hours and stored overnight in 30% saccharose in 0.1 M PBS. Coronal (60 μm) sections were made from the medulla oblongata with a freezing microtome. The sections were then incubated overnight with FOS antibody (raised in sheep, OA-11-823 CRB; 1:10.000 in 0.1 M PBS containing 0.5% Triton X-100 and 1% bovine serum albumin (solution A); free floating immersion). Thereafter the sections were rinsed in PBS and incubated for one hour with donkey anti-sheep Immunoglobulin (Jackson; 1:400 in solution A). Finally, after washing, the sections were incubated for two hours in an ABC solution (1:800 in 0.1 M PBS). The FOS antibody-ABC complex was visualized using DAB to which Ni-ammoniumsulfate was added. The sections were mounted on slides and analyzed with light microscopy.

RESULTS

Control Animals

The only site in the medulla consistently showing immunoreactive neurons was the dorsal cochlear nucleus. Regions sometimes containing weakly labelled cells were the nucleus tractus solitarius (NTS), the medial vestibular nuclei, nucleus raphe pallidus and the spinal nucleus of the trigeminal nerve.

Hypercapnic Animals

In the dorsal medulla a large amount of cells immunoreactive to FOS was present within the gelatinous, commissural and medial subnuclei of the nucleus tractus solitarius. Most cells were found at the level of, and caudally to area postrema. In the ventral medulla, the following regions showed a consistent pattern of FOS-LI: 1) a longitudinal cell column within the ventral part of the medullary reticular nucleus, approximately 2 mm laterally from the midline, extending from the pyramidal decussation to the level of the caudal limits of the lateral part of the paragigantocellular nucleus. Possibly this cell column, which was located very close to the ventral part of the nucleus ambiguus, corresponds with the C1/A1 adrenergic/noradrenergic cell group. The distance of the immunoreactive cell group to the ventral medullary surface was 300-400μm; 2) within the lateral part of the paragigantocellular nucleus: a column of many labelled cells extending caudorostrally approximately from the level of the rostral pole of the inferior olive to the caudal limits of the nuclei of the trapezoid bodies. Caudally, most of these immunoreactive cells were located at a depth varying between 250 and 500 μm, approximately 1.5 mm from the midline. Rostrally, at a level where the facial nucleus occupies its largest cros-sectional area (levels 56-55 in Swanson (18)), the labelled cells were located medially to the facial nuclei, on average at a larger distance from the ventral surface (0.1-1mm), but closer to the midline, immediately laterally from the pyramids.

    In fig 1 an example is presented of neurons showing FOS-LI at this level. In the ventrolateral medulla, our sections also contained a few labelled nuclei of very superficial

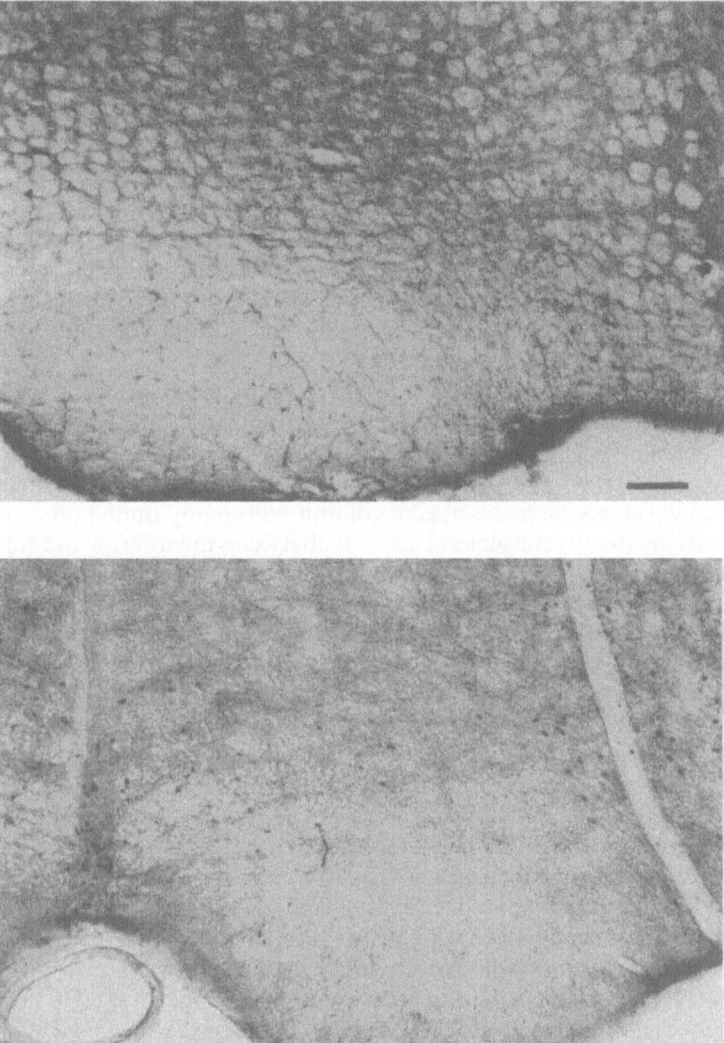

Figure 1. FOS-LI in a hypercapnic rat (bottom photograph) compared with a control animal (top). The level of section is approximately level 56 in Swanson (18). The hypercapnic animal shows cells with FOS-LI both in the ventral raphe nuclei as well as laterally from the pyramid in the lateral paragigantocellular nucleus. Bar: 72 μm.

cells, but in the present study the majority of FOS-LI was found in neurons located deeper and more dispersed. 3) the nuclei raphe pallidus and magnus consistently contained more labelled cells than control animals did.

DISCUSSION

The locations of neurons with FOS-LI within the commissural and (dorso)medial subnuclei within the NTS in the present study correspond roughly with those described by Erickson and Millhorn (4) after hypoxia and carotid sinus nerve (CSN) stimulation. Most labelled cells were present at and behind the obex, in the same areas where chemoreceptor afferents terminate (17). Further studies are needed to investigate whether these immunoreactive neurons within the NTS are specifically involved in respiratory or in cardiovascular regulation, or in both.

In rats, hypoxia can result in hypotension (e.g. see 4), so that vaso(pre)motoneurones may be activated. It is possible that the number and distribution pattern of immunoreactive neurons within the NTS differ during hypoxia and hypercapnia. This is currently under investigation in our lab.

A column of cells with FOS-LI in the ventrolateral medulla, corresponding to the A1/C1 adrenergic and noradrenergic cell groups, was also found by Erickson and Millhorn (4) after hypoxia and CSN stimulation, but not by Sato et al (13) after hypercapnia. A possible explanation may be that the latter authors exposed their rats to hypercapnia only for one hour. It is known that the time course of c-fos expression not only depends on the strength of the applied stimulus, but also on its duration (3).

The column of labelled cells that we found in the lateral part of the paragigantocellular nucleus was neither described by Erickson and Millhorn (4) after hypoxia, nor by Sato et al (13) after hypercapnia. The latter authors described a longitudinal column of immunoreactive neurons located very close to or even lining the ventral surface only. In cats exposed to hypercapnia, however, we have descibed a column with many immunoreactive cells at a comparable level in the rostroventrolateral medulla, between the inferior and superior olives (19). These cells with FOS-LI were concentrated in the retrotrapezoid nucleus (RTN), which has been shown to possess anatomical connections with the NTS and with the ventral respiratory group (16), and the integrity of which is necessary for a normal hypercapnic ventilatory response in the cat (10). The region within the lateral paragigantocellular nucleus where we found the labelled neurons in the present hypercapnic rats is probably at least partly analogous to the RTN in cats. Furthermore, the location of the immunoreactive cells corresponds with the region where Pearce et al (11) found tracer molecules which were retrogradely transported from the ventral respiratory group, and where they recorded neurons with a respiratory rhythm. Recently, Nattie and Li (9) have shown in the rat that lesions in the same area reduce phrenic activity and CO_2 sensitivity.

It is possible that the labelled neurons within or in the vicinity of the RTN that we found in the present hypercapnic rats specifically respond to hypercapnia. However, the fact that they are apparently not activated during hypoxia (4; also personal observation) does not mean that they represent central chemoreceptors. To our knowledge, within the central nervous system as yet FOS-LI has been demonstrated in second - and higher order neurons only. Additional studies are needed to prove that a given neuron showing immunoreactivity to FOS after exposure to hypercapnia, is a chemoreceptor cell. A further complication arising with both hypoxia and hypercapnia is that their physiological effects are not limited to the respiratory system. For example, both activate the hypothalamo-pituitary-adrenal axis and elicit vasomotor responses. The high CO_2 concentration (>10%) which apparently is needed to obtain FOS-LI in the rat medulla (13), is remarkable. Further studies are needed to document the quantitative relationship between FOS-LI in the brain stem and the concentration of CO_2 in the inspiratory air, as well as the inluence of the duration of the hypercapnic exposure on this relationship.

The present study showed that neurons in the raphe pallidus and raphe magnus nuclei were activated during hypercapnia. Because the same nuclei are also activated during hypoxia (4; also personal observation), they may play an integrative role in the responses to hypercapnia and hypoxia. The possible significance of both raphe nuclei in the control of breathing has been the subject of several recent studies (e.g. see 5).

Although interpretation of data obtained from studies using FOS immunohistochemistry is sometimes difficult, the technique could be a valuable tool to gain more insight into the neuro-anatomical substrates for the peripheral and central chemoreflex loops. By combining FOS studies with injection of tracers and by applying double-immunostaining techniques, the physiological role and neurochemical properties of activated pathways can be investigated. This should be the objective of future studies.

REFERENCES

1. Bullit, E. Expression of c-fos-like protein as a marker for neuronal activity following noxious stimulation in the rat. *J. Comp. Neurol.* 296: 517-530, 1990.
2. Ceccatelli, S., M.J. Villar, M. Goldstein and T. Hökfelt. Expression of c-Fos immunoreactivity in transmitter characterized neurons after stress. *Proc Natl. Acad. Sci. USA* 86: 9569-9573, 1989.
3. Dragunow, M. and R. Faull. The use of c-fos as a metabolic marker in neuronal pathway tracing. *J. Neurosci. Meth.* 29: 261-265, 1989.
4. Erickson, J.E. and D.E. Millhorn. Fos-like protein is induced in neurons of the medulla oblongata after stimulation of the carotid sinus nerve in awake and anesthetized rats. *Brain Res.* 567: 11-24, 1991.
5. Lindsey, B.G., Y.M. Hernandez, K.F. Morris and R. Shannon. Functional connectivity between midline brain stem neurons with respiratory modulated firing rates. *J. Neurophysiol.* 67: 890-904, 1992.
6. Morgan, J.I. and T. Curran. Role of ion flux in the control of c-fos expression. *Nature* 322: 552-555, 1986.
7. Morgan, J.I. and T.Curran. Stimulus-transcription coupling in neurons: role of cellular immediate-early genes. *TINS* 12: 459-462, 1989.
8. Morgan, J.I. and T. Curran. Proto-oncogene transcription factors and epilepsy. *TIPS* 12: 343-349, 1991.
9. Nattie, E.E and A. Li. Retrotrapezoid nucleus lesions decrease phrenic activity and CO_2 sensitivity in rats. *Respir. Physiol.* 97: 63-77, 1994.
10. Nattie, E.E., A. Li and W.M. St. John. Lesions in retrotrapezoid nucleus decrease ventilatory output in anesthetized or decerebrate cats. *J. Appl.Physiol.* 71: 1364-1375, 1991.
11. Pearce, R.A., R.L. Stornetta and P. Guyenet. Retrotrapezoid nucleus in the rat. *Neurosci. Lett.* 101: 138-142, 1989.
12. Sagar, S.M., F. R. Sharp and T. Curran. Expression of c-fos protein in brain: metabolic mapping at the cellular level. *Science* 240: 1328-1331, 1988.
13. Sato, M., J. W. Severinghaus and A.I. Bausbaum. Medullary CO_2 neuron identification by c-fos immunocytochemistry. *J. Appl. Physiol.* 73: 96-100, 1992.
14. Schläfke, M.E. Central chemosensitivity: a respiratory drive. *Rev.Physiol.Biochem.Pharmacol.* 90: 171-244, 1981.
15. Sheng, M. and M.E. Greenberg. The regulation and function of c-fos and other immediate early genes in the nervous system. *Neuron* 4: 477-485, 1990.
16. Smith, J.C., D.E. Morrison, H.H. Ellenberger, M.R. Otto and J.L. Feldman. Brainstem projections to the major respiratory neuron populations in the medulla of the cat. *J. Comp. Neurol.* 281: 69-96, 1989.
17. Spyer, K.M. Central nervous mechanisms contributing to cardiovascular control. *J. Physiol. (London)* 474: 1-19, 1994.
18. Swanson, L.W. Brain Maps: Structure of the Rat Brain. Elsevier, Amsterdam 1992.
19. Teppema, L.J., A. Berkenbosch, J.G. Veening and C.N. Olievier. Hypercapnia induces c-fos expression in neurons of retrotrapezoid nucleus in cats. *Brain Res.* 635: 353-356, 1994.

10

TWO DISTINCT DESCENDING INPUTS TO THE CRICOTHYROID MOTONEURON IN THE MEDULLA ORIGINATING FROM THE AMYGDALA AND THE LATERAL HYPOTHALAMIC AREA

H. Arita, I. Kita, and M. Sakamoto

Department of Physiology
Toho University School of Medicine
5-21-14 Omori-nishi, Ota-ku, Tokyo 305, Japan

INTRODUCTION

The purpose of this study was to obtain functional and morphological information concerning neuronal connections between the respiratory center in the ventral medulla and the areas of the limbic system such as the amygdala and the hypothalamus.

It is well established that the ventral medulla possesses components essential to respiration. For example, it contains a neuronal network generating the central respiratory rhythm, bulbospinal premotor neurons projecting to inspiratory or expiratory motoneurons, and vagal motoneurons innervating the upper airway (8). In addition, the rostral ventrolateral medulla plays an important role in providing a tonic drive for the central respiratory rhythm generator (5,6) as well as serving as an integration site for a variety of inputs from the peripheral receptors (1).

However, respiration is known to be influenced by emotion. Vocalization, an emotional expression, involves a coordinated motor activity of the expiratory pump muscles and laryngeal muscles, which interrupts ongoing respiratory rhythm. This interruption suggests direct or indirect connections between the medullary respiratory-related neurons and the limbic system which is concerned with emotional expressions.

Such neuronal connections have been evaluated in this study by focusing on the activity of the cricothyroid (CT) muscle of the larynx. Note that CT motoneurons are located in the nucleus ambiguus of the ventral medulla (2). The CT muscle has respiratory as well as behavioral functions. For example, the CT muscle is active only in inspiration during anesthesia and sleep, when its activity is primarily related to the dilation of the vocal cords during inspiration. In contrast, the CT muscle becomes active in expiration during wakeful-

Modeling and Control of Ventilation, Edited by S. J. G. Semple, L. Adams, and B. J. Whipp
Plenum Press, New York, 1995

ness, when it provides tonic activity to the vocal cords so as to produce phonation or vocalization (7).

Taking the respiratory and behavioral functions of the CT muscle into consideration, the present study focused on how descending inputs originating in the limbic system modify the inspiratory CT activity generated in the medulla.

MORPHOLOGICAL STUDY

In the morphological experiment, we used a retrograde labelling technique with unconjugated cholera toxin subunit B (UCTB). The UCTB injection was made into the nucleus ambiguus and an adjacent area in the rat. For this purpose, we used a two-barrel glass micropipette: one barrel was filled with 1% solution of UCTB and the other was filled with 3M KCl for unit recording. The micropipette was inserted from the dorsal surface toward the nucleus ambiguus through the foramen magnum in an anesthetized (sodium pentobarbital i.p.) rat. When we identified a site where inspiratory activity was recorded, we injected 3 μl of UCTB solution by means of a picopump. The incisions in the skin and the resected parts of the muscles were then sutured. Seventy-two hours after the UCTB injection, the brain was removed for immunohistochemistry.

The use of UCTB immunohistochemistry has already been reported in our previous study (2). In brief, serial transverse sections of the brain were incubated in goat antiserum to UCTB, then in biotylated rabbit antigoat immunoglobulin and finally in avidin-biotin conjugate with horseradish peroxidase.

Retrogradely labelled cells were found in a variety of regions within the limbic system, primarily ipsilaterally. At the posterior hypothalamic level, retrogradely labelled somata were present in the area lateral to the fornix throughout the entire longitudinal extent of the lateral hypothalamic area (LH), predominantly ipsilateral (Fig. 1). In the amygdala, a considerable amount of labelled cells were packed in the medial subdivision of the central nucleus (CEA). Retrogradely labelled cells were also present in the paraventricular hypothalamic nucleus and in the perirhinal cortex. At the anterior hypothalamic level, the labelled somata were detected in the ipsilateral bed nucleus of the stria terminalis.

We also examined the injection site of UCTB in the medulla, showing that the UCTB solution was spread in and around the nucleus ambiguus of the ventral medulla. We realized that the injection site was not restricted to the nucleus ambiguus.

Horst et al. (10) have previously reported a descending pathway from the hypothalamus to the ventral medulla of the rat, i.e., the hypothalamo-medullary pathway, using horseradish peroxidase. This hypothalamo-medullary pathway is supported in this study. Furthermore, the present results reveal a direct descending pathway from the central nucleus of the amygdala to the ventral medulla, namely the amygdalo-medullary pathway, independent from the hypothalamo-medullary pathway. These morphological findings indicate that there are at least two distinct descending pathways from the limbic system to the ventral medulla. Based on this morphological evidence, we then performed physiological experiments to clarify how these two distinct descending pathways act on or interrupt the respiratory activity generated in the medulla.

FUNCTIONAL STUDY

In the physiological experiments, we evaluated the effect on CT activity of electrical stimulation at the lateral hypothalamic area (LH) or the central nucleus of the amygdala in an anesthetized (dial-urethane i.p.) rat. It has been well established that the amygdala and

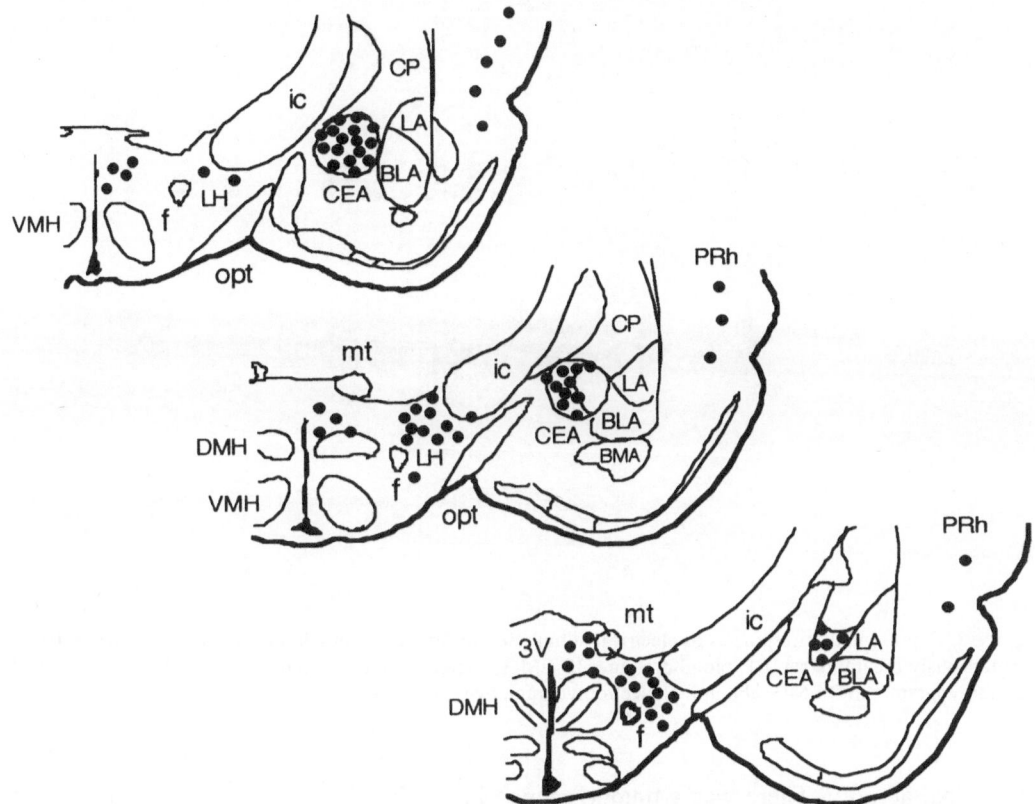

Figure 1. Schematic drawings of transverse sections in the brain, illustrating the distributions of cells retrogradely labelled following injection of cholera toxin B at the nucleus ambiguus and the adjacent region of the ventral medulla in the rat. One symbol represents five labelled cells per section. 3V; 3rd ventricle; BLA, basolateral amygdaloid nucleus; CEA, central nucleus of amygdala; CP, caudate putamen; DMH, dorsomedial hypothalamic nucleus; f, fornix; ic, internal capsule; LA, lateral amygdaloid nucleus; LH, lateral hypothalamic nucleus; mt, mammillothalamic tract; opt, optic tract; PRh, perirhinal cortex; VMH, ventromedial hypothalamic nucleus.

the hypothalamus are responsible for autonomic, somatic and behavioral responses associated with emotion (8). Thus, autonomic responses were examined by measuring blood pressure, heart rate, and intercostal EMG in this experiment. The behavioral response associated with vocalization was evaluated by measuring abdominal EMG and intercostal EMG along with CT EMG. Facial expressions were observed directly.

Figure 2 shows typical responses to electrical stimulation in LH that consists of constant current (100µ) square-wave (0.1 msec in duration) train pulses at the rate of 50Hz for 15 seconds. The LH stimulation evoked a decrease in blood pressure accompanied by bradycardia, indicating an excitation of the parasympathetic nervous system. The responses were essentially the same as those reported previously by Spencer et al. (9), who made a microinjection of L-glutamate in the LH of the rat. Thus evidence suggests that the electrical stimulation applied in this study acts on the cell bodies rather than on the passing fibers located in LH. The change in facial expression was characterized by closure of eyelids and tearing.

As for the respiratory activity, the tonic electrical stimulation of LH produced an enhanced inspiratory activity of the intercostal EMG accompanied by a recruitment of

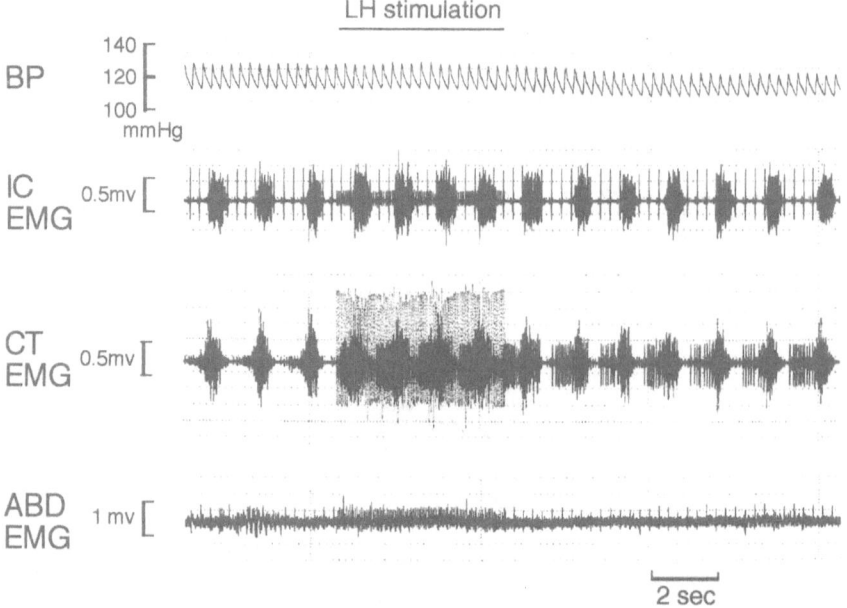

Figure 2. Representative responses to electrical stimulation of lateral hypothalamic area (LH) in anesthetized, spontaneously breathing rat. BP, blood pressure; IC EMG, intercostal electromyogram; CT EMG, cricothyroid electromyogram; ABD EMG, abdominal electromyogram. See text for details.

abdominal activity. There was a unique change in CT EMG, as follows. CT activity was observed only in the inspiratory phase before LH stimulation, as was expected during spontaneous breathing under anesthesia (7). LH stimulation evoked a recruitment of expiratory unit, whose activity was restricted to the late expiratory phase. The inspiratory CT activity was preserved during LH stimulation. The waveform of the recruited expiratory unit was significantly different from that of the ongoing inspiratory unit. This indicates that the CT muscle consists of two distinct motor units controlled by distinguished neuronal systems. In a previous study (2), we have demonstrated morphological evidence for a dual innervation of CT muscle in terms of the locations of the motoneurons and the axonal pathways. The present data provide physiological evidence for the dual innervation in terms of afferent inputs from respiratory or behavioral control system.

The coordinated motor activities producing vocalization are generally described as activation of expiratory pump muscles and enhanced activity of inspiratory pump muscles along with a development of tension in the vocal cords. This stereotypic manifestation has been previously shown in conscious animals that produce vocalization in response to chemical stimulation of the periaqueductal gray (3). All these characteristic changes were observed during LH stimulation in this study, though the animal was under anesthesia.

Responses to electrical stimulation in the amygdala (Fig. 3) were completely different from those to LH stimulation described above. The amygdala stimulation evoked an increase in blood pressure, indicating an excitation of the sympathetic nervous system. The changes in facial expression were also different, i.e., wide opening of the eyelids with protrusion of the eyeballs. The respiratory pump muscles also showed distinct responses, characterized by a decrease in the inspiratory activity and an increase in the expiratory duration, namely an expiratory shift. The CT response was also distinguished, as follows. The stimulation of the amygdala evoked a recruitment of the expiratory unit, whose activity occurred predomi-

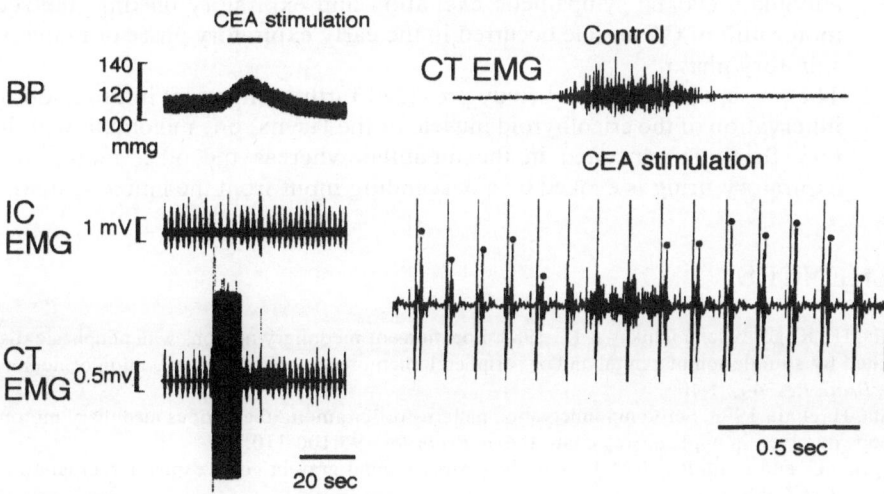

Figure 3. Representative responses to electrical stimulation in the amygdala. The stimulation consisted of constant current (100μA) train pulses at the rate of 10 Hz for 10 seconds: the stimulation period is shown at the top in the left panel (CEA stimulation). BP, blood pressure; IC EMG, intercostal electromyogram; CT EMG, cricothyroid electromyogram. Right panel shows CT EMG recorded at a faster paper speed before (control) and during the stimulation (CEA stimulation). Vertical lines on the CT EMG trace exhibit artifacts due to the stimulation, and each evoked response is pointed by a filled circle. See text for details.

nantly in the early expiratory phase or the post-inspiratory phase, rather than in the late expiratory phase as described above in the case of LH stimulation. It seems unlikely that such coordinated motor activities of respiratory pump muscles and CT muscle produce vocalization. The post-inspiratory activity of CT muscle is considered to brake the expiratory flow, indicating a gentle and prolonged expiration. This type of response might be related to flight or fear rather than fight.

We made a systematic search for the sites where the above responses were elicited. The effective sites were identified in the medial part of the amygdala (1.5-2 mm lateral to the midline and 2.5-3.2 mm posterior to Bregma). A variety of distinct responses were observed in the neighboring parts including several amygdaloid nuclei located laterally or ventrally.

Another important point of the present results is that, although the electrical stimulation was given in a tonic fashion, the evoked response in the CT activity was phasic or gated to the expiratory phase. This implies that the descending inputs from the limbic system act indirectly on the CT motoneurons in the ventral medulla, probably mediated by some interneurons with a respiratory gate mechanism.

SUMMARY

1. The retrograde labelling study revealed that there are at least two independent descending pathways from the limbic system to the ventral medulla, i.e., the hypothalamo-medullary pathway and the amygdalo-medullary pathway.
2. The stimulation in the lateral hypothalamic area produced parasympathetic excitation and vocalization response: the recruited motor unit of CT muscle occurred in the late expiratory phase. By contrast, the stimulation at the medial part of the

amygdala evoked sympathetic excitation and expiratory braking: the recruited motor unit of CT muscle occurred in the early expiratory phase or in the post-inspiratory phase.

3. The present physiological study provided further important information on dual innervation of the cricothyroid muscle of the larynx: one motor unit with inspiratory firing is generated in the medulla, whereas the other motor unit with expiratory firing is evoked by a descending input from the limbic system.

REFERENCES

1. Arita, H., Kogo, N. and Ichikawa, K., 1988, Locations of medullary neurons with nonphasic discharges excited by stimulation of central and/or peripheral chemoreceptors and by activation of nociceptors in cat, *Brain Res.* 442:1-10.
2. Arita, H. et al., 1993, Serotonin innervation patterns differ among the various medullary motoneuronal groups involved in upper airway control, *Exp. Brain Res.* 95:100-110.
3. Jurgens, U. and Pratt, R., 1979, Role of the periaqueductal gray in vocal expression of emotion, *Brain Res.* 167:367-378.
4. Isaacson, R.L., 1974, The limbic system, Plenum Press, New Tork.
5. Kita, I., Sakamoto, M. and Arita, H., 1994, Adrenergic cell group in rostral ventrolateral medulla of cat: its correlation with central chemoreceptors, *Neurosci. Res.* 20:265-274.
6. Loeschcke, H.H., 1982, Central chemosensitivity and the reaction theory, *J. Physiol. Lond.* 332:1-24.
7. Mathew, O.P. et al., 1988, Respiratory activity of the cricothyroid muscle, *Ann. Otol. Rhinol. Laryngol.* 97:680-687.
8. Richter, D.W., 1982, Generation and maintenance of the respiratory rhythm, *J. Exp. Biol.* 100:93-107.
9. Spencer, S.E., Sawyer, W.B. and Loewy, A.D., 1989, Cardiovascular effects produced by L-glutamate stimulation of the lateral hypothalamic area. *Am. J. Physiol.* 257:H540-H552.
10. ter Horst, G.J., Luiten, P.G.M. and Kuipers, F., 1984, Descending pathways from hypothalamus to dorsal motor vagus and ambiguus nuclei in the rat, *J. Auton. Nerv. Syst.* 11:59-75.

11

TRIGEMINAL MOTOR NUCLEUS AND PONTILE RESPIRATORY REGULATION

M. Pokorski and H. Gromysz

Department of Neurophysiology
Polish Academy of Sciences Medical Research Center
00-784 Warsaw, 3 Dworkowa St., Poland

INTRODUCTION

The pontile mechanisms underlying the inspiratory off-switch are not understood in full. Recently, a role in shaping the respiratory timing has been suggested for the expiratory neuronal pool of the motor nucleus of the trigeminal nerve (NVmt) in the rabbit (3). In the present study we set out to further scrutinize the apparently novel respiratory role of the NVmt in another species, the cat. We hypothesized that if the NVmt played a role in terminating the central inspiratory activity then (i) its removal ought to extend the inspiratory phase leading to an apneustic pattern of respiration and (ii) its function could be influenced by the arterial chemoreceptor input, an important regulator of the timing mechanism. We addressed these problems by comparing the effects on neural inspiration (T_I) and expiration (T_E) of a block of the NVmt and the nucleus parabrachialis medialis (NPBM) - hitherto the most considered locus of the off-switch and of arterial chemoreceptor excitation by cyanide before and during the block of each structure.

METHODS

Eleven cats of either sex anesthetized with Nembutal (35 mg/kg, i.p.) were used. The cats were tracheotomized, vagotomized, paralyzed, and artificially ventilated with room air. The right C_4 root of the phrenic nerve was cut and placed on bipolar recording electrodes. The sinus nerve was cut on the same side. The details of the surgical approach to the dorsal brain stem surface have been described previously (3). The stereotaxic search for the pontile nuclei was based on Berman's atlas (2).

Tracheal PCO_2 and phrenic neurogram from which T_I and T_E were measured were recorded simultaneously. Arterial blood pressure was monitored. The reversible block of each nucleus was achieved with ~0.5 µL of a 2% xylocaine and Pontamine Sky Blue mix extruded from the tip of a penetrating glass microelectrode by applying manual pressure.

Modeling and Control of Ventilation, Edited by S. J. G. Semple, L. Adams, and B. J. Whipp
Plenum Press, New York, 1995

The animal was sacrificed by perfusion of the brain with 4% formaldehyde. The brain stem was removed and stored for histological examination of the microelectrode tip's position.

The block of either nucleus was always unilateral, on the side contralateral to the phrenic and sinus nerve sections.

The protocol consisted of measuring the T_I and T_E changes in response to the block of the NVmt and NPBM and to cyanide injection (100 μg i.v. bolus of NaCN) before and on a background of the block. Data were presented as mean ±SE percentage changes over the control values. The corresponding pre- and postblock mean values were compared with a t-test.

RESULTS

Effect of Block of the Trigeminal Motor Nucleus on Neural Respiration

An original record of the respiratory effects of xylocaine block of the NVmt is shown in Fig. 1 A (upper half). The block resulted in a 4-fold prolongation of T_I with nearly unchanged T_E under otherwise stable conditions. The pattern of apneustic respiration developed with an accompanied decrease of the peak phrenic amplitude. The dye traces in this cat's brain stem (not shown) were found in the central part of the NVmt. On average, the block prolonged T_I by 265 ± 44% (Table 1), which was ~12 times more than the concurrent small lengthening of T_E.

Effect of Block of the Nucleus Parabrachialis Medialis on Neural Respiration

An illustration of the respiratory effects of block of the NPBM is shown in Fig.1 B (lower half). The block caused, in contrast to that of the NVmt, rather modest and about equal prolongations of both T_I and T_E; 1.62-fold and 1.55-fold, respectively. Simultaneous and similar changes of both phases of the respiratory cycle are also evident when one compares the average values shown in Table 1. Such changes do not conform to Lumsden's, classic definition of apneustic respiration (4) but rather represent a pattern of slow respiration.

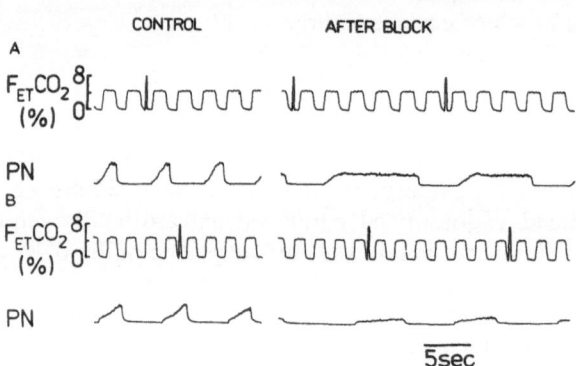

Figure 1. Experimental recordings showing the effects of xylocaine block of the trigeminal motor nucleus (Panel A) and the nucleus parabrachialis medialis (Panel B) in two separate cats. Traces from top in each horizontal panel: end-tidal fractional CO_2 concentration ($F_{ET}CO_2$) with automatic calibration marks and phrenic neurogram (PN).

Table 1. Changes in neural inspiratory time (T_I), expiratory time (T_E), and peak phrenic amplitude (A_P) due to xylocaine block of the trigeminal motor nucleus (NVmt) and the nucleus parabrachialis medialis (NPBM), and to cyanide injection before and during the block of each structure

| | Pre-block n = 11 | Post-block | | | |
| | | NVmt (n = 7) | | NPBM (n = 4) | |
	CN^-	Block alone	CN^-	Block alone	CN^-
T_I (Δ %)	7 ± 4	$265 \pm 44^{*\dagger}$	$84 \pm 18^{*\dagger}$	$49 \pm 12^*$	-9 ± 16
T_E (Δ %)	2 ± 6	$31 \pm 10^*$	-2 ± 5	$41 \pm 11^*$	3 ± 32
A_P (Δ %)	$63 \pm 8^*$	$-32 \pm 6^*$	$46 \pm 11^*$	$-36 \pm 11^*$	$84 \pm 24^*$

Values are mean ± SE percentage changes over the corresponding control state, which is the preceding state in each case.
*P<0.05 vs. the preceding state.
†P<0.05 for the comparison of T_I and T_E changes in each state.

Despite the differential changes in the length of T_I and T_E in response to NVmt and NPBM blocks, a common feature for both nuclei was a sharp decline in the peak height of the phrenic discharge.

Effect of Arterial Chemoreceptor Excitation on Neural Respiration

The effects of cyanide injection in each experimental setting are depicted in Table 1. In the control state, arterial chemoreceptor activation had a variable and, on average, inappreciable effects on both T_I and T_E. The main response was one of an increased peak phrenic amplitude by ~60%.

Chemoexcitation performed during the block of the NVmt caused a significant, further prolongation of inspiratory holdings by 84 ±18% with no changes in T_E and thus potentiated the apneustic type of respiration. The amplitude response was dampened.

In the setting of NPBM block, the cyanide's effect matched that of the control state, i.e., consisted of variable and small T_I and T_E changes with the dominating increase in phrenic amplitude.

DISCUSSION

We assessed the role of the trigeminal motor nucleus that is made up chiefly of expiratory-related neurons (5) in the pneumotaxic mechanism in the cat. We found that inactivation of this nucleus advanced the phrenic discharge and delayed its termination with a nearly unchanged expiratory pause. The inference is that the neurons of the nucleus might be responsible for the inspiratory off-switch. The selective prolongation of the inspiratory phase underlies the apneustic pattern of respiration (4). This pattern was distinctly linked with the NVmt, as was the prolongation of both T_I and T_E with the NPBM. The latter change is likely to represent a mere slowing of respiration. Further prolongation of the apneustic inspiratory holding due to the arterial chemoreceptor excitation during the NVmt block strengthens the suggestion that the expiratory neurons of this nucleus are concerned with the termination of inspiration. When these neurons are inactivated, the inspiratory phase is prolonged. The results therefore point to the NVmt as a structure involved with the respiratory phase-switching. The results also stress the difference between the apneustic and slow respiration, based on T_E changes. Indeed, the lack of T_E evaluation in some previous work seems confounding (1,6).

This study has limitations. The respiratory effects of xylocaine might be due at least partly to its inhibitory action on membrane ionic fluxes, and therefore the results might be different if other blocking agents were used. The tissue spread of xylocaine has not been assessed and is largely unknown. The NPBM and the NVmt lie in the vicinity, therefore some overlap of this spread onto the neighboring area during block of each structure is likely. Since these limitations would apply to the block of both structures, they are unlikely to have been responsible for the differential respiratory effects observed. Finally, xylocaine applied to the NVmt could block the passing fibers on their way down from the NPBM to the medullary centers. But then the block of the very perikarya of these axons in the NPBM would be expected to have even greater apneustic-like changes, which was not the case.

Despite these limitations we believe we have shown that the trigeminal motor nucleus has a part in the inspiratory off-switch effect. The nature of the interaction between this nucleus and the medial parabrachial nuclear complex requires further exploration.

REFERENCES

1. Berger, A.J., D.A. Herbert, and R.A. Mitchell. Properties of apneusis produced by reversible cold block of the rostral pons. Respir. Physiol. 33: 323-337, 1978.
2. Berman, A.L. The brain stem of the cat. Madison, WI: Univ. of Wisconsin Press, 1968.
3. Gromysz, H., W.A. Karczewski, and U. Jernajczyk. Motor nucleus of the Vth nerve and the control of breathing (Breuer-Hering reflexes and apneustic breathing). Acta Physiol. Pol. 41: 4-6, 1990.
4. Lumsden, T. Observations on the respiratory centres in the cat. J. Physiol. London 57: 153-160, 1923.
5. St. John, W.M., and T.A. Bledsoe. Comparison of respiratory-related trigeminal, hypoglossal and phrenic activities. Respir. Physiol. 62: 61-78, 1985.
6. St. John, W.M., R.L. Glasser, and R.A. King. Apneustic breathing after vagotomy in cats with chronic pneumotaxic center lesions. Respir. Physiol. 12: 239-250.

12

AXON BRANCHING OF MEDULLARY EXPIRATORY NEURONS IN THE SACRAL SPINAL CORD OF THE CAT

S.-I. Sasaki[1] and H. Uchino[2]

[1] Department of Physiology
[2] Department of Anesthesiology
Tokyo Medical College
6-1-1, Shinjuku, Shinjuku-ku, Tokyo 160, Japan

INTRODUCTION

It has been described that bulbospinal expiratory neurons are located in the region of the caudal nucleus retroambigualis (7,9). Expiratory (E) neurons in the caudal nucleus retroambigualis have axon collaterals in the upper lumbar segments (8), and they have respiratory synaptic effects on internal intercostal motoneurons (6) and abdominal motoneurons (9). Our previous study has shown that the majority of E neurons extend their descending spinal axons to the lower lumbar (L6-L7) and the sacral spinal cord (S1-S3) (12). Anatomical study has shown the projection from the caudal nucleus retroambigualis to the nucleus of Onuf using autoradiographic techniques (5). In the present study, we investigated the location of axon collaterals in the sacral spinal gray matter of single E neurons. For this purpose, a microstimulation technique was employed and the distribution of effective sites of antidromic activation in axon collaterals was mapped (12,13,14).

METHODS

Experiments were performed on adult cats, anesthetized with sodium pentobarbital. Animals were paralyzed by intravenous administration of pancuronium bromide and kept on artificial ventilation, which was adjusted to maintain the end-expired CO_2 level between 4-6%. The phrenic nerve was dissected free, ligated, and cut distally. In three experiments, after removal of the gluteus muscle, the pudendal nerve was exposed and dissected free and prepared for stimulation to identify the location of the nucleus of Onuf (11,15,16). The sacral spinal cord was exposed by a laminectomy. The dura was opened, mineral oil pools were formed over the exposed brainstem and spinal cords. Dorsal roots were cut to expose the lateral surface of the spinal cord. Glass micropipettes filled with 2 M NaCl solution saturated with Fast Green FCF dye were used for extracellular recordings of single E neurons. In order

Modeling and Control of Ventilation, Edited by S. J. G. Semple, L. Adams, and B. J. Whipp
Plenum Press, New York, 1995

to map effective sites of antidromic activation in axon collaterals, glass insulated tungsten stimulating microelectrodes (exposed tip diameter: 5-10 μm and tip length: 10-20 μm) were used. The duration of cathodal stimulus pulses was 0.15 ms. To search for the rostrocaudal distribution of axon collaterals belonging to single E neurons, the spinal gray matter was microstimulated at 100 μm intervals during advancement of the electrode from the dorsal to the ventral sites with a current of 150-250 μA, and each stimulus track was separated by 1 mm or 500 μm along the spinal cord rostrocaudally. To determine the detailed trajectory of axon collaterals, systematic mapping was performed in and around axon collaterals at a matrix of points 100 or 200 μm apart in the mediolateral direction and 500 μm in the rostrocaudal direction along the spinal cord using a stimulus current under 50 μA in steps of 100 μm on stimulus tracks from the dorsal to the ventral sites, and the trajectory of axon collaterals was reconstructed on the basis of the location of low-threshold foci and the latency of antidromic spikes. A stimulus microelectrode was connected to the recording amplifier at each stimulus track and the location of the nucleus of Onuf was identified by antidromic field potentials following stimulation of the ipsilateral pudendal nerve.

RESULTS

Unit spikes were recorded extracellulary from 45 E neurons in the caudal nucleus retroambigualis. E neurons were characterized on the basis the expiratory discharges during respiratory cycles (7,8,12,13). The most caudal level of spinal projection of descending spinal axons was examined in the sacral segments by antidromic activations following stimulation of the surface of the ventrolateral funiculus of the spinal cord with a small Ag-AgCl ball electrode. Nine E neurons with descending stem axons in the sacral spinal cord were selected for further analysis.

An example of axon collaterals in sacral segment is shown in Fig. 1. In this E neuron, microstimulation was performed from S1 and S2 segments. Axon collaterals were found in the dorsal region to the nucleus of Onuf at the rostral S1. The descending axon terminated at the S3 spinal level.

S1 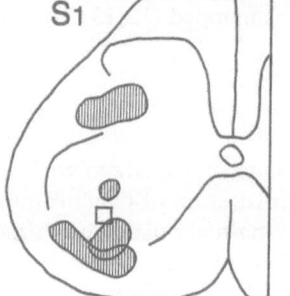   S2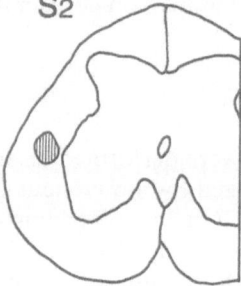

Figure 1. An example of results of electrode tracking of antidromic activation in the sacral spinal cord. A: dorsal view of distribution of effective sites. Filled circles indicate stimulus tracks where antidromic activation was elicited. B: latencies of antidromic spikes relative to the distance along the spinal cord. Open triangles and circles indicate the latencies in the stem axon and the axonal branches, respectively. Open circles below the abscissa show the location of stimulus tracks where antidromic spikes were evoked in axonal branches.

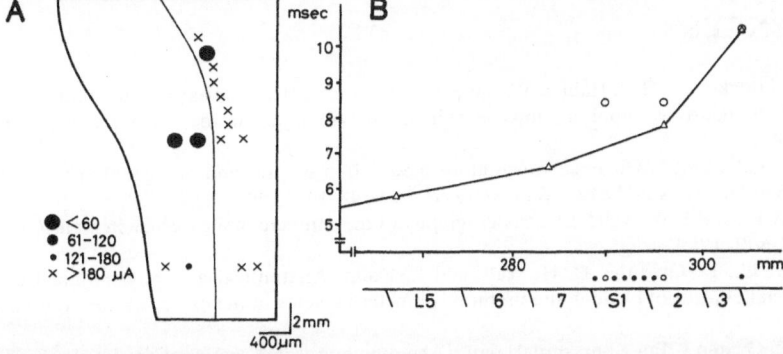

Figure 2. Summary of axon collaterals in S1 and S2 segments. Locations of axon collaterals are shown by hatched areas.

E neurons gave collaterals in the nucleus of Onuf (n=1), the region dorsal to the nucleus of Onuf (n=1), the intermediate region in S1 (n=1), the lateral funiculus at the spinal level of the caudal S1/rostral S2 segments (n=1) and in the marginal region between the lateral funiculus and the gray matter at the caudal S1 spinal level (n=1). In the remaining 4 E neurons, no positive evidence for axon arborizations in the sacral gray matter and in the ventrolateral funiculus was obtained, although descending stem axons could be localized in the ventrolateral funiculus.

DISCUSSION

The present results are the first electrophysiological demonstration that respiratory neurons in the brainstem project to the sacral regions involved in the autonomic functions. Axon collaterals in the nucleus of Onuf were found in one of 9 E neurons, but axon collaterals were limited to the region of the nucleus. Axon collaterals were also found in the region dorsal to the nucleus of Onuf and in the lateral funiculus. It is likely that some axon collaterals of E neurons terminate on the dendrites of motoneurons, since morphological study has shown that the dendrites of motoneurons in the nucleus of Onuf extended in a dorso-lateral direction (15). In an electromyographic study of the sphincter ani externus in man, it has been shown that raised intra-abdominal pressure is accompanied by an increase in sphincter tone (3). These results might support in part that E neurons play a functional role in to co-ordinating sphincter muscles and abdominal muscles during increased intra-abdominal pressure (4), since E neurons have axon collaterals in the upper lumbar (9,13) and the sacral spinal cord.

The existence of a temporal relationship between the activity of sympathetic nerves and that of phrenic nerves has been described (1,2). This evidence indicates the existence of some descending spinal systems of coupling between the respiratory rhythm generator and the sympathetic nervous system. We found axon collaterals in the lateral border of the gray matter and the base of the dorsal horn in the sacral spinal cord. These projection areas seem to correspond with the regions of parasympathetic preganglionic neurons that were stained after the application of HRP to the pelvic nerve (10). The parasympathetic preganglionic neurons in the sacral spinal cord innervate various visceral organs and seem to integrate descending spinal inputs (4). Thus respiratory input may converge to parasympathetic preganglionic neurons.

REFERENCES

1. Boczek-Funcke, A., H.-J. Häbler, W. Jänig, and M. Michaelis. Respiratory modulation of the activity in sympathetic neurones supplying muscle, skin and pelvic organs in the cat. *J. Physiol. London* 449:333-361, 1992.

2. Darnall, R.A., and P. Guyenet. Respiratory modulation of pre- and postganglionic lumbar vasomotor sympathetic neurons in the rat. *Neurosci. Lett.* 119:148-152, 1990.

3. Floyd, W.E., and E.W. Walls. Electromyography of the sphincter ani externus in man. *J. Physiol. London* 122:599-609, 1953.

4. Holstege, G., D. Griffiths, D. H. Wall, and E. Dalm. Anatomical and physiological observations on supraspinal control of bladder and urethral sphincter muscles in the cat. *J. Comp. Neurol.* 250:449-461, 1986.

5. Holstege, G., and J. Tan. Supraspinal control of motoneurons innervating the striated muscles of the pelvic floor including urethral and anal sphincters in the cat. *Brain.* 110 :1323-1344, 1987.

6. Kirkwood, P.A., and T.A. Sears. Monosynaptic excitation of thoracic expiratory motoneurones from lateral respiratory neurones in the medulla of the cat. *J. Physiol. London* 234:87-89, 1973.

7. Merrill, E.G. The lateral respiratory neurones of the medulla: Their associations with nucleus ambiguus, nucleus retroambigualis, the spinal accessory nucleus and the spinal cord. *Brain Res.* 24:11-28, 1970.

8. Merrill, E.G. Finding a respiratory function for the medullary respiratory neurons. In:*Essays on the Nervous System,* edited by R. Bellairs, and E.G. Gray. Oxford, UK, Clarendon, 1974, p. 451-486.

9. Miller, A.D., K. Ezure, and I. Suzuki. Control of abdominal muscles by brain stem respiratory neurons in the cat. *J. Neurophysiol.* 54:155-167, 1985.

10. Nadelhaft, I., W.C. de Groat, and C. Morgan. Location and morphology of parasympathetic preganglionic neurons in the sacral spinal cord of the cat revealed by retrograde axonal transport of horseradish peroxidase. *J. Comp. Neurol.* 193:265-281, 1980.

11. Onuf, B. On the arrangement and function of the cell groups of the sacral region of the spinal cord in man. *Arch. Neurol. Psychopathol.* 3:387-412, 1900.

12. Sasaki, S.-I., H. Uchino, M. Imagawa, T. Miyake, and Y. Uchino. Lower lumbar branching of caudal medullary expiratory neurons of the cat. *Brain Res.* 553:159-162, 1991.

13. Sasaki, S.-I., H. Uchino, and Y. Uchino. Axon branching of medullary expiratory neurons in the lumbar and the sacral spinal cord of the cat. *Brain Res.* 648:229-238, 1994.

14. Shinoda, Y., T. Yamaguchi, and T. Futami. Multiple axon collaterals of single corticospinal axons in the cat spinal cord. *J. Neurophysiol.* 55:425-448, 1986.

15. Thor, K., C. Morgan, I. Nadelhaft, M. Houston, and W.C. de Groat. Organization of afferent and efferent pathways in the pudendal nerve of the cat. *J. Comp. Neurol.* 288:263-279, 1989.

16. Ueyama, T., N. Mizuno, S. Nomura, A. Konishi, K. Itoh, and H. Arakawa. Central distribution of afferent and efferent components of the pudendal nerve in cat. *J. Comp. Neurol.* 222:38-46,1984.

13

VAGAL COOLING AND THE ORIGIN OF PULMONARY REFLEXES IN CATS

C. P. M. van der Grinten, N. E. L. Meessen, and S. C. M. Luijendijk

Department of Pulmonology
University Hospital Maastricht
P.O. Box 5800, 6202 AZ Maastricht, The Netherlands

INTRODUCTION

Cooling the vagus nerve is often used to study respiratory reflexes mediated by pulmonary receptors (2,5). Its use is based on the finding that at low temperatures (e.g. 4°C) unmyelinated fibres still transmit impulses whereas transmission in myelinated fibres is fully blocked. This allows distinguishing reflex effects caused by C-fibres from those caused by slowly and rapidly adapting pulmonary receptors (SAR and RAR) (2). The aim of the present study was to investigate whether vagal cooling can also be used to distinguish between reflex effects caused by SAR and RAR. Cooling a nerve fibre increases its refractory period in the cooled area. During this period an afferent impulse in this fibre cannot pass the cold block. As a consequence, the frequency of impulses arriving at the brain stem respiratory centres is reduced. SAR have higher discharge frequencies than RAR. Therefore, the relative reduction in impulse frequency by cooling is larger for SAR than for RAR (5). Our hypothesis is that during vagal cooling the effects of stimulating SAR start to disappear at a higher temperature than the effects of stimulating RAR. In the present study two clearly distinct temperature ranges were identified. It is argued that in the first range (14-10°C) the reflex effects caused by SAR were strongly reduced and in the second (8-4°C) the reflex effects caused by RAR.

METHODS

Data for this investigation were collected from previous studies of ours in anaesthetized, spontaneously breathing cats (6,7). After initial anaesthesia with ketamine (10 mg/kg) cats (3.8-7.0 kg) were anaesthetized by intravenous injection of a mixture of urethane (12.5 mg/kg) and chloralose (62.5 mg/kg). Anaesthesia was maintained by hourly injections of this mixture using 10% of the initial dose. Cats were kept in the supine position on a warmed operating table. A tracheal tube was inserted just below the larynx. Both vagi were freed of their surrounding sheaths and carefully placed in the grooves of two cooling devices. Vagal

Modeling and Control of Ventilation, Edited by S. J. G. Semple, L. Adams, and B. J. Whipp
Plenum Press, New York, 1995

67

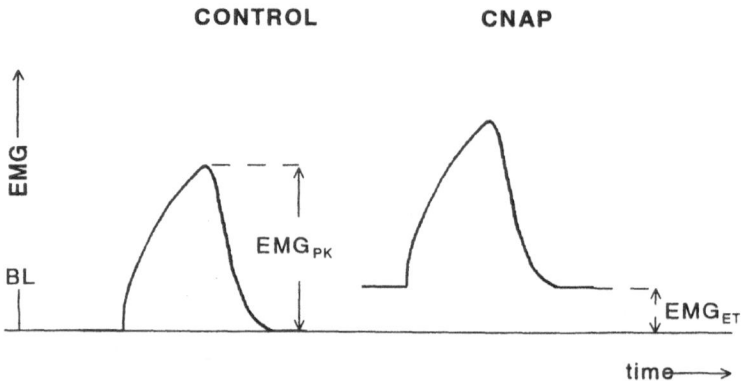

Figure 1. Schematic diagram showing how end-tidal inspiratory activity (ETIA) was calculated: ETIA = $EMG_{ET}/\overline{EMG}_{PK,CTRL}$ x 100% . EMG_{ET} is electrical activity at the end of expiration during CNAP. $\overline{EMG}_{PK,CTRL}$ is the average value of 5 consecutive breaths during control conditions. BL is base line.

temperature (T_v) could be set between 38 and 0 (±0.2) °C. A pair of electrodes was placed in the costal part of the diaphragm (Dia) and another pair in a parasternal intercostal muscle (ICM), aimed at recording inspiratory activity. The electrical activity from the electrodes was amplified, filtered between 150 and 3000 Hz, rectified and passed through a 'leaky' integrator with a time constant of 50 ms. The tracheal tube was connected to a breathing circuit containing a Fleisch-pneumotachograph for measuring air flow and allowing quick steps to positive and negative airway pressures as described previously (6).

All signals measured were sampled by a computer and stored on hard disk for off-line analysis. Altogether, ten signals were A/D converted: integrated activity of Dia and ICM, flow, tracheal and transpulmonary pressure, partial pressure of CO_2 in inspired and expired air, blood pressure, rectal temperature and the measured temperatures of the two cooling devices.

The following tests were used: steps to continuous positive airway pressure (CPAP, 0.9 kPa) mainly stimulating SAR, steps to continuous negative airway pressure (CNAP, -0.9 kPa) stimulating RAR and inhibiting SAR, intravenous injections of histamine-di-phosphate (300 µg) stimulating RAR and intravenous injections of capsaicin (50 µg/kg) stimulating C-fibres. The effects of these stimuli were evaluated at T_v of 37, 22, 14, 12, 10, 8, 6 and 4 °C. The parameters evaluated were: durations of inspiration (t_I) and expiration (t_E) determined from the flow pattern and end-tidal inspiratory activities (ETIA) of Dia and ICM. ETIA was expressed as a percentage of the amplitude of the EMG at the end of expiration divided by the average peak activity of 5 control breaths preceding the test (Fig. 1): ETIA = $EMG_{ET}/\overline{EMG}_{PK,CTRL}$ x 100% . This relative measure was chosen to correct for the inevitable changes in the quality of the EMG signal in the course of the day.

RESULTS

CNAP increased ETIA in Dia and ICM at all T_v tested though the amplitudes and significancies were much reduced at 6 and 4 °C (Fig. 2). Accordingly, the largest decrease in the effect of CNAP on ETIA was found between 8 and 4 °C. During control conditions and during CNAP cooling the vagal nerves increased both t_I and t_E (Fig. 3). The largest changes were found between 14 and 10 °C for t_I and between 8 and 6 °C for t_E.

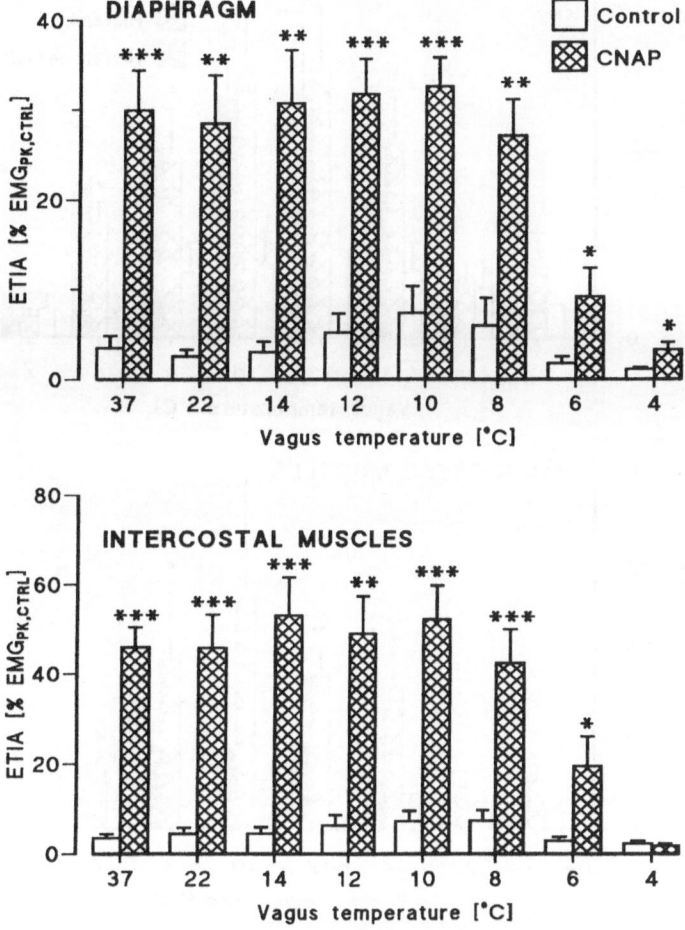

Figure 2. Effects of vagal cooling on ETIA in Dia and ICM in response to CNAP (mean ±SEM). Note that panels have different scalings. Asterisks indicate significant differences from control values obtained at the same T_v. (Student's paired t-test, n=9, *** P<0.001; ** P<0.01; * P<0.05) [*from Meessen et al. (6)*]

Intravenous injection of histamine increased ETIA in both Dia and ICM at all T_v but 4 °C (Fig. 4, cross hatched bars). Only small increases were found at $T_v = 37$ and 22 °C, whereas the largest increases were observed between 14 and 8 °C. ETIA diminished between 8 and 4 °C. Simultaneous application of CPAP-shortly after a histamine induced increase in EMG_{ET} was observed-reduced ETIA in Dia at $T_v = 37$ and 22 °C (Fig. 4, solid bars). At lower T_v this reduction was no longer found. In Dia at 8 °C and especially at 6 °C CPAP seemed to increase histamine-induced ETIA, which was clearly observed in some cats. The results for CPAP in ICM were less pronounced probably due to a load compensating mechanism in these muscles which is likely absent in Dia (4).

Intravenous injection of capsaicin induced apnoea followed by an increase in breathing frequency, which is the normal response to this drug at high doses (2). Nevertheless, capsaicin failed to induce ETIA in both Dia and ICM at all T_v applied.

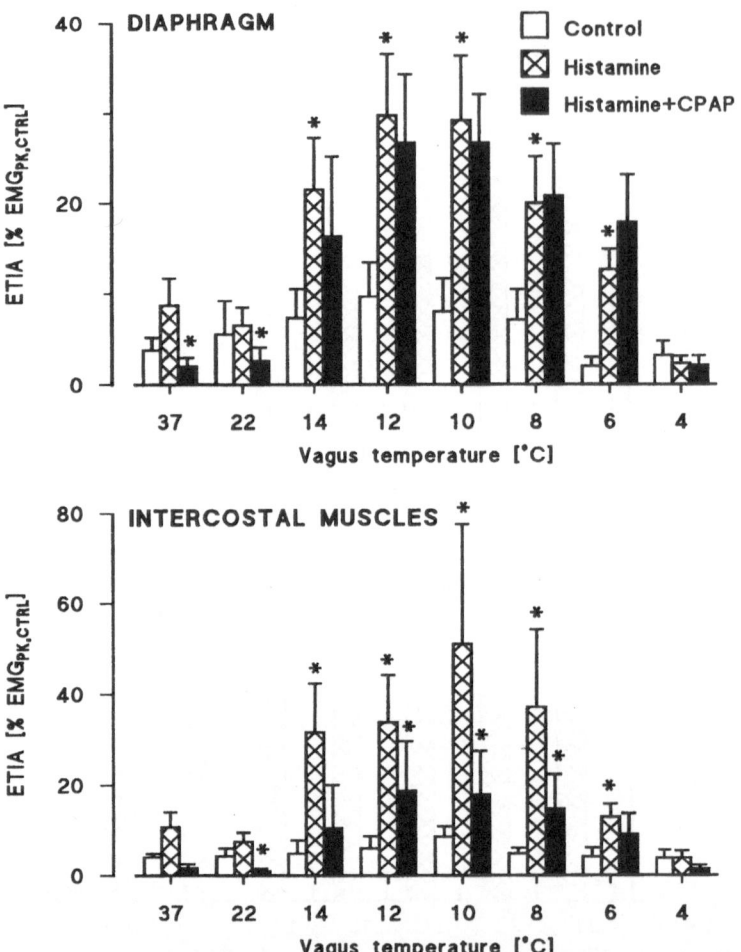

Figure 3. Effects of vagal cooling (mean ±SEM) on t_I and t_E during control and during CNAP. Asterisks indicate significant differences from the corresponding results obtained at 37 °C (Student's paired t-test, n=11, *** P<0.001; ** P<0.01; * P<0.05). [*from Meessen et al. (6)*]

DISCUSSION

In the vagus nerve three types of afferents are generally discerned: SAR, RAR and C-fibres. Since ETIA was not evoked by stimulation of C-fibres, not even by high doses of capsaicin, the results obtained for ETIA should be explained on the basis of the properties of SAR and RAR. From our results two clearly distinct temperature ranges can be discerned: the first between 14 and 10 °C and the second range between 8 and 4 °C, which are referred to as range I and range II, respectively.

Both histamine and CNAP stimulate RAR (1). CNAP simultaneously decreases the activity of SAR, whereas with histamine SAR activity is likely not changed from that during control conditions. Activity of SAR may even be increased due to the concomitant histamine-induced hyperinflation. The decrease in the effects of stimulation of RAR by CNAP and histamine on ETIA was found in range II (Figs. 2 and 4). T_E , which is at least partly

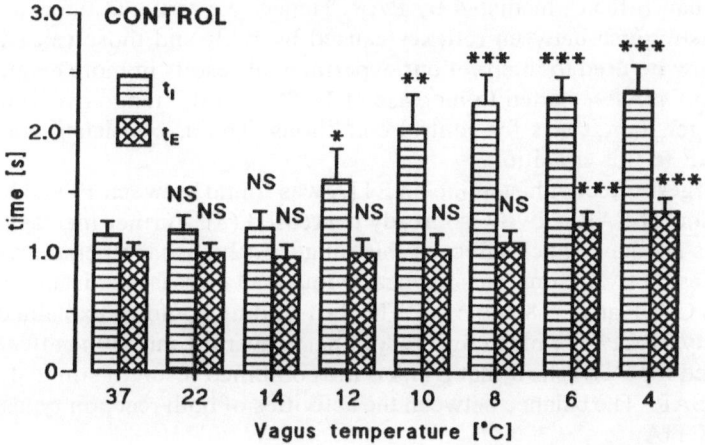

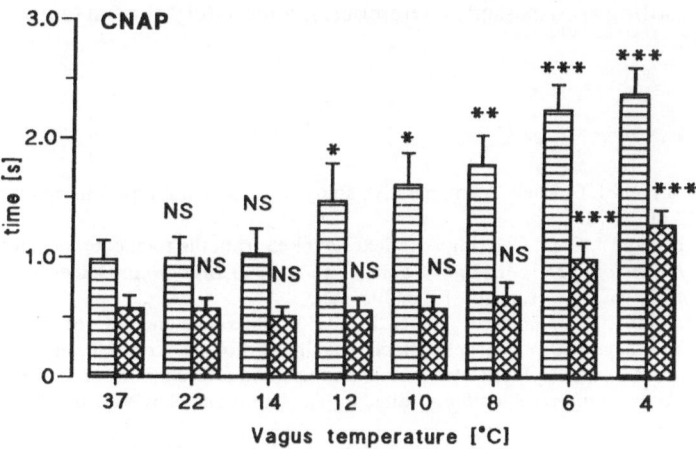

Figure 4. Effects of vagal cooling (mean ±SEM) on ETIA in Dia and ICM during control, after intravenous injection of histamine and during subsequent application of CPAP (Histamine + CPAP). Note that panels have different scalings. Asterisks for histamine bars and histamine + CPAP bars indicate significant differences from ETIA during control conditions and after histamine injection, respectively, obtained at the same T_v (Wilcoxon's test for paired observations, n=7, *** P<0.001; ** P<0.01; * P<0.05).

determined by RAR activity (2,3), also changed in range II. This was especially clear during CNAP where t_E was considerably reduced compared to control (Fig. 3, lower panel).

CPAP strongly stimulates SAR and to a smaller extend also RAR if changes in lung volume are large enough. CPAP suppressed the histamine induced increase in ETIA at 37 and 22 °C. This effect was diminished at 14 °C and completely disappeared at lower T_v (range I). During cooling t_I , which is mainly controlled by SAR activity (off-switch mechanism), also increased in range I.

The above discussed results show that reflex effects that may be ascribed to stimulation of SAR were attenuated in range I whereas the reflex effects originating from stimulation of RAR were attenuated in range II. This sequence is in agreement with our hypothesis, namely that during vagal cooling reflexes mediated by SAR would be affected at higher T_v than reflexes mediated by RAR. Hence, we arrive at the conclusion that it is possible to distinguish between reflexes caused by SAR and those caused by RAR. This finding can now be used to interpret our experimental results in more detail.

During CNAP t_I started to increase at 14 °C, but the range over which t_I increased was much larger than it was for control conditions. This is consistent with the decreased activity of SAR in this condition.

The largest effect of histamine on ETIA was found between 14 and 8 °C. At these T_v the transmission of SAR activity is already decreased (5). Further increasing SAR activity by CPAP does not have much effect in this range. At higher temperatures (22 and 37 °C) CPAP decreases ETIA. In some cats we clearly found an increase in histamine-induced ETIA in response to CPAP at T_v = 8 and 6 °C. This last finding could be explained by an increase in RAR activity by CPAP, which now predominates over the inhibiting effect of SAR. These results obtained for ETIA show that ETIA is the combined result of stimulation by RAR and inhibition by SAR. The balance between the activities of both receptor types determined the amplitude of ETIA.

The results discussed above demonstrated that vagal cooling can be a useful tool to study the origin of vagal reflexes. However, this tool can only be applied successfully if the effects of vagal cooling are assessed at a number of adequately chosen temperatures between 37 and 0 °C.

REFERENCES

1. Armstrong, D. J., and J. C. Luck. A comparative study of irritant and type J receptors in the cat. *Respir. Physiol.* 21: 47-60, 1974.
2. Coleridge, H. M., and J. C. G. Coleridge. Reflexes evoked from the tracheobronchial tree and lungs. In: *Handbook of Physiology. The respiratory System. Control of Breathing.* Bethesda, MD: Am. Physiol. Soc., 1986, sect. 3, vol. II, pt 1, chapt. 12, p. 395-429.
3. Davies, A., M. Dixon, D. Callanan, A. Huszcuk, J. G. Widdicombe and J. C. M. Wise. Lung reflexes in rabbits during pulmonary stretch receptor block by sulphur dioxide. *Respir. Physiol.* 34: 83-101, 1978.
4. Duron, B. Intercostal and diaphragmatic muscle endings and afferents. In: *Lung Biology in Health and Disease, Vol. 17: Regulation of Breathing*, edited by T.F. Hornbein, New York and Basel: Marcel Dekker, 1981, p. 473-540.
5. Jonzon, A., T. E. Pisarri, A. M. Roberts, J. C. G. Coleridge and H. M. Coleridge. Attenuation of pulmonary afferent input by vagal cooling in dogs. *Respir. Physiol.* 72: 19-34, 1988.
6. Meessen, N. E. L., C. P. M. van der Grinten, H. Th. M. Folgering and S. C. M. Luijendijk. Tonic activity in inspiratory muscles during continuous negative airway pressure. *Respir Physiol.* 92: 151-166, 1993.
7. Meessen, N. E. L., C. P. M. van der Grinten, H. Th. M. Folgering and S. C. M. Luijendijk. Histamine-induced end-tidal inspiratory activity and lung receptors in cats. *Europ. Resp. J.* (In press).

14

APNEIC SNOUT IMMERSION IN TRAINED PIGS ELICITS A "DIVING RESPONSE"

Erika Schagatay and Marja van Kampen

University of Lund
Department of Animal Physiology
Helgonavägen 3B, 223 62 Lund, Sweden

INTRODUCTION

Diving mammals and birds exhibit drastic cardiovascular adjustments to restrict oxygen consumption during diving. The response involves selective peripheral vasoconstriction and heart rate reduction (reviews 1-3). Similar but less pronounced adjustments occur at apneic face immersion in man and dog, but data from voluntary diving in other terrestrial species are scarce (review 4). In some species, including man, the response is temperature dependent, indicating the involvement of thermoreceptors in its elicitation (5-7). The aim of the present study was to quantify some circulatory adjustments occuring at voluntary apneic snout immersion in trained pigs, and to reveal if temperature receptors contribute to the triggering of this response in the pig.

METHODS

Four newly weaned piglets (Sus scrofa; Swedish land race) were trained to perform apenic snout immersion by feeding them milk formula from a nipple at the bottom of a large water container. At the time for experiments the pigs were about 3 months of age and weighed approximately 20 kg. Heart rate, skin capillary blood flow and breathing movements of the pigs were registered continously by non-invasive methods over the series of dives. Repeated dives at different water temperatures were performed by one pig.

RESULTS AND DISCUSSION

Maximum apneic time varied between 15 and 53 sec among the individual pigs. All pigs reacted with prompt and pronounced heart rate- and skin blood flow reductions, but the magnitude and rate of onset of the response varied among the individuals. The final level of

Modeling and Control of Ventilation, Edited by S. J. G. Semple, L. Adams, and B. J. Whipp
Plenum Press, New York, 1995

73

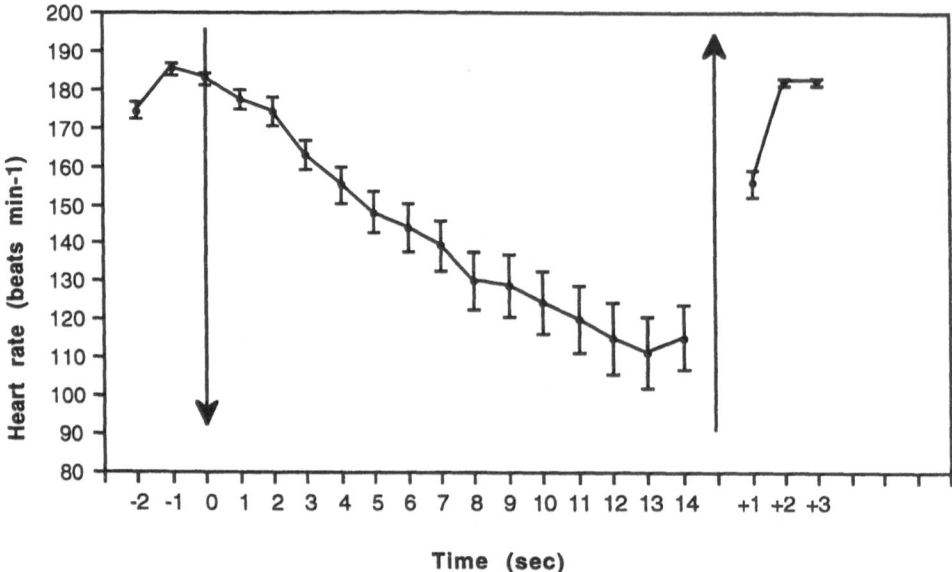

Figure 1. Bradycardia response at apneic snout immersion. Mean values (with SD) of second by second registrations from 10 "dives" of at least 14 sec duration (5 from each of two pigs). Arrows indicate onset and end of dives. "+" time indicates absolute time after resumed breathing.

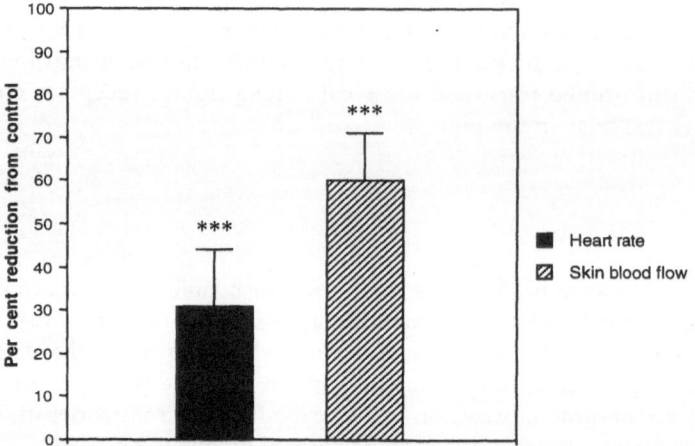

Figure 2. Mean (with SD) heart rate and skin blood flow reductions from 23 dives of at least 10 sec duration (from 3 pigs). HR values were obtained by measuring the 5 last r-r immediately preceding immersion (control) and the lowest value of 5 consecutive r-r intervals during diving. For skin blood flow values the control used was average flow during 30 sec preceding each series of apneic episodes and as dive value the lowest value at each dive was used. Values presented are mean per cent reductions from control. Significant diference from control at the 1% level is indicated by ***.

heart rate reduction was reached within 10-12 sec and the blood flow reduction within 5 sec of apneic immersion (Fig 1).

For 23 dives of at least 10 sec duration (from 3 pigs) the mean heart rate reduction was 31% and the skin blood flow reduction was 60% from control values prior to immersion (Fig 2). As there was no change in the response when water temperatures of 5, 10 , 20 and 35°C were used there seems to be no influence of temperature receptors in eliciting the response in pigs, contrary to in man. The primary sensory input seems to be sessation of respiration, but changes in neural activity from other facial or upper airway receptors may contribute to initiate the response.

Pigs thus share the defence mechanism against asphyxia brought to its extreme in mammalian divers. The magnitude of the heart rate reduction in pigs is less than that found at similar experiments in habitual divers like seals, sealions and dolphins, (about 80%: 8-10), and in trained human divers (50%: 11 if data are treated as in the present study) and dogs (43-48%:12, 13) but similar to that of humans unexperienced in diving (about 30%: 11 if data are treated as in the present study). In addition to being a necessary qualification for extensive diving, a well developed diving response may faciliate other behaviour where tolerance to apnea is advantageous, like the grubbing in the ground for food of the pig, and the use of water or mudbaths for thermoregulation. The pig diving response may be a remnant from possible semiaquatic ancestors shared with the peccary and the hippopotamus. Future research will hopefully yield data from voluntary apneic immersion of strictly terrestrial species.

Acknowledgements

This study was supported by the Hierta Retzius Foundation, the Royal Physiographic Society, Lund and the Faculty of Science, University of Lund, Sweden.

REFERENCES

1. Andersen, H. T. Physiological adaptations in diving vertebrates. Physiol Rev 46, 212-243, 1966.
2. Elsner, R. and B. Gooden. Diving and Asphyxia: A comparative study of animals and man. Physiological Society Monograph 40. Cambridge, Cambridge University Press, 1983, 175 pp.
3. Kooyman, G. L. Diverse Divers: Physiology and behavior. Zoophysiology vol 23. Berlin Heidelberg New York, Springer-Verlag, 1989, 201 pp
4. Lin, Y. C. Breath-hold diving in terrestrial mammals. Exercise and Sports Sciences Reviews 10, 270-307, 1982.
5. Hong, S. K., S. H. Song, P. K. Kim and C. S. Suh. Seasonal observations on the cardiac rhythm during diving in the Korean Ama. J Appl Physiol 23, 18-22, 1967.
6. Corriol, J. and J. J. Rohner. Role de la temperature de l'eau dans la bradycardie d'immersion de la face. Arch Sci Physiol 22, 265-274, 1968.
7. Furedy, J. J. and J. W. Morrison. Effects of water temperature on some noninvasively measured components of the human dive reflex: An experimental response-topography analysis. Psychoohysiology 20, 569-578, 1983.
8. Elsner, R. Heart rate response in forced versus trained experimental dives in pinnipeds. Hvalrad Skr 48: 24-29, 1965.
9. Elsner, R., D.L Franklin and R.L. Van Citters. Cardiac output during diving in an unrestrained sea lion. Nature 202, 809-810, 1964.
10. Elsner, R., D.W. Kenney and K. Burgess. Diving bradycardia in the trained dolphin. Nature 212, 407-408, 1966.
11. Schagatay, E. The significance of the human diving reflex. In: M. Roede, J. Wind, J. Patrick and W. Reynolds (eds), The aquatic ape: Fact or fiction? London, Souvenir Press Ltd, 1991, pp 247-254.
12. Lin, Y. C., E. L. Carlsson, E. P. Mc Cutcheon and H. Sandler. Cardiovascular functions during voluntary apnea in dogs. Am J Physiol 245, 143-150, 1983.

13. Gooden, B. A., H. L. Stone and S. Young. Cardiac responses to snout immersion in trained dogs. J Physiol 242, 405-414, 1974.

15

INTERACTION BETWEEN EXPIRATORY TIME AND INSPIRATION IN CONSCIOUS HUMANS

G. F. Rafferty and W. N. Gardner

Department of Physiology
Biomedical Sciences Division
King's College, London
Campden Hill Road, London W8 7AH, United Kingdom

A breath can be divided into inspiratory and expiratory drive and timing variables. The regulation of expiratory time (T_E) in conscious humans is important in speech and many non-metabolic respiratory functions. T_E is linked to inspiratory variables via mechanical, vagal reflex, chemical and central neuronal mechanisms. However, the mechanisms of these linkages are poorly understood in conscious humans and very little is known about the influence that T_E has on the following breath, especially inspiratory time (T_I). To study these linkages, respiratory patterning variables were measured for each breath in real time by computer in various combinations of 17 normal awake humans breathing mildly hyperoxic ($P_{I,O2}$ 35%) and hypercapnic gas mixtures via a pneumotachograph into an open circuit. Computer controlled auditory feedback allowed gradual and imperceptible 3 minute step alterations at a constant end-tidal PCO_2 ($P_{ET,CO2}$) over 45-60 minutes of (1) T_I over a range of (800 msec at constant inspired tidal volume (V_{TI}), (2) V_{TI} up and down in repeated steps of 200 ml at constant T_I, and (3) T_E over a range of (2000 msec at constant V_{TI}. In each case, the timing of the uncontrolled half of the respiratory cycle was free to be determined by the subjects' automatic respiratory control mechanisms. Variables were averaged over the final minute of each step, and in protocol 2 the average was obtained of 8 up and down steps respectively. In protocol 1, T_E changed significantly ($P<0.05$) from 1.94 to 3.10 sec in parallel with a mean achieved change of T_I from 1.48 to 2.91 sec ($N = 10$) despite constant V_{TI}. In protocol 2, T_E and time for expiratory flow did not change significantly in response to a mean step change of V_{TI} of 150 ml at constant T_I averaged over 10 steps in 7 subjects. In protocol 3, change of T_E from 1.63 to 4.87 sec ($N = 8$) had no significant influence on the subsequent T_I, but V_{TI} increased slightly by a mean of 240 ml ($P<0.01$) as T_E lengthened despite attempts to clamp using the auditory feedback. Thus, in conscious humans, inspiratory timing has a direct influence on expiratory timing independent of volume change and chemical drive, but expiratory timing has no influence on the inspiratory timing of the subsequent breath but has a small influence on tidal volume.

Modeling and Control of Ventilation, Edited by S. J. G. Semple, L. Adams, and B. J. Whipp
Plenum Press, New York, 1995

16

CONTROL OF THE RESPIRATORY CYCLE IN CONSCIOUS HUMANS

G. F. Rafferty and W. N. Gardner

Department of Physiology
Biomedical Sciences Division
King's College, London
Campden Hill Road, London W8 7AH, United Kingdom

There is no simple model for control of breathing in man. There are a number of degrees of freedom in the respiratory cycle. In inspiration, these include tidal volume (V_{TI}), inspiratory time (T_I) and mean inspiratory flow (MIF or V_{TI}/T_I), and in expiration, expiratory time (T_E). We have developed a new technique to clamp and vary these variables individually or in combination to determine the strength of the mechanisms limiting change of each variable, and the hierarchy of control between variables. Subjects breathed into and from an open circuit via a mouthpiece and pneumotachograph and variables were measured breath-by-breath on-line by computer. Subjects coincided two bleeps of different pitches activated by the computer, one triggered at a fixed V_{TI} and one at a fixed time after the start of inspiration (i.e. T_I), expiration being initiated at this point. The threshold for each bleep could be varied from the keyboard without the subjects' knowledge and were slowly applied to reduce subject awareness of the induced change. End-tidal PCO_2 ($P_{ET,CO2}$) was held constant by manipulation of the inspired gas. We performed 4 sets of experiments in various combinations of 17 normal subjects in mild hyperoxia at a constant $P_{ET,CO2}$ slightly above resting. Each lasted 3/4-1 hour. The first experiment examined the range over which V_{TI} and MIF could be increased and decreased by changing the threshold for the V_{TI} bleep in 50-100 ml steps every 3 minutes at constant T_I. Subjects tracked the bleeps without difficulty as V_{TI} increased by 500 ml but were not able to reduce V_{TI} by more than 20 ml below the resting value, associated with a reduction in MIF of 12%. In a second series of experiments, T_I was changed by ñ 800 msec at constant V_{TI} using a similar technique; no difficulty was experienced in either direction, T_I changing from, on average, 1.48 to 2.91 sec. In a third series of experiments, V_{TI} and T_I were reduced together from starting values at inspired PCO2 of 1, 3 and 5% to keep MIF constant while reducing V_{TI}. At each level of chemical drive, V_{TI} overshot the bleeps and could not be reduced below the free breathing resting value. These experiments show that timing can be changed over a wide range, that V_{TI} and MIF can increase without difficulty, but that it is impossible to reduce V_{TI} by more than a small amount below the resting values as dictated by the chemical drive. In a fourth series of experiments, we studied rapid shallow breathing without feedback control at constant $P_{ET,CO2}$ in 4 subjects. V_{TI} fell without difficulty for prolonged periods to about 1/2 resting but

Modeling and Control of Ventilation, Edited by S. J. G. Semple, L. Adams, and B. J. Whipp
Plenum Press, New York, 1995

79

end-expiratory volume as measured by a respiratory inductive plethysmograph increased to keep peak absolute tidal volume constant. In summary, these results suggest that in conscious humans, the major role of the control mechanism is to prevent a fall of tidal volume below that dictated by the chemical drive, presumably to ensure the maintenance of metabolic requirements. Mechanisms controlling timing are of lesser importance, consistent with the needs for non-metabolic functions of breathing such as speech. When V_{TI} is forced to decrease by a programmed manoeuvre such as panting, end-expiratory volume increases to keep absolute volume constant.

17

INTELLECTUAL WORK USING A VIDEO GAME INHIBITS POST HYPERVENTILATION HYPERPNOEA FOLLOWING VOLUNTARY HYPERVENTILATION WHILE IT STIMULATES BREATHING AT REST

K. Chin, M. Ohi, M. Fukui, H. Kita, T. Tsuboi, N. Otsuka, H. Hirata,
T. Noguchi, M. Mishima, and K. Kuno

Department of Clinical Physiology
Chest Disease Research Institute
Kyoto University
Shogoin-Kawaharacho 53, Sakyo-Ku, Kyoto 606, Japan

INTRODUCTION

It has been reported that visual and auditory stimuli significantly increase respiratory frequency (f) and ventilation ($\dot{V}_E$) relative to resting levels (8). However, the effects of intellectual work, which would require widespread parallel activation of central nervous system (CNS) in motor, sensory and associated integrative areas of the cortex, on $\dot{V}_E$ and breathing patterns have not been well documented. Post hyperventilation hyperpnoea (PHH) (9, 10) or ventilatory afterdischarge (2) is the time-dependent continued hyperventilation after the abrupt termination of a ventilatory stimulus. Plum et al. (6) showed that PHH was frequently impaired in patients with brain disorders. These results have led to the hypothesis that intellectual work could have effects on resting ventilation and PHH in man. To test this hypothesis we evaluated $\dot{V}_E$, breathing patterns and the changes in $PaCO_2$ using inductive plethysmography and transcutaneous PCO_2 ($PtcCO_2$) measurements during intellectual work in normal subjects. We also investigated the effects of intellectual work on PHH following voluntary hyperventilation (VHV) for 3 minutes.

METHODS

Subjects and Monitoring

5 normal subjects (Mean (+/- S.D.) age = 27.2 +/- 2.2 yr) were studied; all were male doctors in the supine position. Intellectual work was initiated using a video game

Modeling and Control of Ventilation, Edited by S. J. G. Semple, L. Adams, and B. J. Whipp
Plenum Press, New York, 1995

on a Family Computer^{TM} (Nintendo®). The five subjects were unfamiliar with the video game. They practised the game for up to 15 minutes before the experiments. Ventilation was monitored noninvasively with inductive plethysmography. Calibration factors were obtained by multiple linear regression methods (4). Surface electrodes to monitor the electroencephalogram, electromyogram of the chin, electrocardiogram, and an electrooculogram were applied using standard techniques. A PtcCO$_2$ electrode was applied and heated to 45°C during the study. The concentration of inspired O$_2$ and the end-tidal PCO$_2$ were sampled by a nasal prong and were measured with a mass spectrometer (Airspec 2000 SP, Biggin Hill. Kent).

Protocol

The subjects participated in two sets of activities. First, they watched television (TV) and then played the video game for 10 minutes. Next, they watched TV, voluntarily hyperventilated for 3 minutes and then played the video game or watched TV. The ventilation immediately following VHV was measured for 30 seconds.

VHV was performed for 3 minutes without a mouthpiece. The subjects were instructed to hyperventilate voluntarily as much as they could, while keeping in mind that they had to be able to continue for 3 minutes. When PtcCO$_2$ was not lower than 35 mmHg after two minutes of hyperventilation, they were urged to hyperventilate more vigorously. The subjects were permitted to increase either their tidal volume or their respiratory frequency. In a previous study (5), we ascertained that 3 minutes of VHV in the normal subjects was well tolerated and the subjects decreased their PtcCO$_2$ by more than 10 mmHg when compared with their eucapnic state. The degree of the lowered PtcCO$_2$ levels during each VHV was not significantly different.

Data Analysis

The data are presented as means ± S.D. The differences between the values obtained for any two conditions were tested by the two tailed paired t test. Differences between values obtained for more than two conditions were tested for significance using a one-way analysis of variance (ANOVA) for repeated measures. If a significant difference was found by ANOVA, the difference between those two conditions was re-tested by a multiple comparison method (Fisher's protected least significant difference). A p-value less than 0.05 was considered to be statistically significant.

RESULTS

The intellectual task had a stimulatory effect (p<0.05 by paired t test) on $\dot{V}_E$ at normocapnia via an increase in respiratory output (mean inspiratory flow rate: VT/TI) (p<0.01) and respiratory frequency (f) (p<0.01) while tidal volume (V$_T$) and relative duration of inspiration per breath (T$_I$/T$_{TOT}$) did not change significantly. Their PtcCO$_2$ levels also decreased significantly (p<0.01) when they played the video game (Table 1).

Performance of the task immediately following VHV significantly (p<0.01) attenuated PHH ($\dot{V}_E$ while watching television: 4.90±0.56 l/min; the first 30 seconds of $\dot{V}_E$ while watching television following VHV: 8.12±3.06 l/min; the first 30 seconds of $\dot{V}_E$ while playing a video game following VHV: 2.67±1.57 l/min) (Table 2).

Table 1. Breathing patterns during watching television and playing the video game

Period	$\dot{V}_E$ (l/min)	V_T (ml)	f (1/min)	V_T/T_I (ml/min)	T_I/T_{ToT}	PtcCO$_2$ (mmHg)
Control	4.90	312	16.3	186	0.44	43.7
	(0.56)	(58)	(2.5)	(16)	(0.03)	(3.6)
Game (0-1)	5.88*	287	20.6†	226†	0.43	44.0
	(1.19)	(24)	(3.9)	(25)	(0.04)	(3.9)
Game (7-10)	6.06*	293	21.3†	230†	0.44	42.5†
	(0.56)	(32)	(3.9)	(18)	(0.02)	(3.7)

Mean(SD). *$p<0.05$, †$p<0.01$ vs. Control. Control, breathing during watching television; Game(0-1) (7-10), breathing during the first 1 minute (from minute 7 to minute 10) of ventilation while playing the game; PtcCO$_2$, transcutaneous PCO$_2$.

DISCUSSION

We found that the intellectual work using a video game inhibited PHH following voluntary hyperventilation while it stimulated breathing at rest. We also found that the effects of the task depended on the PaCO$_2$ level. Namely, with normocapnia, the intellectual task was a stimulant for ventilation via an increase in the respiratory frequency or inspiratory flow rate, and produced slight hypocapnia. However, with severe hypocapnia the intellectual task had a depressant effect on ventilation. Therefore, it is suggested that breathing is an obstacle to performance of an intellectual task when there is no chemical stimulation for ventilation.

Shea et al. showed that playing a video game caused normal children to breathe faster as compared to their resting ventilation, but this result was not statistically significant (7). Their results were almost the same as our results under the normocapnic conditions. Younes (11) said that abnormalities in the PHH mechanism are a prerequisite for the development

Table 2. Minute ventilation during control conditions and after voluntary hyperventilation with and without a video game

Subjects No.	Control (l/min)	VHV-TV (l/min)	VHV-Ga (l/min)
1	4.68	5.38	0.00
2	4.50	7.34	3.65
3	5.52	5.79	2.82
4	4.32	12.90	2.77
5	5.47	9.20	2.14
Mean	4.90	8.12*	2.67†
SD	0.56	3.06	1.57

*$p<0.05$ vs. Control; †$p<0.01$ vs. Control and VHV-TV.

Control, the minute ventilation during watching television; VHV-TV (Ga), the first 30 seconds of ventilation while watching televsion (playing the video game) following the voluntary hyperventilation.

of sustained periodic breathing in the clinical setting. It has been reported that PHH is impaired by the depressant effects on the CNS of brain infarction (6) and by sustained hypoxia, which induces hypoxic depression (13). During non-rapid eye movement sleep, PHH occurs as long as hypoxia is brief and arterial PCO_2 is maintained (1). We showed that an intellectual task requiring widespread parallel activation of motor, sensory (visual and auditory) and associated integrative areas of the cortex, had an inhibitory effect on PHH, while watching TV did not. From our study and those of others (1, 3,6, 10), it is suggested that in humans PHH following hyperventilation requires the intact higher centres in the CNS or no depression of the CNS, although intense activity such as that needed to accomplish an intellectual task impairs PHH. These facts indicate that the cerebral cortex can have several effects on PHH according to its intensity of activity.

In conclusion, the intellectual task had a stimulatory effect on breathing at normo-capnia via respiratory output, and respiratory frequency. However, the intellectual task depressed the post hyperventilaton hyperpnoea following the voluntary hyperventilation. Because the numbers of our subjects in this study were relatively small, the results mentioned above await confirmation from further studies.

REFERENCES

1. Badr, M.S., J.B. Skatrud, And J.A. Dempsey. Determinants of poststimulus potentiation in humans during NREM sleep. *J. Appl. Physiol.* 73: 1958-1971, 1992.
2. Eldridge, F.L., And P. Gill-Kumar. Central neural respiratory drive and afterdischarge. *Respir. Physiol.* 40: 49-63, 1980.
3. Georgopolos, D., Z. Bshouty, M. Younes, And N.R. Anthoniesen. Hypoxic exposure and activation of the afterdischarge mechanism in conscious humans. *J. Appl. Phsiol.* 69: 1159-1164, 1990.
4. Loveridge, B., P. West, N.R. Anthonisen, And M.H. Kryger. Single-position calibration of the respiratory inductance plethysmograph, *J. Appl. Physiol.* 55: 1031-1034, 1983.
5. Ohi, M., K. Chin, M. Hirai, T. Kuriyama, M. Fukui, Y. Sagawa, And K. Kuno. Oxygen desaturation following voluntary hyperventilation in normal subjects. *Am. J. Crit. Care. Med.* 149: 731-738, 1994.
6. Plum, F., H.W. Bown, And E. Snoep. The neurological significance of post-hyperventilation apnea, *JAMA.* 181: 1050-1055, 1962.
7. Shea, S.A., L.P. Andres, D. Paydarfar, R.B. Banzett, And D.C. Shannon. Effect of mental activity on breathing in congenital central hypoventilation syndrome. *Respir. Physiol.* 94: 251-263, 1993.
8. Shea, S.A., J. Walter, C. Pelley, K. Murphy, And A Guz. The effect of visual and auditory stimuli upon resting ventilation in man. *Respir. Physiol.* 68: 345-357, 1987.
9. Swanson, G.D., D.S. Ward, And J. W. Bellville. Post-hyperventilation isocapnic hyperpnea. *J. Appl. Physiol.* 40: 592-596, 1976.
10. Tawadrous, F.D., And F.L. Eldridge. Posthyperventilation breathing patterns after active hyperventilation in man. *J. Appl. Physiol.* 37: 353-356, 1974.
11. Younes. M. The physiology basis of central apnea and periodic breathing. *Curr. Pulmonol.* 10: 265-326, 1989.

Pathophysiology of Breathing Control and Breathing Awake and Asleep

18

INTRODUCTION TO SESSION ON THE PATHOPHYSIOLOGY OF BREATHING CONTROL AND BREATHING: AWAKE AND ASLEEP

N. S. Cherniack

Case Western Reserve University
School of Medicine
Cleveland, Ohio 44106-4915

INTRODUCTION

From a modelling perspective the respiratory system is often considered to function as two interactive parts. First, there is a controller made up of chemoreceptors and the respiratory neurones in the pons and the medulla which acts on the second part, the respiratory plant which consists of the lung, the respiratory muscles, and the body tissues connected to the lung by the circulation. The assumption is made that ventilation is regulated so as to keep arterial levels of PCO_2 and PO_2 constant. Over the years this basic concept has been amplified with more elements included in the model.

The body tissues have been divided into compartments and modelled individually in more detail. Since the chemoreceptors are within the tissues, a particularly important advance has been to specifically model the brain compartment in which the central chemoreceptors reside with greater accuracy, and to include the variations in cerebral blood flow that occur with changes in PCO_2 and PO_2 levels. More recent models attempt to reproduce patterns of breathing as well as actual ventilation levels and have included in some form, the mechanical properties of the lung and upper airways.

While these models are able to reproduce many of the features of acute breathing responses to chemical stimuli, it is clear that they are simplistic and fail to account for significant features of breathing responses observed in patients. Two sorts of deficiencies are particularly significant: First the absence of models that take into account the important effects of higher brain centres on breathing, and second, the absence of models that can describe long-term changes in breathing, since it is likely that breathing responses to respiratory disease adapt over time.

The bulbo-pontine neurons in existing models remain pretty much as a black box. No general model of respiration has attempted to include any of the known information

Modeling and Control of Ventilation, Edited by S. J. G. Semple, L. Adams, and B. J. Whipp
Plenum Press, New York, 1995

concerning pattern generation. This indeed may require too complex a model while not adding appreciably to improved simulations of ventilatory response.

However, these neurons receive inputs from higher centres of the brain that have strong effects on respiratory behaviour. For example, in recent years the clinical significance of apnoeas, hypopnoeas and other disorders of breathing during sleep has become apparent. Models of breathing that include the interaction of the respiratory control system with the system regulating the sleep-waking cycle are only beginning to be developed. One example of such a model is contained in the contribution by Gottshalk which appears in this section. In addition, these interactions are complicated by differences in the effects of the various sleep stages on the relative contribution of the respiratory muscles. The extensive data which has been gathered on loaded breathing in humans, which is known to have a significant central component, are yet to be included in models of sleep. The respiratory muscles themselves are not homogeneous and develop forces during contraction that are quite complex even in the awake state. Some examples of the patterns of intercostal and diaphragm responses in disease are illustrated in the paper by M. D. Goldman. The translation of the output of the respiratory pattern generator to the action of specific thoracic and upper airway muscles has not appeared in models even though it has significant effects on ventilation and is clearly influenced by sleep stage.

Besides the effect of the sleep-waking cycle, various other forms of inputs from higher levels of the brain impinge on respiratory neurones. Some of the inputs arise from voluntary activity and can themselves be divided into novel voluntary acts (eg. breath holding) as opposed to automatic voluntary acts (voluntary breathing changes that occur for example in the trained athlete during exercise).

One of the characteristic features of the breathing system is its ability to vary and perhaps to adapt as a constant stimulus is maintained. Hypoxia, for example, is an instance where breathing waxes and wanes over time as stages of hypoxic excitation, hypoxic depression and acclimatization succeed one another. These fluctuations in breathing arise from changing cellular phenomena incited by hypoxic interaction with the pattern generator and are probably influenced as well by hypoxic actions on suprapontine neuronal collections. Similar adaptations are likely to occur with chronic respiratory illness and further experimental data is needed to include these adaptations in models.

It is clear that respiration is far from stereotyped and is not just the result of the algebraic summation of the action of fixed reflexes. Short and long-term changes in respiration can occur and at every instance in time it is possible that the respiratory system makes choices as to breathing pattern, muscles used and even level of ventilation achieved. It may be that there are considerations other than adequacy of gas exchange that allow the respiratory system to select among options.

The idea that respiratory behaviour is optimized in some way and this could be involved in adaptive processes is not new. Several papers in this conference, not this section, address this possibility (eg. Benchetrit). What is being optimized is less clear and only a few mathematical models have attempted to use optimizing criteria in model construction. One possibility is that the energy cost of breathing as well as satisfactory gas exchange is considered by the respiratory system in making choices. Another possibility, particularly in disease states, is that minimizing discomfort is a goal sought by the respiratory system in addition to stability of arterial gas tensions.

Further experimental exploration of the possibility of optimization could be a rewarding direction of respiratory research and may lead to a better quantitative understanding of breathing.

19

POSSIBLE GENOMIC MECHANISM INVOLVED IN CONTROL SYSTEMS RESPONSES TO HYPOXIA

N. S. Cherniack, Nanduri Prabhakar, and Musa Haxhiu

Case Western Reserve University
School of Medicine
Cleveland, Ohio 441064915

INTRODUCTION

It has been increasingly evident that the response of the respiratory control system (i.e. the controller and its interactions with the controlled system) is not invariant but changes over time. This is particularly obvious in the effects of hypoxia on the control system. For example, ventilation increases sharply, decreases within a few minutes and then increases over time even though constant levels of hypoxia are maintained [1]. Controlled system reactions to signals from the controller also vary as hypoxia is maintained and are modulated by the release of neurotransmitters such as NO from the vascular endothelium, enhanced secretion of catecholamines, and ultimately by changes such as smooth muscle hypertrophy and increased red cell mass [2].

While many of these changes in response characteristic are caused by cytoplasmic events which lead to neurotransmitter release, there is increasing evidence that changes in gene expression triggered and sustained by relatively brief exposures to hypoxia play a significant role in the adjustments of the control system [2]. These adjustments may be of particular importance during disease, when hypoxia cannot be avoided.

POTENTIAL EFFECTS OF HYPOXIA ON GENE EXPRESSION

Cells seem to be able to sense and respond to hypoxia [3,4]. In cell culture systems, the endogeneous EPO gene can be induced by hypoxia [3,4]. A hypoxia inducible factor (HIF-1) which binds to DNA has been identified in Hep 3 B cells cultured in 1% O_2 for 12 hours. HIF-1 binds with the EPO gene to produce erythropoetin [4]. It has been demonstrated that HIF-1 is generated in many other cells by hypoxia in which the EPO gene is not transcribed [3]. Atrial natriuretic peptide is formed in cardiac myocytes and tryosine hydroxylase formation initiated in PC-12 cells exposed to hypoxia [5,6]. Endothelin and platelet-de-

Modeling and Control of Ventilation, Edited by S. J. G. Semple, L. Adams, and B. J. Whipp
Plenum Press, New York, 1995

rived growth factor (PDGF) are formed in pulmonary arterial endothelial cells made hypoxic [7,8].

There are likely to be many other examples of the induction of gene expression by hypoxia. It is also conceivable that hypoxia may repress the expression of some genes. Whether or not gene expression is altered probably depends on the interplay of many factors in which one or more elements exhibit sensitivity to oxidation - reduction states [3,9].

IMMEDIATE EARLY RESPONSE GENES

A recent formulation by Morgan and Curran [10] proposed that gene expression can be affected by extra cellular stimuli acting in three different ways: 1) by direct binding of the stimulating agent to nuclear protein; 2) by a cascade of secondary intracellular messengers acting on target genes which is started by the binding of extracellular signaling chemicals to membrane receptors, and 3) via intracellular cascades of proteins which bind to the promoter and enhancer regions of immediate early response genes (IERG) and which, in turn, causes the formation of so called third messengers proteins that then act on target genes producing increased transcription and/or increased stability of messenger RNA [11]. Immediate early response genes seem play a crucial role in many biological processes including cell division, differentiation and aptosis.

The most extensively studied of the IERG belongs to the Fos-Jun family. But other families of IERG include the CREB/ATF family and the C/EBP family [12]. These have been reported to interact with DNA and with each other thereby potentially allowing specific extracellular stimuli to produce unique patterns of effects and cell specificity. The basis of this interaction may be in part due to the overlapping enhancing elements present in the promoter regions of IERG and other genes which allows them to respond to different degrees to a broad range of response element binding proteins and in part by a variety of complexes formed by the protein binding elements themselves which connect to DNA with different affinities and transcription activating capacities [13].

THE Fos/Jun/IERG FAMILY

Both Fos and Jun are members of a family of genes which are excited within a short time after the occurrence of a stimulus (within minutes), give rise to messenger RNA's and produce different protein products which then by binding to DNA at an AP-1 site influence cell phenotype [10,14,15]. C-Fos and to a degree c-Jun expression are induced by a wide variety of transmembrane signaling agents via a mechanism independent of new protein synthesis. The intracellular factors that participate in the triggering of IERG include increases in calcium concentration or enhanced activity of protein kinase C or A. Agents which increase cyclic GMP activity also can trigger IERG expression possibly indirectly by mechanisms that alter levels of cAMP [16].

Studies of the promoter region 5' flanking region of the c-fos gene reveals both so called serum response elements (which can be activated by protein kinase C) and calcium response elements (which can be activated by protein kinase A and cAMP) of known nucleotide structure. These response elements are located -60 to -300 base pairs from the transcription initiation site of c-fos and in roughly corresponding regions of C-jun [10]. The translation products (Fos and Jun protein) form heterodimers through leucine zipper binding to generate an AP-1 factor which attaches to DNA. The DNA binding sites are on either side of the leucine being oxidized or reduced, and it has been demonstrated in vitro that reduction of cysteine enhances binding to DNA [17,18] suggesting one possible site of action of hypoxia on IERG.

The AP-1 binding factor which activates target genes (which are largely unknown) can be formed by *c-Jun* homodimers as well as by *c-Fos/c-Jun* heterodimers. *C-Jun* may also form heterodimers with other Fos-like members (less well studied) like Fra-1 and Fra B as well as joining with proteins of the CREB/ATF families to bind with variable affinities to AP-1 and CRE regions in target genes resulting either in stimulation or repression of transcription. Protein products of other jun-like IERG (i.e. Jun B and Jun D) can also join with Fos-like genes and c-Jun to form AP-1 binding factors, which vary in their capacity to activate transcription [6,17,19,20]. The diversity of potential DNA transactivating factors which arise from expression of IERG of the Fos-Jun type could allow cells to modulate their response to hypoxia according to its severity, duration, or according to the mechanism producing hypoxia [6,10]. Moreover, the complexity of the cascades which can trigger IERG action and their effect on target genes could make it possible to finely tune the response to hypoxia in specific cells.

Fos-Jun EXPRESSION IN INTACT ANIMALS

Hypoxia as well as a wide range of other stimuli including seizures, peripheral nerve stimulation, water deprivation, and photic excitation have been shown to cause the appearance of Fos-like immunoreactive proteins in intact animals [6, 21].

Stimulation of the carotid sinus nerve and hypoxia cause C-Fos expression in the Nucleus Tractus Solitarius (NTS) [24]. Our own studies demonstrate that in rats and cats hypoxia can induce the appearance of Fos-like immunoreactivity not just in the NTS but in other areas of the medulla even after peripheral chemo denervation. We have also observed that 8% O_2 breathing in intact rats for 30 minutes induces Fos-like protein expression in higher brain regions. Our studies show that the N Methyl D-Aspartate (NMDA) antagonist dizoclipine attenuates the formation of Fos-like protein by hypoxia in the NTS. Thus, it is clear that in intact animals hypoxia can potentially act at critical sites in the brain known to influence respiratory responses by inducing the expression of IERG.

The specificity of these effects of hypoxia in the intact animals is not clear. Fos-like proteins also appear in cats and rats exposed to high levels of CO2. With both hypoxia and hypercapnia and Fos expression is observed in many of the same regions of the brain. Since it is known that Fos expression can be caused by stress by itself, and because analgesics and anesthetics depress Fos expression, it is extremely difficult to evaluate the mechanism or importance of IERG expression in whole animals whether they are conscious or anesthetized. Hence, studies of the mechanism responsible for IERG expression by hypoxia require an initial examination in cell culture systems, although it is crucial that the cell culture results be accompanied by correlative studies in the whole animal.

CELL CULTURE SYSTEMS

The effects of hypoxia on gene expression have been studied in several cell types such as hepatoblastoma cells and cardiac myocytes. PC-12 cells have been extensively used to examine the effects of various stimuli on IERG [10, 16, 20]. These cells have secretory properties like the Type 1 cells of the carotid body. Both PC-12 and Type 1 cells produce tyrosine hydroxylase and form catecholamines [5]. Like Type 1 carotid body cells but unlike smooth muscle cells (A_7r5) hypoxia increases intracellular calcium in PC-12 cells and inhibits an outward potassium conductance [23, 24, 25]. Recently, we observed PC-12 release catecholamines in response to hypoxia [26]. PC-12 cells can be made to form neurites and develop neuronal-like structures by nerve growth factor (NGF) [16]. This offers the possibility of using the same cell before and after differentiation by NGF to examine the effects of hypoxia on IERG and their target genes.

In PC-12 cells IERG has been induced by a number of different stimuli including depolarization, changes in extracellular calcium, phorbol esters, and by a number of neurotransmitters including glutamate and NO [by exposing the cells to nitroprusside] [10,16,20]. We have been able to show that brief and moderate hypoxia causes IERG expression in PC-12 cells.

EFFECTS OF HYPOXIA ON IERG EXPRESSION IN PC-12 CELLS

We have examined the effects of hypoxia on expression of *c-fos*, Jun-B, Jun-C, Jun-D mRNA's in PC-12 cells. Cells were maintained in cultures as described above. Cells were exposed for one hour either to normoxia (pO_2 = 149 mmHg; pCO_2 = 40 mmHg; or to hypoxia (pO_2 = 33 mmHg; pCO_2 = 42 mmHg) and processed for RNA extraction and Northern blot hybridization assay. Hypoxia stimulated expression of mRNA's for *c-fos*, Jun B, Jun C, Jun D. However, relative increases amongst IERG's varied considerably. For instance, *c-fos* expression increased by 12 fold, whereas increases in Jun B, C, D ranged between 1.5 to 3 fold. The effects of hypoxia could be seen with low pO_2 exposure as short as 15 min. Moreover, mRNA levels were greater with one compared to 10 hrs of hypoxia. These results demonstrate a) that hypoxia induces IERG expression, b) the effects of low pO_2 are not uniform across different IERG's, and c) the effects of low pO_2 depend on duration of hypoxia.

EFFECTS OF HYPOXIA ON IERG EXPRESSION IN DIFFERENT CELL LINES

We have also examined whether hypoxia stimulates IERG expression in propagating cells other than PC-12 cells. Not all cells responded with an increased IERG expression. For example, no induction in IERG's could be seen in neuroblastoma cells by exposing them to pO_2 of 25-35 mmHg. In the same cells IERG expression, however, could be induced by phorbol ester. Hypoxia stimulated IERG's in hepatoblastoma (Hep3 b) cells, but the magnitude of the response was substantially less than PC-12 cells. These results demonstrate that the effect of hypoxia are cell specific, that all cells do not respond with increased IERG expression by hypoxia.

EFFECT OF CARBON DIOXIDE (CO_2) ON IERG EXPRESSION IN PC-12 CELLS

Previous studies from us and others have shown that high levels of CO_2 induces *c-fos* protein in the central nervous system of anaesthetized animals. We examined c-fos expression by CO_2 in PC-12 cells. Exposing the cells to CO_2 (20%; 2-4 hrs) stimulated *c-fos* mRNA. Although hypercapnia induces *c-Fos* in cell culture systems, the levels of CO_2 needed to stimulate IERG expression appear to be rather high.

These results then support the idea that even brief episodes of hypoxia can trigger the expression of genes and produce new protein products. It is of interest that tyrosine hydroxylase gene contains an AP-1 site in its promoter region indicating the possibility that

Fos-Jun dimers may be involved in the hypoxic induction of tyrosine hydroxylase expression [5]. Future studies of the dynamics of the control system may need to take these gnomic changes into account.

REFERENCES

1. Easton PA, Slykerman LJ, Anthonisen NR. Ventiulatory response to sustained hypoxia in normal adults. J. Appl Physiol: 906-911, 1986.
2.. Fanberg B, Massaro D, Cerutti P, Gault D, Berberich M. Regulation of gene expression by O_2 tension. Am. J. Physiol. 262:L235-241, 1992.
3. Wong GL, Semenza GL. General involvement of hypoxia inducible factor 1 in trasncriptional response. Proc. Nat. Acad Sci (USA) 90:4304-4308, 1993.
4. Wong GL, Semenza GL. Characterization of Hypoxia-inducible Factor 1 and Regulation of DNA Binding Activity by Hypoxia J. Biol Chem 268:21513-21518, 1993.
5. Czyzky-Krzeskak MF, Funair BA, Millhorn DE. Hypoxia increases rate of transcription and stability of tyrosine hydroxylase mRNA in pheochromocytoma LPC-12 cells. J. Biol Chem 269:760-764, 1994.
6. Morgan JI, Curran T. Stimulus-transcription coupling in the nervous system: involvement of the inducible proto-oncogenes fos and jun. Ann Rev Neurosci 14:421-51, 1991.
7. Rao UJS Prasadam, Preslow ND, Block ER. Hypoxia induces the synthesis of tropomyosin in cultured porcine pulmonary cells. Am J. Physiol 267: (271-281, 1994.
8. Kourembanas SR, Hannan L, Faller PV. Oxygen tension regulates the expression of platelet derived growthf actor-B chain gene in human endothelial cells. J. Clin Invest 86:670-674, 1990.
9. Kunoh O, Akahori A, Sato H, Xanthoudakis S, Currant & Iba H. Escape from redox regulation enhance the transforming activity of Fos. Oncogene 8:695-701, 1993.
10 Morgan JI, Curran T. Stimulus-transcription coupling in neurons: role of cellular immediate-early genes. TINS 12:459-462, 1989.
11. Herschman HR. Primary response genes induced by growth factors and tumor promoters. Annu Rev. of Biochem 60:281-319, 1991.
12. Hai T, Curran T. Cross-family dimerization of transcription factors Fos/Jun and ATF/CREB alters DNA binding specificity. Proc Nat Acad Sci 88:3720-3724, 1991.
13. Dynan WS. Modularity in Promoters and Enhancers Cell 58:1-4, 1989
14. Abate C C, Rauscher III,FJ, Gentz R, Curran T. Expression and purification of the leucine zipper and the DNA binding domains of Fos and Jun: Both Fos and Jun directly contact DNA. Proc Natl Acad Sci (USA) 87:1032-1036, 1990.
15. MacGregor PF, Abate C, Curran T. Direct cloning of leucine zipper proteins: Jun binds cooperatively to the CRE with CRE-BP-1. Oncogene 5:451-458, 1990.
16. Haby C, Lisovoski F, Aunis D, Zwiller J. Stimulation of the cyclic GMP pathway by NO induces expression of the immediate early genes C-fos and Jun B. JourNeurochem 62: 490-501, 1994.
17. Guis D, Cao X, Rauscher III, FJ, Cohen D, Curran T, Sokhatme VP. Transcription activation and repression by Fos are independent functions: The c-terminus represses immediate early gene expression via CArG elements. Mol & Cell Biol 10:4243-4255, 1990.
18. Yanthoudakis S, Curran T. Identification and clearance of Ref-1, a nuclear protein that facilitates DNA binding activity. EMBO 11:653-665, 1992.
19. Angel P, Hattori K, Smeal T, Karin M. The jun proto-oncogene is positively autoregulated by its product Jun/AP-1. Cell 55:875-885, 1988.
20. Bartel DP, Shong M, Lau LF, Greenberg ME. Growth factors and membrane depolarizaiton activate distinct programs of early response gene expression: disssociation of fos and jun induction. Gene Dev 3:309-313, 1989.
21. Dragunow M, Faull R. The use of c-fos as a metabolic marker in neuronal pathway tracing. Neurosci Meth 29:261-265, 1986.
22. Erickson JT, Millhorn DE. Fos-like protein is induced in neurons of the medulla oblongta after stimulation of the caretoid sinus nerve in awake and anesthetized rats. Brain Res 567:11-24, 1991.
23. Lopez-Lopez J, Gonzalers C, Urena J, Lopez-Barnes J. Low pO2 selectively inhibits K^+ channel activity in chemoreceptors of the mammalian carotid body. J. Gen Physiol 93:1001-1015, 1989.
24. Bright G, Agani F, Haxhiu M, Prabhakar NR. hypoxia and intracellular calcium changes in adult rat carotid body cells: Heterogeneity in responses. FASEB J: A 397, 1993.

25. Agani F, Haxhiu M, Bright G and Prabhakar NR.Hypoxia and intracellular calcium changes in PC-12 cells: Comparison withcarotid body cells. FASEBJ 7: A 398, 1993.
26. Kumar G and Prabhakar NR. hypoxia stimulates dopamine release in pheochromocytoma cells. FASEBJ: 1995 (in press).

20

ASYNCHRONOUS THORACOABDOMINAL MOVEMENTS IN CHRONIC AIRFLOW OBSTRUCTION (CAO)

Active Expiration during Spontaneous Breathing in Sleep and Wakefulness

M. D. Goldman,[1] A. J. Williams,[1] G. Soo Hoo,[1] T. T. H. Trang,[2] and C. Gaultier[2]

[1] University of California at Los Angeles
VA Medical Center, West Los Angeles
Wilshire and Sawtelle Blvds, Los Angeles CA 90073
[2] Université de Paris XI
Hopital Antoine Beclere
157 Rue de la Porte de Trivaux, Clamart, France

INTRODUCTION

Thoracoabdominal movements (TAM) represent two parallel pathways for the distribution of body wall volume displacements. Thoracic and abdominal pathways provide two "degrees of freedom" by which body wall movements accommodate lung volume changes (12). During spontaneous breathing in normal humans, inspiration is the "active phase" of respiration, and expiration is normally passive. Respiratory movements of the thorax and abdominal wall are both driven by the actions of the diaphragm, as well as by muscles intrinsic to each structure. To a useful approximation, abdominal movements reflect directly the movements of the diaphragm in quiet breathing. Thoracic movements may be passive (caused by the action of the diaphragm, using the abdominal contents as a fulcrum to lift the lower rib cage) or active (caused by action of the intercostal and thoracic "accessory" inspiratory and expiratory muscles). During normal quiet breathing, thoracic and abdominal movements are in large part, passive, closely synchronized (in phase), reflecting primarily the action of the diaphragm driving both structures (8). It may be said that TAM during resting breathing are relatively unitary: not only are thoracic and abdominal movements in phase, but their relative amplitudes are nearly constant. This constitutes TAM manifesting only one "degree of freedom," (despite the potential for utilizing two degrees of freedom).

Modeling and Control of Ventilation, Edited by S. J. G. Semple, L. Adams, and B. J. Whipp
Plenum Press, New York, 1995

When breathing is stimulated (chemically, voluntarily, or by exercise), TAM usually depart from one degree of freedom (3,7,10). This may be manifest by changes in relative tidal amplitudes of thoracic and abdominal movements or by phase differences between the two (which may change during the respiratory cycle). When mechanical loads are added to normal respiration (obesity, increased airway resistance), amplitude and phase differences of TAM usually occur (1,2,9).

The earliest reports of TAM asynchrony described patients with cardiac or pulmonary disease. Visual observations were interpreted as "dis-coordination" ("paradoxical" or "desynchronized") of respiratory muscle actions. (4,11). In severely ill patients with CAO, biphasic abdominal movements were attributed to increased activity of thoracic inspiratory muscles lifting the rib cage, and extending the spine in the latter part of inspiration (2). Subsequently, abnormal "paradoxical" abdominal movements in patients with paralyzed or paretic diaphrams (13) were reported, associated with normal neural commands to non-diaphragmatic respiratory muscles.

In anesthetized cats, TAM measured along with respiratory muscle electromyographic (EMG) signals showed "paradoxical" nonuniform rib cage movements in association with physiologic, normally synchronized neural respiratory commands and "passive" mechanical resistance (5). This model of nonuniform thoracic movements occurring with normally synchronized neural commands and respiratory muscle efforts was central to further analyses of measurements in both animals and humans. In supine anesthetized dogs paradoxical outward movement of the anterior abdominal wall occurred during expiration in association with EMG activity of internal intercostal muscles, while the anterior abdominal wall moved paradoxically inward during inspiration. Lateral abdominal and thoracic movements were normal (outward during inspiration and inward during expiration). In this preparation during active contraction of thoracic expiratory muscles, the abdomen passively became more circular in cross section, and returned to a more elliptical shape during inspiration. Local abdominal paradoxical movements were due to abdominal shape changes, validated by pressure, volume, and respiratory muscle EMG measurements (14). This description of normally coordinated respiratory muscle commands leading to local "asynchrony," with paradoxical abdominal movements caused by active thoracic expiratory muscle contraction provides the analytical framework for interpretation of the present studies.

Thoracoabdominal Movements in Chronic Airflow Obstruction

A previous study of respiratory muscle EMG activity and intra-abdominal pressure changes in patients with CAO in the supine posture showed that respiratory muscle efforts are "coordinated" or "synchronized" with expiratory muscle activity during spontaneous breathing at rest (6). A prominent feature of resting spontaneous breathing in chronic airflow obstruction is the active participation of expiratory muscles. This alters the normal phasic relationship of thoracic and abdominal movements, with the appearance of local nonuniformities of thoracic and/or abdominal wall movements. While normal subjects show increased relative amplitude of thoracic movements in the upright posture (and increased abdominal amplitudes supine), TAM in the upright and supine postures often manifest comparable thoracic amplitudes in patients with CAO. This may be due to thoracic "emphasis" associated with increased chemical respiratory stimuli in chronic lung disease. However, phase relationships change dramatically in the supine posture (9).

METHODS

The present study was undertaken to make use of a simplified graphical analysis to infer respiratory muscle actions in adult patients with CAO. We measured thoracic and

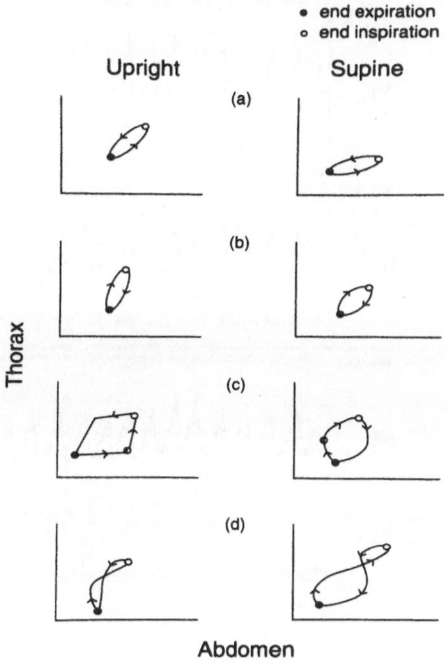

Figure 1. Schematic representation of Konno-Mead diagrams in upright and supine postures during different respiratory muscular efforts. From top to bottom, tracings show thoracic movements on the vertical axis and abdominal movements on the horizontal. a = quiet breathing; b = thoracic inspiratory muscle activity; c = active expiration (thoracic and abdominal muscles); d = thoracic inspiratory and expiratory activity. End expiration is indicated by the filled circle and end-inspiration by the open circle. See text for discussion.

abdominal movements noninvasively, using a respiratory inductive plethysmograph. We displayed TAM in time, and as a Lissajous figure (the Konno-Mead [KM] diagram [12]) with thoracic movements plotted on the vertical axis against abdominal movements on the horizontal axis.

Figure 1 shows a schematic representation of KM diagrams in upright and supine postures during different patterns of breathing. In the uppermost panels (a), resting spontaneous breathing is shown. A counterclockwise loop is formed, with passive thoracic and abdominal movements driven by the action of the diaphragm. Relative amplitude of thoracic movements is greater upright than supine. In panels (b), active thoracic accessory inspiratory muscle contraction increases thoracic amplitude relative to abdominal, and also causes the tidal loop to form in the clockwise direction. In panels (c), active expiration is shown. The half-filled circle indicates mid-inspiration. In the upright posture, active expiration is associated with substantially smaller abdominal dimensions during expiration, reflecting the mechanical advantage of the abdominal wall muscles. In the supine posture, the mechanical advantage of thoracic expiratory muscles is relatively better than that of the abdominal muscles, and active expiration is associated with increased abdominal size during expiration. In constrast, during the first portion of inspiration (up to the half-filled circle), relaxation of expiratory muscles causes an increase of abdominal size in the upright posture, and a decrease of abdominal size ("paradox") in the supine posture. Panels (d) represent combined activity (emphasis) of thoracic inspiratory muscles early in inspiration, diaphragm action late in inspiration and expiratory muscle activity during expiration. This results in a "figure 8" both in the upright and supine postures. In the upright posture, the graphic representation is the equivalent of adding the loop in panel a to the top of the loop in panel b (active lifting of the thorax tends to cancel the effect of abdominal muscle relaxation early in inspiration). In the supine posture, the graphic result is the equivalent of adding the loop in panel a to the top of the loop in panel c.

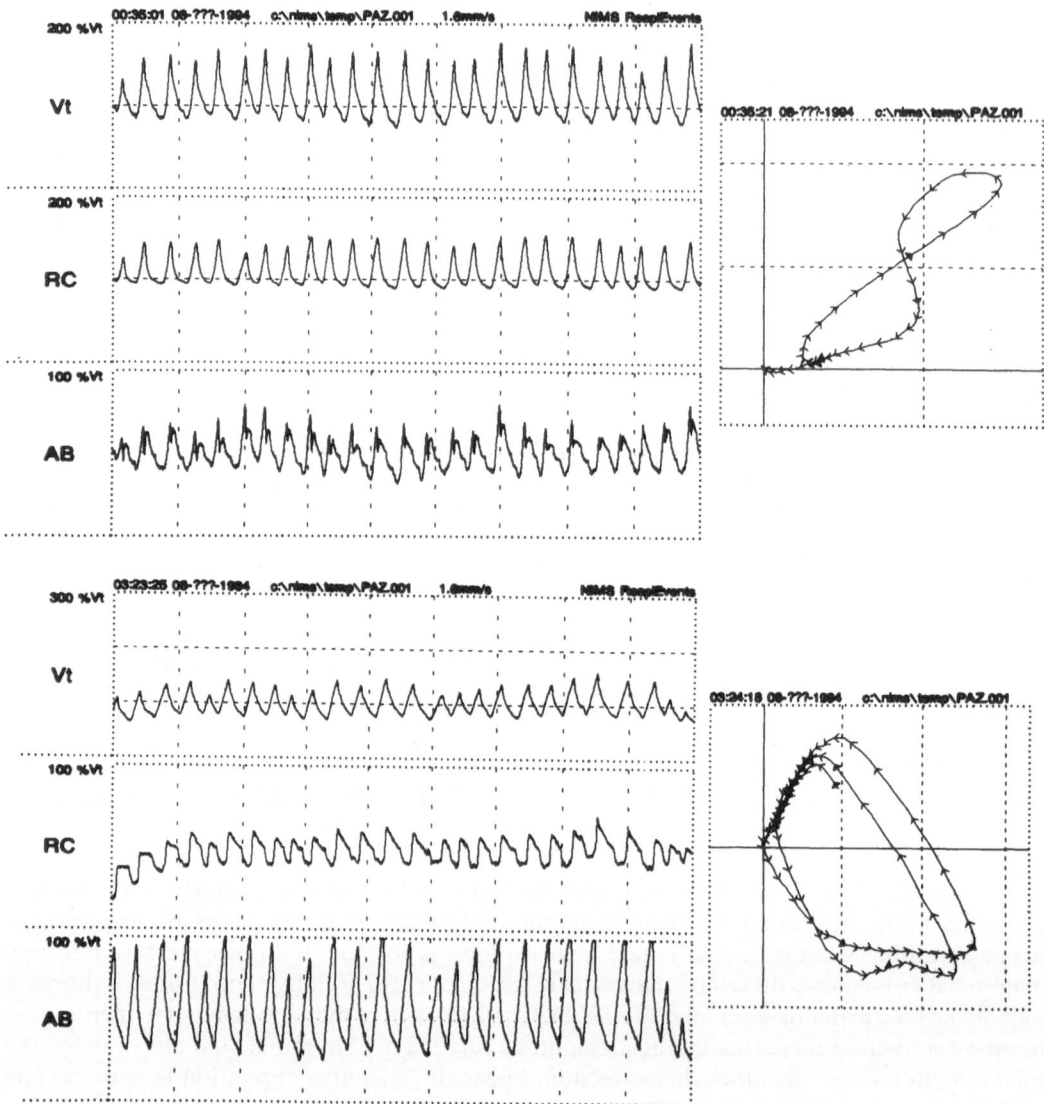

Figure 2. Temporal and Lissajous patterns of thoracic and abdominal motion in a patient with CAO. a = Temporal and KM representations of thoracic and abdominal movements during nonREM sleep in a patient with CAO. b = Temporal and KM representations of thoracic and abdominal movements during REM sleep. Vt = estimated tidal volume, which is given by the sum of RC plus AB displacements. RC = Rib cage (thoracic) displacements. AB = Abdominal displacements. See text for discussion.

RESULTS

A typical study is in a patient with CAO during nonREM and REM sleep is shown in Figure 2.

Figure 2 shows temporal patterns of thoracic and abdominal motion and associated KM diagrams. Fig. 2a and 2b represent nonREM and REM sleep respectively in a patient with CAO. During wakefulness and nonREM sleep, abdominal movements are biphasic. At the onset of inspiration, abdominal movement is paradoxical, but is in phase during the latter portion of inspiration. The abdomen remains in phase with airflow during at the onset of expiration, but subsequently moves paradoxically outward before returning inward again at the end of expiration. This results in an apparent respiratory frequency of abdominal movements twice that of the thorax. During REM sleep, inhibition of non-diaphragmatic respiratory muscles results in paradoxical thoracic movements during both inspiration and the first portion of expiration (similar to children with CAO [9]). Abdominal movement is in phase with airflow at all times, driven by action of the diaphragm. The KM diagrams are shown at right in each figure. End expiration is at the left edge of the loop, at the level of the solid horizontal line. Paradoxical abdominal motion is seen at onset of inspiration in nonREM sleep; and paradoxical thoracic motion in REM sleep.

DISCUSSION

In the examples shown, abnormal abdominal movements occur because of increased accessory inspiratory and expiratory muscle activity in chronic obstructive lung disease. Abdominal movements are not "uncoordinated;" but rather are readily explained by passive and active movments in association with normally coordinated respiratory muscle activation.

In summary, simple graphic analysis of thoracic and abdominal respiratory movements indicates directly the phase relationship between the two parallel pathways for distribution of lung volume change. The X-Y patterns can be used to infer respiratory muscle activity during spontaneous breathing. In normal humans, thoracic and abdominal movements occur passively, as a consequence of action of the diaphragm. Abdominal movements are phase-advanced relative to those of the thorax. In chronic airflow obstruction, TAM patterns are more complex. Action of thoracic inspiratory muscles increases the prominence of thoracic movements, and causes them to be phase-advanced relative to those of the abdomen. Inspiratory and expiratory thoracic muscle activity in CAO results in a phase-advance of thoracic movements early in inspiration and late in expiration. In the latter part of inspiration and early in expiration, action of the diaphragm results in a "normal" phase advance of abdominal movements. Thus, especially in the supine posture, patients with CAO manifest an inconstant phase relationship between thoracic and abdominal movements during the course of the respiratory cycle during wakefulness and nonREM sleep. Under these conditions, determination of a single "phase angle" for the respiratory cycle would be misleading. During REM sleep, however, the marked decrease in accessory inspiratory and expiratory muscle activity results in a relatively constant phase advance of abdomen relative to thorax. Paradoxical thoracic movements do not indicate "discoordination" of respiratory muscle activity; but rather reflect the inability of the thoracic wall to passively resist deformation caused by the decrease in intrathoracic pressure.

REFERENCES

1. Agostoni E & Mognoni P. (1966). Deformation of the chest wall during breathing efforts. J. Appl Physiol 21:1827-32.
2. Ashutosh K, Gilbert R, Auchincloss JH & Peppi D. (1975) Asynchronous breathing movements in patients with chronic obstructive pulmonary disease. Chest, 67, 553-7.
3. Campbell, EJM & Green JH. (1955). The behaviour of the abdominal muscles and the intra-abdominal pressure during quiet breathing and increased pulmonary ventilation: a study in man. J. Physiol, London 127:423-26.
4. Cournand A, Brock HJ, Rappaport J & Richards DW, Jr. (1936). Disturbance of action of respiratory muscles as a contributing cause of dyspnea. (1936) Arch Intern Med. 57: 1008-11.

5. Da Silva KMC, Sayers BMA, Sears TA, Stagg DT. (1977). The changes in configuration of the rib cage and abdomen during breathing in the anaesthetized cat. J. Physiol. London 266; 499-521.

6. Goldman M. (1995). Abnormal thoracoabdominal movements in patients with chronic lung disease. Chap 31, in The Neurobiology of Disease: Contributions from Neuroscience to Clinical Neurology. Ed. Bostock, Kirkwood, & Pullen. Cambridge University Press

7. Goldman MD, Grimby G & Mead J. (1976). Mechanical work of breathing derived from rib cage and abdominal V-P partitioning. J. Appl. Physiol. 41:752-63.

8. Goldman MD & Mead J. (1973). Mechanical interaction between the diaphragm and rib cage. J. Appl. Physiol. 35: 197-204.

9. Goldman MD, Pagani M, Trang H, Praud J, Sartene R & Gaultier C. (1993). Asynchronous Chest Wall Movements during Non-Rapid Eye Movement and Rapid Eye Movement Sleep in Children with Bronchopulmonary Dysplasia. Am Rev Resp Dis 147:1175-84.

10. Grimby G, Bunn J & Mead J. (1966). Relative contribution of rib cage and abdomen to ventilation during exercise. J. Appl. Physiol. 24: 159-66.

11. Hoover CF. (1920) Definitive percussion and inspection in estimating size and contour of heart. J.A.M.A. 75:1626-30.

12. Konno K & Mead J. (1967). Measurement of the separate volume changes of rib cage and abdomen during breathing. J. Appl. Physiol. 22: 407-22.

13. Newsom Davis J, Goldman M, Loh L & Casson M. (1976) Diaphragm function and alveolar hypoventilation. Q. J. Med. 45: 87-100.

14. Nochomovitz ML, Goldman MD, Mitra J & Cherniack NS. (1981) Respiratory responses in reversible diaphragm paralysis. J Appl Physiol: Respirat Environ Exercise Physiol 51:1150-56

21

BREATHING PATTERNS UNDER ENFLURANE, HALOTHANE AND PROPOFOL SEDATION IN HUMANS

B. Nagyova, K. L. Dorrington, and P. A. Robbins

University Laboratory of Physiology
Parks Road
Oxford OX1 3PT, England

INTRODUCTION

Goodman and collegues (4) reported a fall in tidal volume (V_T), ventilation ($\dot{V}_E$) and the inspiratory time:total time ratio (T_I:T_{tot}) during propofol anaesthesia compared with the awake state. On the other hand, Rosa and coworkers (8) were unable to show any significant changes in respiratory frequency (f), $\dot{V}_E$, V_T, inspiratory and expiratory time (T_I, T_E) and T_I:T_{tot} during conscious sedation with low doses of propofol when compared with the awake state. Their results indicate that propofol at repeated small boluses of 0.3 and 0.6 mg. kg^{-1} does not produce significant alterations in inspiratory drive, respiratory pattern, gas exchange, or minute volume. Similar findings to those of Goodman (4) were published by Reich et al. (6) for the ventilatory effects of halothane at doses of 1.2 minimum alveolar concentration (MAC) and 1.7 MAC. They concluded that halothane caused a dose-dependent decrease in inspiratory drive and an increase in breathing frequency. The aim of our study was to investigate the effect of sedative doses of enflurane, halothane and propofol on breathing pattern, while maintaining end-tidal P_{CO2} 1-2 Torr above the subjects' natural resting value.

METHODS

Each agent was studied in 12 adults. Each adult was studied without any anaesthetic being administered and with three different low doses of anaesthetics. The measured doses used were: 0.04 MAC, 0.07 MAC and 0.13 MAC for enflurane, 0.05 MAC, 0.11 MAC and 0.2 MAC for halothane; and 0.06 effective plasma concentration$_{50}$ (EC$_{50}$), 0.13 EC$_{50}$ and 0.26 EC$_{50}$ for propofol. For enflurane 1 MAC = 1.7% end-tidal, for halothane 1 MAC = 0.77%, and for propofol 1 EC$_{50}$ = 8.1 μg.ml^{-1} (1). The doses were given in random order and each anaesthetic was studied on a different day. A mass spectrometer was used to measure inspired and expired enflurane and halothane concentrations. The infusion regime for

Modeling and Control of Ventilation, Edited by S. J. G. Semple, L. Adams, and B. J. Whipp
Plenum Press, New York, 1995

propofol was derived using the pharmacokinetic model of Gepts *et al.*(3). Blood samples were collected and propofol concentrations were measured after the experiment using high performance liquid chromatography. During the experiment subjects were seated and breathed through a mouthpiece connected to a turbine volume measuring device and pneumotachograph. The end-tidal values for CO_2 and O_2 were controlled and ventilation was measured using our automated laboratory technique (5, 7).

STATISTICAL ANALYSIS

The values for each respiratory variable for each subject with each anaesthetic dose were first expressed as a percentage of the control (no anaesthetic) condition. Analysis of variance was then used to assess the significance or otherwise of the various anaesthetics and doses on each respiratory variable. The analysis for each respiratory variable was based on the same linear model for variance of the form:

$$X_{ij} = \mu + A_i + D_j + AD_{ij} + e_{ij}$$

where X_{ij} is the variable being analysed ($\dot{V}_E$, V_T, f, T_{tot}), μ is a mean value, A_i is the anaesthetic (a factor), D_j is the dose (a co-variate), and e_{ij} is an error term. Anaesthetics were compared by splitting the anaesthetics' term into a "propofol vs other" term, and an "enflurane vs

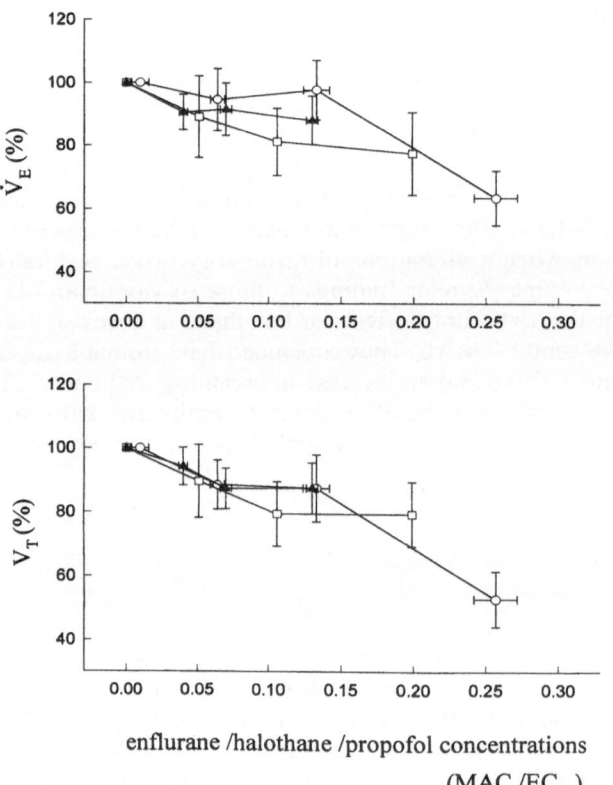

enflurane /halothane /propofol concentrations

(MAC /EC_{50})

Figure 1. Effects on $\dot{V}_E$ and V_T of enflurane (▲), halothane (□) and propofol (○).

halothane" term. Probability values <0.05 were considered to indicate statistical significance.

RESULTS

The relationships between doses of enflurane, halothane and propofol (MAC/EC_{50}) and respiratory variables describing the pattern of breathing ($\dot{V}_E$, V_T, f, T_{tot}) are shown in Figures 1 and 2.

There were significant declines in both $\dot{V}_E$ ($p<0.001$) and V_T ($p<0.001$) for all three drugs with increasing dose. V_T but not $\dot{V}_E$ decreased more with increasing dose of propofol in comparison with enflurane and halothane ($p<0.019$). There were no differences between the effects of enflurane and halothane on $\dot{V}_E$ and V_T. There were no significant dose-dependent effects on T_{tot} or f.

CONCLUSION

In contrast to Rosa and coworkers (8), we were able to detect dose-related decreases in $\dot{V}_E$ and V_T during sedation with low-doses propofol. The depression of V_T was more pronounced during sedation with propofol than with the volatile anaesthetics. However, this

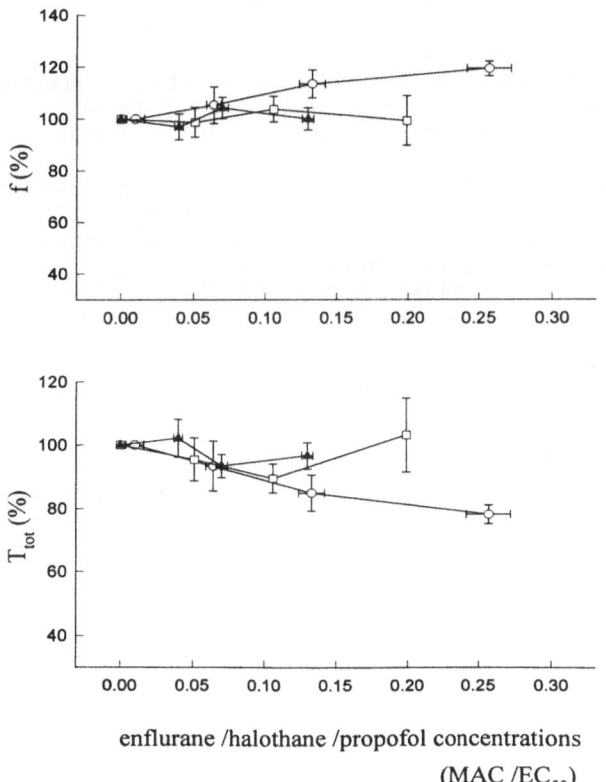

enflurane /halothane /propofol concentrations

(MAC /EC_{50})

Figure 2. Effects on f and T_{tot} of enflurane (▲), halothane (□) and propofol (○).

conclusion has to be treated with some care, because the concentrations of propofol reached were somewhat higher than intended. The statistical analysis employed provides a linear correction for variations in dose, but cannot control for any element which is not linearly related to dose.

The ventilatory effects of propofol anaesthesia are qualitatively the same (4) as those seen with other i.v. agents (2) and halothane in anaesthetic doses (6): a decrease in tidal volume, an increase in frequency, and a reduction in $T_I:T_{tot}$ ratio. It is likely therefore, that at least some of the changes are a consequence of the change in the state of the central nervous system between wakefulness and anaesthesia, rather than being a quantitative dose-related effect of a particular anaesthetic agent. This conclusion for propofol and volatile anaesthetic anaesthesia seems to be in accord with the results of our study.

ACKNOWLEDGMENTS

Our laboratory is supported by Wellcome Trust. Dr. Dorrington is supported by the Dunhill Medical Trust. Dr. Nagyova is supported by Balliol College, Oxford. The authors wish to thank Mr. David O'Connor for technical assistance.

REFERENCES

1. Davidson, J.A.H., Macleod, A.D., Howie, J.C., White, M. and Kenny G.N.C.,1993, Effective concentration 50 for propofol with and without 67% nitrous oxide. *Acta. Anaesth. Scand.* 37:458-464.
2. Gautier, H. and Gaudy, J.H., 1978, Changes in ventilatory pattern induced by intravenous anaesthetic agents in human subjects. *J. Appl. Physiol.* 45:171.
3. Gepts, E., Camu, F., Cockshott, I.D. and Douglas, E.J., 1987, Disposition of propofol administered as constant rate intravenous infusions in humans. *Anesth. Analg.* 66:1256-1263.
4. Goodman, N.W., Black, A.M.S. and Carter, J.A., 1987, Some ventilatory effects of propofol as sole anaesthetic agent. *Br. J. Anaesth.* 59:1497-1503.
5. Howson, M.G., Khamnei, S., McIntrye, M.E., O'Connor, D.F. and Robbins, P.A., 1987, A rapid computer-controlled binary gas-mixing system for studies in respiratory control. *J. Physiol.* 394:7P.
6. Reich, O., Brown, K. and Bates, J.H.T., 1994, Breathing patterns in infants and children under halothane anaesthesia: effect of dose and CO_2. *J. Appl. Physiol.* 76:79-85.
7. Robbins, P.A., Swanson, G.D. and Howson, M.G., 1982, A prediction correction scheme for forcing alveolar gases along certain time-courses. *J. Appl. Physiol.* 52:1353-1357.
8. Rosa, G., Conti, G., Orsi, P., D'Alessandro, F., La Rosa, I., Di Giugno, G. and Gasparetto, A., 1992, Effects of low-dose propofol administration on central respiratory drive, gas exchanges and respiratory pattern. *Acta Anaesthesiol. Scand.* 36:128-131.

22

MULTIPLE MODES OF PERIODIC BREATHING DURING SLEEP

A. Gottschalk,[1] M. C. K. Khoo,[3] and A. I. Pack[2]

[1] Department of Anesthesia
[2] Department of Medicine
  University of Pennsylvania
  Philadelphia, Pennsylvania 19104
[3] Biomedical Engineering Department
  University of Southern California, Los Angeles, California 90089

INTRODUCTION

Periodic breathing (PB) during sleep is commonly observed and may be accompanied by frequent arousals. The full extent of the associated public health issues and the roles of medical and surgical treatment have yet to be fully elucidated. An appreciation of the underlying mechanism(s) would aid the discussion by focusing on specific physiological differences which could predict and discriminate clinically significant PB. One means of clarifying the mechanism(s) of PB is through the formulation of quantitative models of respiratory control during wakefulness and sleep which also consider the transitions between these states. We recently demonstrated that a model with otherwise normal parameters which also incorporated a wakefulness drive and the capacity to arouse from sleep could produce PB if the wakefulness drive is of sufficient magnitude and is withdrawn sufficiently rapidly with sleep onset (5). We now examine such a model in greater detail to reveal the multiple modes by which PB can emerge during sleep.

METHODS

Respiratory Model

A respiratory system model similar to that employed previously (5) was the starting point for the current study. The plant consisted of a single alveolar chamber whose contents are in equilibrium with the blood, and this is connected in series with a fixed dead space. Blood reaches the central compartment and peripheral chemoreceptor after an appropriate transport delay and passage through a second order mixing function. The blood gas tensions at the central chemoreceptor are a function of blood flow to and the metabolic rate of the

Modeling and Control of Ventilation, Edited by S. J. G. Semple, L. Adams, and B. J. Whipp
Plenum Press, New York, 1995

central brain compartment. The rate of gas movement through the alveolar chamber is proportional to the integrated ventilatory drive.

The integrated ventilatory drive is the product of the algebraic sum of the peripheral, central, and wakefulness drives, and a sleep state dependent coefficient. The wakefulness drive is responsible for the "dogleg" in the minute ventilation versus arterial CO_2 curve, and is fully withdrawn with the achievement of stage 1 NREM sleep (2). In addition to withdrawing the wakefulness drive with sleep onset, the slope of the CO_2 response curve is progressively decreased with increasing depth of NREM sleep (2). The option to include an exponential respiratory after discharge function (3) is included in the equations governing the integrated ventilatory drive.

Arousal from the sleep state is possible if the combined output of the chemical drives exceeds a predetermined threshold. This arousal drive may be different from the integrated drive to the alveolar compartment, because it does not include the wakefulness drive, and can also include the effects of abnormal levels of sleep state dependent airway obstruction. Should ventilatory drive fall below an obstructive threshold, the airway remains obstructed until arousal occurs. Should an arousal occur, ventilatory parameters revert to those of the awake state, any airway obstruction is terminated, and sleep onset begins when the arousal drive becomes less than the arousal threshold.

Analysis

The properties of the model were studied with a combination of simulation and local analysis of the linearized model. At each equilibrium point of interest, the linearized model was examined to determine its frequency response and the local stability of the equilibrium point. The frequency response can be computed directly from the resulting linear differential-difference equations (1). The roots of the corresponding characteristic equation can be used to determine the local stability of the system. However, the exponential terms in the characteristic equation, which arise because of the time delay in the system, make this impractical. Consequently, we perform a Taylor expansion about the time delays, and convert the problem of finding the roots of the characteristic equation into a well defined eigenvalue problem. The local stability of the system can then be determined by examining the eigenvalues obtained in this manner. The presence of a positive real component implies localized instability and oscillation, whereas the presence of negative real components for all of the eigenvalues indicates localized stability. Where present, the imaginary components of complex pairs is closely related to the natural frequency of the system.

RESULTS

Frequency Response of the Model

The frequency response of the normally configured system to alterations in arterial CO_2 in wakefulness and in stages 1-4 NREM sleep is shown in Fig. 1. Here, the curve for wakefulness is normalized with respect to the response at the lowest frequency shown, and each of the curves during sleep is normalized with respect to this same value in order to facilitate direct comparisons between the responses at each state. Here, note the similarity of the peak response in wakefulness and light sleep. Calculations for perturbations in the arousal state (not shown) produce similar results in wakefulness and stage 1. However, in stages 2-4, the responsiveness is dramatically attenuated because there is no contribution of the wakefulness drive to ventilation in these stages.

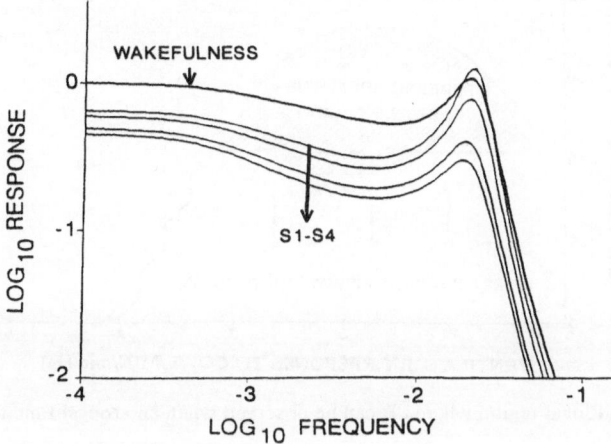

Figure 1. Frequency response of the normally configured model to alterations in arterial CO_2 in wakefulness and stages 1-4 NREM sleep.

The Stability Diagram

To examine the stability of the many possible configurations which could be generated by the model, we studied the stability of each configuration as a function of its combined chemoreceptor gain (normal: 2.25 l/min/torr CO_2), and the magnitude of the wakefulness drive. Here, the wakefulness drive represents a quantity which is subtracted from the integrated drive with sleep onset. The normal value of the wakefulness drive is chosen so that loss of the wakefulness drive with sleep onset results in an increase in arterial CO_2 of 3 torr. Stability is determined for the equilibrium points obtained when an arousal function is not present. Fig. 2 illustrates the stability boundary for the normally configured system in stage 1 sleep. A dot marks the location of the normal values of gain and wakefulness drive. Note the instability of the system at very low gains. Deeper sleep stages increase the stable region, whereas decreased lung volumes, increased transport delays, and an increased participation of the peripheral chemoreceptor all reduce the stable region. Importantly, these different factors do not alter the qualitative nature of the instability or the oscillations which

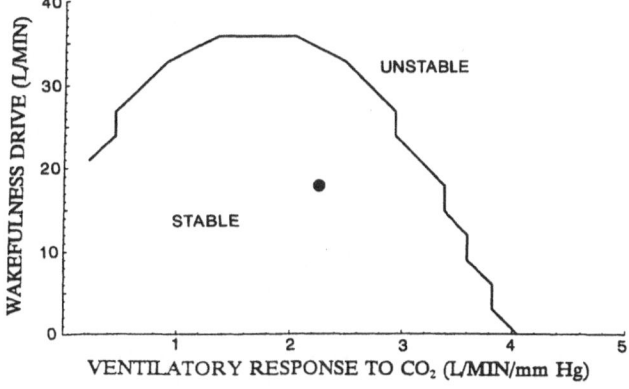

Figure 2. Stability diagram for the normally configured system in stage 1 NREM sleep.

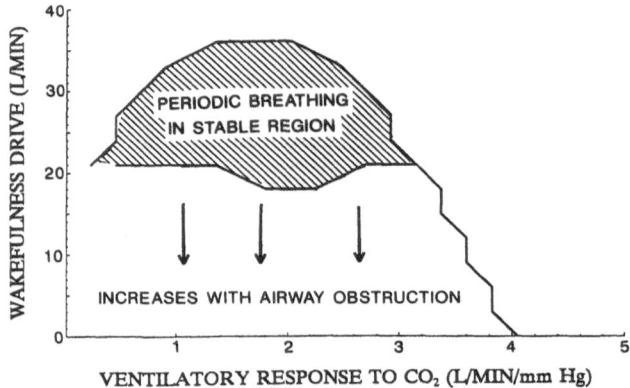

Figure 3. Additional region where PB can be observed when an arousal function is present.

result, since only the location of the stability region is modified. Although stable, configurations at the edge of the stable region are poorly damped. Consequently, although not sustained, the oscillations produced by the wake-to-sleep transition could persist sufficiently long to be of clinical significance.

Other Types of Oscillation

When an arousal function is present, several additional types of oscillations become possible, and, as shown in Fig. 3, make PB observable in a large portion of the stable region of Fig. 2. If the transition to sleep is sufficiently rapid, and the magnitude of the wakefulness drive sufficiently great, then the resulting oscillations trigger the arousal function, and this process repeats itself. This type of oscillation coexists with a stable equilibrium point (see next section), and can be more difficult to observe in the model if respiratory after discharge (3) is included in the model. More complex oscillations of this basic type are possible in and near the unstable region of the stability diagram.

The presence of an arousal function can also convert a previously stable center to one which is unstable. This occurs when the equilibrium blood gas values during sleep without an arousal function present are sufficient to trigger an arousal when that mechanism is intact. These conditions only arise in the model when there is a substantial difference between the magnitudes of the integrated ventilatory drive and the arousal drive. This occurs in the model when there is a marked sleep state dependent decrease in alveolar ventilation (as in airway obstruction), but the sensitivity of the arousal mechanism to chemical stimuli is affected less by the sleep state. For this type of PB, the size of the crosshatched area of Fig. 3 varies with the level of airway obstruction. Again, more complex oscillations of this basic type can be observed in and near the unstable region of the stability diagram.

Resetting the Oscillator

When oscillations occur about a stable equilibrium point, it is theoretically possible to drive the oscillating system to this stable equilibrium point with a pulse of energy of the appropriate magnitude if it is applied in an appropriate manner. The results of this computational experiment are illustrated in Fig. 4. Here, once sustained PB is present, a well chosen arousal pulse is applied, terminating the cycle of PB and leaving only damped oscillations at the system's natural frequency.

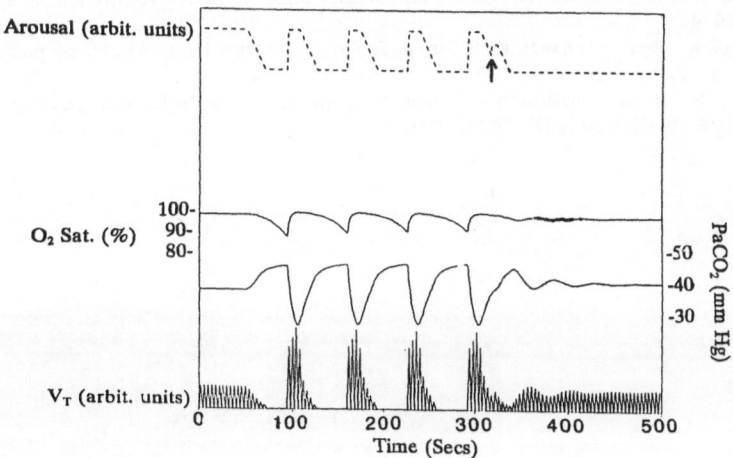

Figure 4. Termination of PB with a well chosen arousal pulse (arrow).

DISCUSSION

Clearly, many factors influence respiratory system behavior at sleep onset. This degree of complexity offers the possibility that PB could originate from many different mechanisms. Earlier theoretical studies have stressed enhanced chemoreceptor gain and increased lung-to-chemoreceptor transport delays as the mechanism of PB (4). However, recent theoretical studies (5) have shown how PB can be generated in a model that would otherwise be considered stable. This led to speculation that multiple modes of PB may exist and motivated the current theoretical study, in which we observed at least three distinct modes of sustained PB.

One important attribute of each mode of PB is the extent to which arousals are a component of the pattern of PB. In these studies, it is possible to observe PB where arousals are absent, incidental, facilitatory, or essential. A similar observation can be made for the variable role of airway obstruction in PB in this model. Airway obstruction can be absent, incidental, facilitatory, or essential to the generation of PB during sleep. The existence of multiple interacting modes of PB may help to explain why correcting single factors, such as airway obstruction, does not always eliminate PB, or why pure central or obstructive apnea patterns are not usually observed.

ACKNOWLEDGMENTS

Supported in part by the American Lung Association of Pennsylvania, the Hartford Foundation, National Institutes of Health Grant RR-01861, Specialized Center of Research Grant HL-42236, Teaching Nursing Home Award AG-03934, and National Institutes of Health Research Career Development Award HL-02536.

REFERENCES

1. Bellman, R., and K. L. Cooke. *Differential-Difference Equations.* New York: Academic Press, 1963.
2. Bulow, K. Respiration and wakefulness in man. *Acta Physiol. Scand. Suppl.* 59(209): 1-110, 1963.

3. Eldridge, F. L. Central neural respiratory stimulatory effect of active respiration. *J. Appl. Physiol.* 37: 723-735, 1974.

4. Khoo, M. C. K., R. E. Kronauer, K. P. Strohl, and A. S. Slutsky. Factors inducing periodic breathing in humans: A general model. *J. Appl. Physiol.* 53: 644-659, 1982.

5. Khoo, M. C. K., A. Gottschalk, and A. I. Pack. Sleep induced periodic breathing and apnea: A theoretical study. *J. Appl. Physiol.* 70: 2014-2024, 1991.

23

VOLUME HISTORY RESPONSE OF AIRWAY RESISTANCE

H. R. Ahmad, M. A. Khan,[1] M. Memon,[1] and M. N. Khan

Department of Physiology
[1] Department of Medicine
Faculty of Health Sciences
The Aga Khan University Medical Centre
Stadium Road, P.O. Box 3500, Karachi-74800, Pakistan

INTRODUCTION

Airways exhibit pressure volume hysteresis. The degree of airway hysteresis is directly proportional to smooth muscle tone. The lung parenchyma also exhibits similar pressure volume hysteresis. The effect on airway size depends on the relative difference between parenchymal and airway hysteresis as demonstrated by Froeb and Mead [1]. Hysteretic behaviour of airways and lung can be demonstrated by an act of deep inhalation, which is called a volume history (VH) manoeuvre. The model of Mead and Froeb has been analysed by Burns and Taylor [2]. These authors have shown that if airway volume represented by dead space (V_D) is plotted against lung volume and the subject is asked to breath in from FRC to TLC and back, three different pattern of responses emerge. First, an anticlockwise loop represents bronchodilation in response to VH; second, a clockwise loop will indicate bronchoconstriction; third, no loop will show absence of VH response. When airway hysteresis exceed parenchymal hysteresis, airway size is greater on the deflation limb than it is at any given lung volume along the inflation limb. Airway hysteresis less than parenchymal hysteresis leads to the opposite volume history response. Finally, if airway and parenchymal hysteresis are equal, volume history has no effect on airway size. It has been shown that this volume history response is maximal almost immediately after a deep inhalation, decays rapidly during the first minute and usually is completely reversed within 2-3 minutes following inspiration in normal subjects.

The response of airway resistance to β_2-agonist may depend on the relationship between the airway and lung parenchymal hysteresis in several conditions [2]. Since airway responsiveness to a bronchodilator is also influenced by a volume history (VH) induced bronchodilation or bronchoconstriction, this study is aimed

a) to dissect out the β_2-agonist response from VH response in asthmatics and patients with COPD, b) to find out whether or not change in VH response of airway resistance could detect early changes in structural functional relationship of lung parenchyma in respiratory disease.

Modeling and Control of Ventilation, Edited by S. J. G. Semple, L. Adams, and B. J. Whipp
Plenum Press, New York, 1995

SUBJECTS AND METHODS

25 asthmatics, 25 COPD patients and 25 normal volunteers (controls) were studied. Patients were selected on the basis of criteria of the American Thoracic Society (3). All patients were in the stable phase of their disease and the intervals between the last medication and the time of recording was 12 hours or more depending on the duration of action of the bronchodilator taken by the patients. Patients also avoided tea and coffee for atleast the past 12 hours. Non were currently taking cortico-steroids or chromolyn sodium.

The oscilloresistometer (Siregnost, Siemens) was used to continuously record the airway resistance as an indictor of airway tone in response to a VH in the form of deep inhalation before and 10 minutes after inhalation of 200 µg salbutamol (β_2-agonist) in metered aerosol.

Protocol

The experimental protocol was as follows. All subjects were comfortably seated and were asked to avoid deep inhalation for 5 minutes. During this period their baseline airway resistance (Raw) was continuously recorded at functional residual capacity (FRC) level. Subjects were then asked to inhale to total lung capacity (TLC) and back to FRC. The response of airway resistance to this deep inhalation (VH) was recorded over the subsequent 3 minutes and the mininum valve of Raw was observed. Salbutamol was then administered by aerosol inhaler (200 µg) and the response of airway resistance to bronchodilation was recorded after 10 minutes. Subjects then performed the VH manouvres to yield the combined effect of both bronchodilation and VH on airway resistance (Raw).

Mean (SD) pre and post VH, β_2 and VH + β_2 values of Raw were analyzed by means of paired t-test. The change in Raw was compared with baseline Raw values and all p-values obtained were with reference to this base-line value for each group.

RESULTS

Volume history (VH) alone reduced significantly the average Raw from 5.9±0.8 to 4.3±0.5 mbar.l⁻1.s⁻¹ in 25 asthamatics, whereas inhalation of β_2 agonist (β_2) decreased Raw to 3.6±0.6 mbar.l⁻¹.s⁻¹. However, it should be noted as shown in table 1 that VH response of Raw was reduced after β_2 inhalation. 6 out of 25 asthmatics showed a constrictor response to VH, which was observed to be reduced after β_2 inhalation.

In patients with COPD, the average Raw dropped from 6.8±0.9 to 5.5±0.5 and 6±0.5 in response to VH and β_2 inhalation respectively (table 1). However, Raw response to VH was absent after β_2 inhalation in contrast to asthmatics. 9 out of 25 patients with COPD showed a constrictor response to VH. There was significant VH response of Raw in 25 health volunteers both before and after β_2 inhalation, while salbutamol itself did not have any effect.

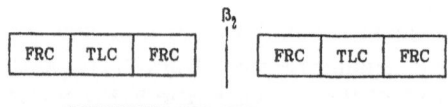

CONTINUOUS RECORDING OF Raw

Figure 1. Schematic representation of experimental protocol.

Table 1. Comparison of response of airway resistance (Raw) to volume history (VH) before and after B2 administration, p-values were obtained after applying paired t-test with reference to * baseline value of Raw at FRC

Group	Pre-β_2 inhalation				Post-β_2 inhalation			
	*Raw mbar.l^{-1}.s^{-1} ($\bar{x} \pm$ SD)	VH	Raw mbar.l^{-1}.s^{-1} ($\bar{x} \pm$ SD)		Raw mbar.l^{-1}.s^{-1} ($\bar{x} \pm$ SD)	VH	Raw mbar.l^{-1}.s^{-1} ($\bar{x} \pm$ SD)	
Asthma								
Group mean n=25	5.9 + 0.8	VH	4.3 = 0.5 P=0.0007		3.6 + 0.6 P=0.0002	VH	2.9 + 0.4 P=0.0001	
Constrictors n = 6	5.4 + 0.9	VH	6.0 + 0.7 P=0.002		3.8 + 0.7 P=0.0008	VH	4.2 + 0.8 P=0.001	
COPD								
Group mean n = 25	6.8 + 0.9	VH	5.5 + 0.5 P=0.001		6.0 + 0.5 P=0.002	VH	6.0 + 0.4 P=0.002	
Constrictors n = 9	5.7 + 0.4	VH	6.1 + 0.5 P=0.01		5.6 + 0.5 P=0.06	VH	5.8 + 0.6 P=0.06	
Controls								
Group mean n = 25	2.4 + 0.4	VH	1.3 + 0.4 P=0.0006		2.3 + 0.5 P=0.1	VH	1.2 + 0.4 P=0.0005	

Fig 2 shows comparison of the percentage reduction in Raw in response to VH, β_2 and VH + β_2. In controls, Raw response to VH, β_2 and VH + β_2 has been found to be 45±8.2, 5±3 and 48±9.5 percentage respectively. In asthmatics, VH alone decreased the average Raw by 28±9.7%. β_2 decreased it by 39±6.4% while VH + β_2 decreased Raw by 50±11.3%. In patients with COPD, Raw response to VH, β_2 and VH + β_2 was observed to be 20±9.4, 12±3 and 12±6.3 percentage respectively.

DISCUSSION

This study has shown the VH response of Raw can be considered as an indicator of relative airway and lung parenchymal hysteresis determing the degree of airway tone. The effect of VH & β_2 seems to be additive in asthmatics; while Raw response to VH + β_2 has been shown to be markedly reduced in patients with COPD when compared with asthmatics.

Pre-β_2 response of airway resistance to volume history in the asthmatic group can be explained by the greater airway hysteresis relative to that of the parenchyma [1]. It has been observed that β_2 inhalation significantly reduced VH response of airway resistance by 17% as shown in Fig. 2. This effect is assumed to be due to the reducation in airway hysteresis caused by β_2 inhalation [4,5,6]. This response can be both dilation as well as constriction of airways depending upon how much the airway hysteresis is reduced in relation to the parenchyma hysteresis. This study has shown that 6 asthmatics out of 25 had a bronchoconstrictor response to VH. It means their airway hysteresis was reduced more than that of parenchyma in response to VH. It may be assumed that factors such as local release of prostaglandins, reflex bronchon-striction and "contractile interstitial cells" of Kapanci could alter lung parenchymal hysteresis in addition to VH and β_2 [7,8,9].

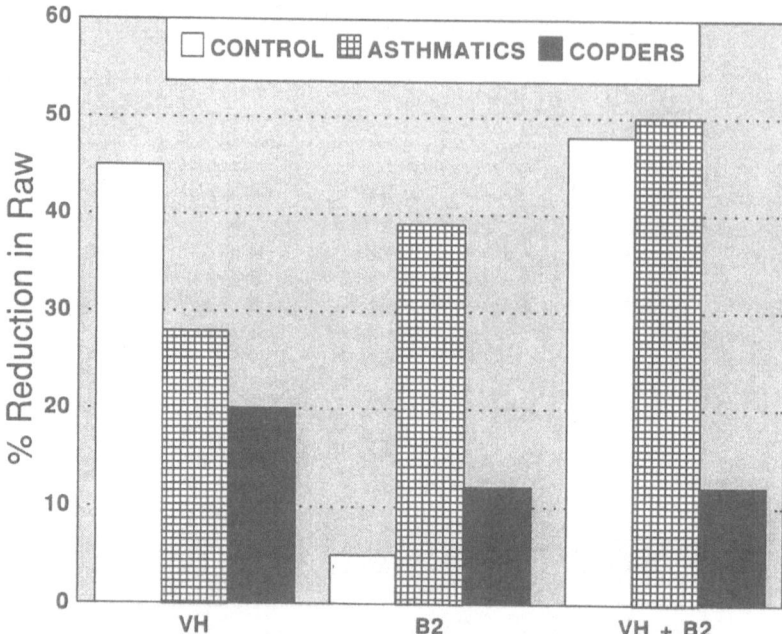

Figure 2. Comparison of the percentage reduction in airway resistance in response to volume history (VH), inhalation of β_2-agonist (β_2) and VH+β_2

Volume history as well as β_2 response of Raw in patients with COPD has been shown to be markedly reduced when compared with asthmatics. The VH response of Raw in these patients will critically depend on the intackness of morphology of lung parenchyma and/or airways. This study has shown that the VH response in COPD is demonstrable (table 1). It means the parenchymal support action or radial traction is operating. However, it should be mentioned that 9 out of 25 patients with COPDers showed a constrictor response to VH indicating probably loss of parenchymal integrity, which in turn can impair airways hysteresis as demonstrated by Fairshter [4]. It means that the difference in airway and parenchymal hysteresis in patients with COPD seems to be reduced.

The post β_2 VH response of Raw was reduced in asthmatics and was absent in patients with COPD (table 1). Since β_2 inhalation reduces the airway tone and thereby airway hysteresis in relation to that of parenchyma, the dilatory response of VH is reduced in asthmatics [6]. Lack of VH response of Raw after β_2 inhalation indicates that the airway and parenchymal hysteresis are equal in patients with COPD [1,2].

The VH response of Raw has been shown by this study to be 48, 28 and 20% in controls, asthmatics and patients with COPD respectively (fig 2). It also shows that both VH and β_2 responses of Raw are additive in controls as well asthmatics. However, such an effect is absent in patients with COPD. This study raises a question whether or not patients with COPD sigh less than control and asthmatics? However, this hypothesis and its consequence for patients with COPD has to be experimentaly established.

ACKNOWLEDGMENTS

Dedicated to the memory of Professor Dr.med. Dr.h.c. Hans H. Loeschcke (1912-86).

REFERENCES

1. Froeb HF and Mead J. Relative hysteresis of the dead space and lung in vivo. J. Appl. Physiol. 25: 244-248, 1968.
2. Burns CB, Taylor WR and Ingram RH Effect of deep inhalation in asthma; relative airway and pareveliymal hysteresis. J. Appl. Physiol. 59: 1590-1596, 1985.
3. American Thoracic Society. Chronic bronchitis, asthma, and pulmonary emphysema. Am. Rev. Respir. Dis. 1962; 84: 762-768.
4. Fairshter RD. Airway hysteresis in normal subject and individuals with chronic airflow obstruction. J. Appl. Physiol 1985; 58(5): 1505-1510.
5. Kariya ST, Thompson LM, Ingenito EP and Ingram RH. Effect of lung volume, volume history and methacholine on lung tissue viscance. J. Appl. Physiol 1989: 66(2): 977-982.
6. Wang YT, Thompson LM, Ingenito EP and Ingram RH. Effects of increasing doses of ß-agonists on airway and parenchymal hysteresis. J. Appl. Physiol 1990; 68(1): 363-368.
7. Gayrand P, Ovehek J, Grimand C and Charpin J. Bronchoconstriction effects of a deep inspiration in patients with asthma. Am. Re. Resp. Dis. 111:933-439, 1975.
8. Beaspre A and Orehek J. Factors influencing the bronchodilation effects of deep inspiration in asthmatic patients with provoked bronchoconstriction. Thorax 37: 124-128, 1982.
9. Kapanci Y, Assimacopoulos A, Irelec, Zwahlen A and Gabbiani G. "Contractile interstitial cells" in pulmonary alveolar septa: a possible regulator of ventilation/perfusion ratio. J. Cell Biol. 60: 375-392, 1974.
10. Weir DC, and Sherwood BP. Measures of reversibility in response to bronchodilators in chronic airflow obstruction: relation to airway calibre. Thorax 1991; 46:43-45.

24

NON-STATIONARITY OF BREATH-BY-BREATH VENTILATION AND APPROACHES TO MODELLING THE PHENOMENON

Pei-Ji Liang, Jaideep J. Pandit, and Peter A. Robbins

University Laboratory of Physiology
Parks Road, Oxford OX1 3PT, United Kingdom

INTRODUCTION

In resting human subjects, successive breaths are correlated (12). In order to investigate the relationship between the present breath and previous ones, time-series analysis (autoregressive modelling) has been applied to study the breath-by-breath variations in ventilation during steady breathing under different conditions for different species (4, 8, 9, 10).

As an alternative approach, noise processes in state-space form can be used to model the correlation between the residuals, and this has been applied in combination with dynamic models to describe the ventilatory response to changes in end-tidal P_{CO_2} (3, 5, 6).

Part of the purpose of the current study was to compare a range of simple time-series models to determine which might form an adequate description of the breath-by-breath correlation present in human ventilatory data. The study also investigates the use of a simple state-space model without an input function to provide a description of the correlation present in these data.

Theoretically, in simple time-series models and simple state-space models, stationarity is required for the sequences analysed. However, it has been suggested that respiratory data may often be non-stationary (1). Therefore a second part of the study involved testing the data for stationarity. Finally, a possible way of handling time-dependent respiratory data was investigated.

METHODS

Experimental Data

The experimental data came from a previous study (11). Experiments were performed on five subjects with two different protocols each lasting 43 min. In the first protocol

Modeling and Control of Ventilation, Edited by S. J. G. Semple, L. Adams, and B. J. Whipp
Plenum Press, New York, 1995

(protocol A), the subject was at rest and the end-tidal P_{CO2} was held constant at 2-5 Torr above the subject's natural resting value. In the second protocol (protocol B), the subject undertook exercise at 70 W, and the end-tidal P_{CO_2} was held at 2-5 Torr above the subject's natural value during exercise. During both protocols, end-tidal P_{O_2} was held at 100 Torr. Each protocol was repeated six times for each subject.

Model Fitting and Statistical Analysis

Ventilatory data during steady-state breathing were used for all the model fitting and statistical analysis. Data collected during the first 50 breaths for protocol A and the first 300 breaths for protocol B were discarded to remove any initial transients in the response.

The mean value for ventilation together with any linear trend were removed from the data before any model fitting and statistical analysis was undertaken.

Arma Model Fitting. The general form for an autoregressive moving average (ARMA) model is:

$$y(t)-a_1y(t-1)-\ldots-a_py(t-p) = \varepsilon(t)-c_1\varepsilon(t-1)-\ldots-c_q\varepsilon(t-q)$$

In this study three different low order models were applied to the data, they are AR_1 (one autoregressive coefficient only), AR_2 (two autoregressive coefficients) and AR_1MA_1 (one autoregressive coefficient and one moving average coefficient). A maximum likelihood technique was used for parameter estimation. A "portmanteau" test was used to test the model adequacy. The fittings were done using S-plus statistical software.

State-Space Model Fitting. As an alternative to ARMA models, the correlation of the successive ventilations during steady-breathing can also be modelled in state-space form with noise processes:

$$x(t+1) = fx(t)+v(t)$$

$$y(t) = x(t)+w(t)$$

where $x(t)$ is system state at time t; $y(t)$ is the observation at time t; f is the system gain; $v(t)$ and $w(t)$ are mutually independent white Gaussian noise processes with means of zero and constant variances of R_V and R_W respectively.

Using the algorithm of Kalman filtering, the system parameter f along with the ratio R_V/R_W can be estimated (2). The parameters of the state-space model were estimated using a least-squares method provided in the Numerical Algorithms Group Fortran library (NAG; Oxford, UK), subroutine E04FDF. Goodness of fit was tested by a "portmanteau" test to determine if the overall residuals sequence can be accepted as white.

Stationarity Test. Stationarity within data sets was tested by using evolutionary spectral analysis (13). With this method, a classification of the non-stationarity (whether or not the sequence is uniformly modulated) is also obtained. The principle of evolutionary spectral analysis is to apply a double window technique in both time and frequency domains to calculate values for evolutionary spectral density corresponding to the selected times and frequencies. A two-way analysis of variance on the logarithmic values of the spectral density using factors of time and frequency is then applied to test whether there is significant non-stationarity in the spectral density with time. If the sequence is non-stationary, the analysis also indicates whether or not the sequence is uniformly modulated.

Table 1. Number of data sets that were white for original data sequences, residuals from ARMA models and state-space model

Number of data sets analysed	Number of originally white sequences	White residuals from ARMA models			White residuals from state-space model
		AR_1	AR_2	AR_1MA_1	
60	1	15	31	46	48

RESULTS

Arma Model Fitting

Three kinds of low order time-series models were applied to each individual data set. They were AR_1, AR_2 and AR_1MA_1. The whiteness of residuals of each model was tested by a "portmanteau" test. The numbers of data sets with white residuals are listed in table 1. The AR_1MA_1 model is the most satisfactory model for both protocols.

State-Space Model Fitting

As an alternative to the approach of using ARMA models, one state-space model of the noise processed was investigated. The residuals were tested for whiteness in the same manner as for the residuals from the ARMA model to examine the model adequacy. The results are shown in table 1.

Stationarity Within Data Sets

The results of stationarity tests are listed in table 2, which show that only a few of the data sets were stationary. However, about half were uniformly modulated.

A uniformly-modulated sequence $X(t)$ can be written as:

$$X(t) = C(t)X_0(t)$$

where $X_0(t)$ is a stationary process and $C(t)$ is the modulating function. In this sequence, the mean value remains constant (equal to zero) and the only factor causing non-stationarity is a changing variance. We chose to try and construct a modulating function $C(t)$ for our data using an autoregressive structure as follows:

$$C^2(t) = 0.95C^2(t-1)+0.05y^2(t)$$

where $y(t)$ is the observation of the present time t.

Demodulating the original sequence by $C(t)$, we obtain the "demodulated" sequence. The stationarity analysis for this demodulated sequence was then repeated as for the original sequence. Following this, many more sequences become stationary as shown in table 2.

DISCUSSION

The major findings of this study are:

Table 2. Numbers of data sets or residuals (from the state-space model) which were either stationary or uniformly modulated before and after demodulation

	Number of data sets	Original		Resid. state-space	
		Stationary	Unif. mod.	Stationary	Unif. mod.
Before demodulation	60	5	31	4	31
After demodulation	60	35	40	45	48

1. Both a simple AR_1MA_1 model and a simple linear time invariant state-space model for the breath-to-breath correlations present in the ventilatory data can produce residuals which in the majority of cases are white overall.
2. Neither the simple AR_1MA_1 models nor the simple linear time invariant state-space model adequately describe the data because the data are non-stationary.
3. Though non-stationary, about half of the data sets (both original data sequences and residuals sequences from the state-space model) appear to be uniformly modulated. Further analysis suggests that the non-stationarity within the data sets can be considered to arise from variations in the ventilatory variance rather than from variations in the correlational structure.

These basic findings suggest that applying a state-space model in a more general form, known as a heteroscedastic form (7), may be appropriate. In this form the system equations can be written as follows:

$$x(t+1) = fx(t)+\zeta(t)$$

$$y(t) = x(t)+\eta(t)$$

where $\zeta(t)$ and $\eta(t)$ are noise processes with time-dependent variances:

$$R_\zeta = g(x(t), t)R_V$$

$$R_\eta = g(x(t), t)R_W$$

R_V and R_W have the same meaning as explained previously.

This heteroscedastic form allows the variance to vary with amplitude of ventilation and with time through the function $g(x(t),t)$. The advantage of this heteroscedastic form is that the same function $g(x(t),t)$ is contained in the variances of both noise sequences R_ζ and R_η so that it enables the same Kalman filtering algorithm to be used to estimate the coefficients as before, since the term in g drops out while estimating the ratio of the variances.

As a result of optimization, the ratio R_ζ/R_η (which equals to the ratio R_V/R_W) is estimated. Thus provided the residuals are white, the heteroscedastic form may well be an adequate form to represent the time-dependent behaviour exhibited by these sequences.

REFERENCES

1. Ackerson, L.M., Jones, R.H., and Bruce, E.N., 1989, Adaptive multivariate autoregressive modelling of respiratory cycle variables, In: *Respiratory Control-A Modeling Perspective.* New York: Plenum, p. 309-316.
2. Anderson, B.D.O., and Moore, J.B., 1979, *Optimal Filtering.* Englewood Cliffs: Prentice-Hall Inc., p. 36-45.

3. Bellville, J.W., Ward, D.S., and Wiberg, D., 1988, Respiratory system: modelling and identification, In: *System & Control Encyclopedia.* Oxford: Pergamon Press, p.4055-4062.

4. Benchetrit, G., and Bertrand, F., 1975, A short-term memory in the respiratory centres: statistical analysis, *Respir. Physiol.* 23: 147-158.

5. Dahan, A., Olievier, L.C.W., Berkenbosch, A., and DeGoede, J., 1989, Modelling the dynamic ventilatory response to carbon dioxide in healthy human subjects during normoxia, In: *Respiratory Control-A Modeling Perspective.* New York: Plenum, p. 265-273.

6. DeGoede, J., and Berkenbosch, A., 1989, Dynamic end-tidal forcing technique: modelling the ventilatory response to carbon dioxide, In: *Modelling and Parameter Estimation in Respiratory Control.* New York: Plenum, p. 59-69.

7. Harvey, A.C., 1989, *Forecasting, Structural Time Series Models and the Kalman Filter.* Cambrige University Press, p.344-345.

8. Jensen, J.I., 1987, *An Analysis of Breath-to-Breath Variability in Steady-States of Breathing in Man,* Ph.D. Thesis, Arahus University, Denmark.

9. Khatib, M.F., Oku, Y., and Bruce, E.N., 1991, Contribution of chemical feedback loops to breath-to-breath variability of tidal volume, *Respir. Physiol.* 83: 115-138.

10. Modarreszadeh, M., Bruce, E.N., and Gothe, B., 1990, Nonrandom variability in respiratory cycle parameters of humans during stage 2 sleep, *J. Appl. Physiol.* 69: 630-639.

11. Pandit, J.J., and Robbins, P.A., 1992, Ventilation and gas exchange during sustained exercise at normal and raised CO_2 in man, *Respir. Physiol.* 88: 101-112.

12. Priban, I. P., 1963, An analysis of some short-term patterns of breathing in man at rest, *J. Physiol.* 166: 425-434.

13. Priestley M.B., and Rao, S.S., 1969, A testing of non-stationarity of time-series, *J. Royal Stat. Society, Ser. B,* 31: 140-149.

25

EFFECT OF REPETITIVE TESTING ON BREATHLESSNESS

James W. Reed and M. M. Feisal Subhan

Department of Physiological Sciences
University of Newcastle, Medical School
Framlington Place, Newcastle upon Tyne, NE2 4HH, United Kingdom

INTRODUCTION

Estimation of breathlessness during exercise by the use of either the visual analogue scale (VAS) or the modified Borg scale is reported to be highly stable over time periods ranging from minutes to weeks (1,6,8,9). Such reproducibility is essential both for clinical studies of the efficacy of pharmacological intervention on the sensations of breathlessness in patients and also in mechanistic studies of the effects of dyspnogenic agents in healthy volunteers. Recent studies however, in both normal subjects and patients with chronic obstructive lung disease, have shown poor reproducibility (4,5,10) and thrown doubt on the basic validity of the measurement.

We have previously demonstrated (7) that although the slope of the VAS/ventilation relationship during progressive exercise is indeed highly reproducible when assessed daily for ten days, the mean threshold increased successively to a plateau value. Using change point regression the plateau was found to occur at approximately the fifth occasion. The underlying mechanism was unclear but changes in the subjective criteria used to estimate breathlessness as a consequence of repetitive exposure could not be ruled out. To investigate the contribution of repetitive testing we have therefore repeated the exercise experiments and compared the breathlessness responses to those reported during repetition of carbon dioxide (CO_2) rebreathing over a similar period. As the major effect appears to take place over the first five occasions, the time course of this study was limited to 5 days.

The underlying premise was that comparable changes in breathlessness during repetitive CO_2 rebreathing and exercise would be likely to be as a result of a change in assessment criteria.

METHODS

The subjects were volunteers, drawn from students and staff of the University. All were healthy non-smokers, naive as to the purpose of the study and with no history of

Modeling and Control of Ventilation, Edited by S. J. G. Semple, L. Adams, and B. J. Whipp
Plenum Press, New York, 1995

123

cardiopulmonary disease. Each undertook either the exercise or the rebreathing procedure daily for 5 days, at the same time of day and at a fixed time after the last meal. Breathlessness was quantified using a 100 mm linear potentiometer which controlled a 100 mm horizontal linear visual analogue scale of light emitting diodes the extremes of which were labelled 'not at all breathless' (0 mm) and 'extremely breathless' (100mm). The output also went to a pen recorder from which the measurements were taken. The study was approved by the local Ethical Committee

Exercise Test

Eight subjects (7 male) took part, mean age was 21 ± 2 years (mean ± S.D., range 19 to 24). The progressive exercise tests were conducted on an electrically braked cycle ergometer (Siemens, 740 Ergomed). Ventilation minute volume was measured on a breath-by-breath basis by means of a pneumotachograph (P.K. Morgan Benchmark Exercise System). Prior to the exercise beginning, the subject was seated on the ergometer and whilst breathing through a mouthpiece, with a nose clip, was allowed a few minutes to become accustomed to the apparatus. The subjects were requested to differentiate between breathlessness and other sensations associated with exercise such as leg fatigue or the awareness of increased breathing. On subsequent questioning all considered that they were able to do this without difficulty.

Data was recorded for two minutes at rest, then the subject started unloaded pedalling at a frequency of between 50-60 revolutions per minute. After one minute, the resistance to pedalling was increased by 20 Watts each minute. The exercise continued to a symptom-limited maximum, with breathlessness being reported each minute. The protocol was repeated exactly on all subsequent occasions, the subjects were told to use the same criteria in estimating their sensations.

CO$_2$ Rebreathing

An additional eight subjects (4 male) undertook the CO$_2$ rebreathing. Their mean age was 26 ± 7 years (mean ± S.D., range 21 to 39). The procedure entailed rebreathing CO$_2$ in oxygen (modified Read rebreathing procedure) from a 6 litre bag-in-box system via a mouthpiece and nose clip. Gas was sampled from the mouthpiece and analysed for CO$_2$ (infra-red) and oxygen (paramagnetic) before being returned to the centre of the rebreathing bag. The ventilation minute volume and respiratory frequency were measured using a rotating vane anemometer and recorded each half minute. The sensations of breathlessness were described in exactly the same manner as for the exercise tests, estimations were recorded at half minute intervals. The experiment was terminated once the CO$_2$ level in the system exceeded 9.0% or was symptom limited. As with the exercise testing, the rebreathing protocol was repeated exactly on all subsequent occasions; the subjects were told to use the same criteria in estimating their sensations.

For both studies, the VAS scores were regressed against the minute ventilation and the relationship expressed in terms of slope and intercept.

RESULTS

Large inter-subject variations in reported breathlessness were observed, which is consistent with previous studies (2, 9). During exercise the slope of the VAS/$\dot{V}_E$ relationship was between 0.11 - 1.75 mm. min. l^{-1}; during CO$_2$ rebreathing the equivalent values were 0.48 - 1.92 mm. min. l^{-1}. The range of inter-subject values for the threshold response was

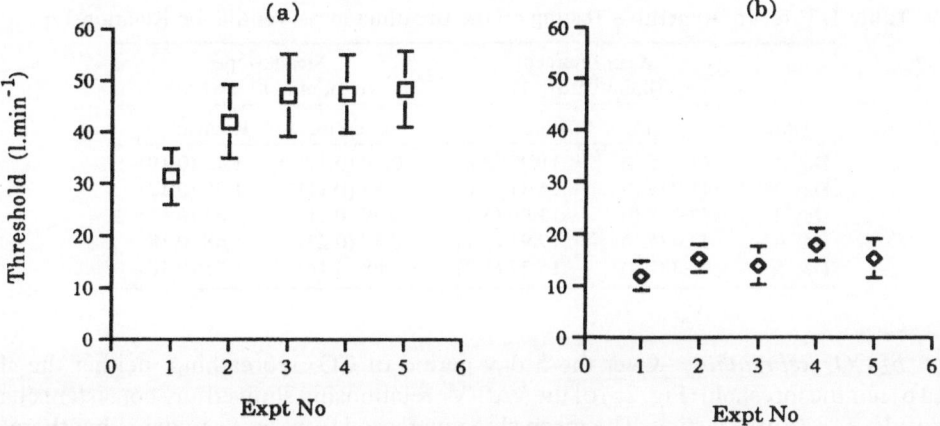

Figure 1. Effect of repetitive exercise (a) and CO_2 rebreathing (b) on the mean slope of the VAS/Ventilation relationship.

for the exercise group from 14.06 - 55.40 l.min^{-1}, and for the rebreathing group 3.52 - 30.19 l.min^{-1}.

There was no difference in response as between the male and female volunteers.

Effects of Repetition

a) Exercise. As found previously, the slope of the VAS/$\dot{V}_E$ relationship during exercise showed no consistent change over the repetition period (Fig. 1a). The threshold, however, progressively increased from an initial value of 31.3 ± 5.4 l.min^{-1} (mean ± SEM) to a maximal 48.3 ± 7.4 l.min^{-1} (mean ± SEM) by day 5 (Fig. 2a). Change point regression confirmed that the threshold did not increase significantly after day 3, resulting in a mean plateau of 47.6 ± 7.7 l.min^{-1} (mean ± SEM). The mean daily values for the exercise threshold and slope are shown in Table 1.

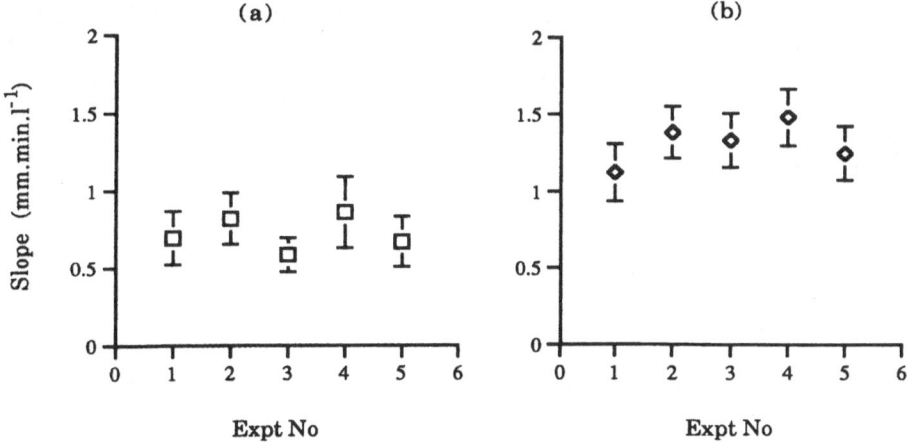

Figure 2. Effect of repetitive exercise (a) and CO_2 rebreathing (b) on the mean threshold of the VAS/Ventilation relationship.

Table 1. Effect of Repetitive Testing on the Breathlessness/Ventilation Relationship

Time	Mean Intercept (l.min^{-1} ± SEM)		Mean Slope (mm. min. l^{-1} ± SEM)	
	Exercise	Rebreathing	Exercise	Rebreathing
Day 1	31.3 (5.4)	11.99 (2.8)	0.70 (0.17)	1.12 (0.19)
Day 2	41.9 (7.2)	15.31 (2.6)	0.83 (0.17)	1.38 (0.17)
Day 3	47.1 (8.0)	13.90 (3.7)	0.59 (0.11)	1.33 (0.17)
Day 4	47.4 (7.6)	17.91 (3.1)	0.87 (0.23)	1.48 (0.18)
Day 5	48.3 (7.4)	15.37 (3.7)	0.68 (0.16)	1.25 (0.18)

b) CO_2 Rebreathing. Over the 5 day period of CO_2 rebreathing, neither the slope (Fig. 1b) nor the threshold (Fig. 2b) of the VAS/$\dot{V}_E$ relationship showed any consistent change as a result of repetitive testing. The mean slope increased from day 1 to day 2 but thereafter was reproducible. The mean intercept also showed a slight increase from the first day to the second, but also then stabilised and showed no consistent effect of repetition. The mean daily values are shown in Table 1.

As expected, the sensation of breathlessness at any given ventilation was materially greater during CO_2 rebreathing than during exercise. This was as a result of both an earlier response and increased gain.

DISCUSSION

For logistical reasons it was not possible to use the same subjects for both exercise and CO_2, but as the response of this exercise group to repeat exercise testing was similar to that of the initial study (7) this is not thought to be a confounding issue. Similarly, the pattern of response of the female subjects was identical to that of the males for both exercise and carbon dioxide rebreathing, so the imbalance between the two test groups was not considered to have influenced the results.

As anticipated, the ventilatory threshold for breathlessness during exercise was found to increase as a consequence of repetitive testing. The underlying mechanism for the observed changes in the threshold are unclear, but if we postulate that breathlessness is dependent on efferent output from some medullary centre (3) then the response could be due to either (a) an alteration in the criteria used by the subjects to estimate the magnitude of the sensation, (b) a reduction in the efferent output for a given input, or (c) a modification of the afferent input.

Although some variation cannot entirely be ruled out, the lack of any change during the rebreathing experiments suggest that changes in the criteria of sensing breathlessness are unlikely to account for the magnitude of the exercise effect.

As an equivalent repetition of CO_2 rebreathing does not have any consistent effect on reported breathlessness, the effect of repetitive testing *per se* is unlikely to have contributed to the effect seen during repetitive exercise testing. The inference therefore is that it is the repeated exposure to exercise itself which is the determining factor. Repeated exercise is commonly associated with a cardiovascular or muscular training effect; such an effect would be consistent with (c) above, a modification of afferent input. This remains to be fully tested.

Whatever the basis, it is of interest that the change only occurs in the threshold, and not the slope, which suggests that at least two mechanisms are involved in the generation of the sensation.

In summary, the slope of the VAS/$\dot{V}_E$ relationship during exercise is highly reproducible on a day-to-day basis. However, the threshold for breathlessness during exercise shows a progressive increase up to a plateau value. The mechanism underlying these changes is unclear but data from the rebreathing study does not support the concept of a change in the assessment criteria as a result of repetitive exposure.

REFERENCES

1. Adams, L., N. Chronos, R. Lane, and A. Guz. The measurement of breathlessness induced in normal subjects: validity of two scaling techniques. *Clin. Sci.* 69: 7-16, 1985.
2. Adams, L., N. Chronos, R. Lane, and A. Guz. The measurement of breathlessness induced in normal subjects: individual differences. *Clin. Sci.* 70: 131-140, 1986.
3. Adams, L., and A. Guz . Dyspnea on Exertion. In: Whipp BJ, Wassermann K, eds. *Exercise, pulmonary physiology and pathophysiology.* New York: Marcel Dekker: 449-494, 1991.
4. Belman, M.j., L.r. Brooks, D.j. Ross, and Z. Mohsenifar. Variability of breathlessness measurement in patients with chronic obstructive pulmonary disease. *Chest* 99: 566-571, 1991.
5. Mador, M.j., and T.j. Kufel. Reproducibility of visual analogue scale measurements of dyspnea in patients with chronic obstructive pulmonary disease. *Am. Rev. Respir. Dis.* 146: 82-87, 1992.
6. O'neill, P.a., R.d. Stark, S.c. Allen, and T.b. Stretton. The relationship between breathlessness and ventilation during steady-state exercise. *Clin. Respir. Physiol.* 22: 247-250, 1986.
7. Reed, J.W., and M.M.F. Subhan. Effect of repetitive testing on breathlessness (Abstract). *J. Physiol.* 480: 54P, 1994a.
8. Silverman, M., J. Barry, H. Hellerstein, J. Janos, and S. Kelsen. Variability of the perceived sense of effort in breathing during exercise in patients with chronic obstructive pulmonary disease. *Am. Rev. Respir. Dis.* 137: 206-209, 1988.
9. Stark, R.D., S.A. Gambles, and J.A. Lewis. Methods to assess breathlessness in healthy subjects: a critical evaluation and application to analyse the acute effects of diazepam and promethazine on breathlessness induced by exercise or by exposure to raised levels of carbon dioxide. *Clin. Sci.* 61: 429-439, 1981.
10. Wilson, R.C., and P.W. Jones. A comparison of the visual analogue scale and modified Borg scale for the measurement of dyspnoea during exercise. *Clin. Sci.* 76: 277-282, 1989.

26

PATHOPHYSIOLOGY OF BREATHING CONTROL AND BREATHING AWAKE AND ASLEEP

Postscript

Richard Casaburi

Division of Respiratory and Critical Care, Physiology and Medicine
Harbor-UCLA Medical Center
1000 W. Carson Street, Torrance, California 90509

> *"The proper study of mankind is man"*
> —Alexander Pope (1688-1744)

The years have made Pope's comment politically incorrect (womankind, too, must be studied), but no less valid. Though important physiological insights have been obtained from studies of experimental animals, most would agree that the final test of relevance depends on studies of human subjects.

There is a long tradition in physiologic investigation of drawing inferences from the responses of individuals in whom one or more potential information pathways have been ablated. These studies fall into two categories: experiments of man and experiments of nature. Experiments of man usually involve (reversible) blockade of neural pathways. Experiments of nature are exploited when we study the response to organ system deficits, e.g. damage to the heart, lung, brain or neural structures.

Most of the studies in this session could be characterized as experiments of man or of nature. In the former category, Drs Kochi and Nishino[*] sought to understand the mechanisms of response to inspiratory resistance by depressing the upper airway muscles in anesthetized subjects. Dr. Nagyova and his colleagues at Oxford studied the genesis of normal breathing patterns by observing the relative effects of two inhaled and one infused anesthetic agents on respiratory volume and timing. In a second study from the Oxford group, Dr. Liang and her colleagues observed the long term variation in breathing pattern in an attempt to gain insight into the algorithms programmed into the respiratory controller. Dr. Datta and his coinvestigators in London sought to determine if the sensation of breathlessness

---

[*] The authors noted with asterisk gave presentations at the meeting but did not submit these for publication.

Modeling and Control of Ventilation, Edited by S. J. G. Semple, L. Adams, and B. J. Whipp
Plenum Press, New York, 1995

that occurs as a consequence of chronic obstructive pulmonary disease can be modified by an anxiolytic agent.

In the category of studies founded on experiments of nature, Heywood et al.* investigated the consequences of lower brainstem abnormalities on inhaled CO_2 sensitivity and other aspects of respiratory control. Kobayashi* and his colleagues sought to explain the increased ventilatory requirement for exercise observed in patients with chronic heart failure. Finally Dr. Murphy* and his coinvestigators sought to define the factors assuring upper airway patency by observing the configuration of the upper airways in laryngectomized patients.

These studies illustrate the value of investigations involving human subjects in improving our understanding of respiratory control mechanisms.

Exercise and Pulmonary Ventilation

27

EXERCISE HYPERPNEA

Chairman's Introduction

J. A. Dempsey

University of Wisconsin-Madison
Department of Preventive Medicine
504 N. Walnut Street, Madison, Wisconsin 53705

HYPERPNEA OF MODERATE EXERCISE

As judged by the number of presentations at this meeting dealing with this topic, exercise hyperpnea continues to be a popular focus of research-although its primary mechanisms remain a dilemma. To summarize recent trends, this reviewer's bias is that little if any support remains for the traditional "CO_2 flow" hypothesis. The normal ventilatory response to steady-state exercise in the lung/heart transplant patient have added the final piece of negative evidence (in humans) against this hypotheses. Similarly, cardiopulmonary afferents from working locomotor muscles probably play only a minor reflex role in hyperpnea. Based on studies of chemoreceptor denervation, it is also unlikely that these types of time-dependent influences play a significant role in hyperpnea. Finally, a role for traditional carotid and/or medullary chemoreceptors in hyperpnea continues to be advanced with the evidence that arterial H^+ is tightly correlated with the hyperpnea during progressive increases in work rate (see W. Stringer); but as has been the case so often in the past, it seems just as likely that the ventilatory response is dictating the change in arterial H^+ rather than vice versa. The kinetics of the ventilatory response to exercise onset may involve mechanisms which are masked during steady-state periods. Several studies in this conference dealt with this topic (see M. Walsh *et al.*, A. Datta *et al.*). Some suggested-based on indirect evidence -that carotid chemoreceptors might play some stabilizing role for ventilation during these transient phases of exercise.

Feed-forward locomotor-linked stimuli originating in the higher CNS may be the primary stimulus to exercise hyperpnea. Beginning with the pioneering studies of Eldridge and colleagues in the early 80's using fictive locomotion in the decorticate cat (1), a few subsequent studies have confirmed the importance of this feed-forward effect. However, these studies are extremely difficult to conduct and the relevance of the decorticate, locomoting cat to physiologic exercise hyperpnea is not entirely certain. Nevertheless, this reviewer's bias is that these locomotor-linked feed-forward stimuli are likely to be a very important player in the hyperpnea mechanism.

Modeling and Control of Ventilation, Edited by S. J. G. Semple, L. Adams, and B. J. Whipp
Plenum Press, New York, 1995

Some work which relates to this mechanism and to the concept of a "central" programmer was presented at this conference. For example, we learned that exercise increased activation of the primary motor cortex in humans (see L. Adams *et al.*) and that at least some of the control of exercise hyperpnea in quadrupeds is malleable and therefore perhaps "learned" in response to repeated, added sensory stimuli (see D. Turner *et al.*). It was also shown that the trotting quadruped commonly breathes at 5-6 Hz frequencies and shows electrical activation and significant shortening of the diaphragm and activation of expiratory muscles of the rib cage and abdominal wall with every breath (see D. Ainsworth *et al.*). This staccato-like locomotor-linked pattern of respiratory muscle recruitment appears very much as if it was "pre-programmed" to suit multifaceted demands of gas exchange, thermoregulation and even assist in locomotion and chest wall stabilization.

The role of supra-medullary, neural feed-forward influences on exercise hyperpnea and its susceptibility to short- and long-term modulation by sensory influences deserves our attention in future research of the mechanisms of hyperpnea.

THE HYPERVENTILATORY RESPONSE TO HEAVY EXERCISE

Hyperventilation usually begins as work rate is increased above about 60-65% of $\dot{V}O_{2max}$. This hyperventilatory response is important, because it helps alleviate some of the concomitant metabolic acidosis. Also in very heavy exercise-especially in the highly trained human-as the alveolar-to-arterial O_2 difference widens to 3-4 times resting levels, this hyperventilatory response is also important in preventing significant arterial hypoxemia. As discussed at this conference, the magnitude of this compensatory hyperventilation was shown to depend on the test protocol (fast *vs* slow "ramp", see B. W. Scheuermann *et al.*). Furthermore, the change in arterial PCO_2 was shown to have far reaching effects, even on O_2 off-loading at the site of the skeletal muscle capillary (see Y. Fukuba *et al.*).

What is the primary mechanism of this hyperventilatory response in heavy exercise? Based on classic studies in carotid body denervated asthmatic humans (2) and the effects of transient hyperoxia in intact humans, it is commonly believed that the carotid chemoreceptors are the key mediator of this hyperventilatory response. What has been debated more recently is exactly what type of humoral stimulus is responsible for this chemoreceptor activation. Based on correlative data in humans, or their demonstrated capability for increasing carotid sinus nerve activity in the anaesthetized animal, increases in circulating H^+, K^+ or norepinephrine are all prime candidates. However, it is debatable whether any of these potential stimuli present sufficient sensitivity to account for the very large amounts of "extra" ventilation being generated in heavy exercise. Two of these potential stimuli, K^+ and norepinephrine, were discussed in the present conference. One correlative study in anaesthetized cats showed the stimulatory effects of plasma norepinephrine and K^+ on carotid sinus nerve discharge (see G. Heinert *et al.*). Previous work in normal humans (with glycogen depletion) has shown that increases in metabolic $[H^+]$ in heavy exercise are not obligatory for the hyperventilatory response.

So the hunt continues for the stimulus or combination of circulating stimuli acting on the carotid chemoreceptors which might act synergistically to explain the hyperventilation of heavy exercise. There seems little doubt that these known chemoreceptor stimuli must change sufficiently in heavy exercise and have sufficient sensitivity to account for at least some portion of the hyperventilatory response. However, these questions have not been answered in the exercising human.

There are two additional considerations that add to the confusion of deciphering the causes of the hyperventilatory response to heavy exercise. First, there has been only one longitudinal experiment conducted (with the appropriate use of controls) to test the role of the carotid chemoreceptors during heavy exercise. These data were obtained in the pony which showed that denervation of the carotid chemoreceptor caused **more**-not less-hyperventilation during heavy exercise when compared to the same animals when intact (3). Perhaps species differences may explain this apparent discrepancy with the human asthmatic data (2). For example, hyperventilation occurs even during moderate exercise in quadrupeds like the pony. On the other hand, the intact pony does show a human-like ventilatory responsiveness to infused lactate and other chemoreceptor stimuli and hyperventilation is enhanced in this animal with the onset of heavy exercise.

So, these data point to an inhibitory role for the carotid chemoreceptor in heavy exercise, and perhaps the source of this inhibition might be found in the arterial hypocapnia incurred during heavy exercise. Recent data in the awake, resting goat and dog with isolated perfusion of the carotid chemoreceptor speak to this question, and some of these data are presented at this conference (see C.A. Smith *et al.*). They show that carotid body hypocapnia in the physiologic range (-3 to -14 mmHg ΔPCB_{CO2}:

a. causes a progressive reduction in VT and $\dot{V}E$;

b. causes reductions in $\dot{V}E$ that are identical to those caused by carotid body hyperoxia (>500 mmHg PCB_{O2});

c. that when hypocapnia was combined with moderate to severe hypoxemia-both changes occurring at the site of the isolated carotid body alone-much of the ventilatory response to hypoxia, was negated by the carotid body hypocapnia.

So, we think that this feedback inhibition may be of significant consequence to the "net" output of the carotid chemoreceptor and to ventilation itself in the heavily exercising animal. In other words, one should not think only of stimulation occurring at the carotid chemoreceptor under these circumstances.

Of course, these suggestions also require that some factor other than carotid chemoreceptor stimulation causes the hyperventilation. We would speculate that the central, locomotor-linked feed-forward stimulus that we favour for the hyperpnea of moderate exercise (please see above) would also account for much of the curvilinear increase in ventilation (with respect to $\dot{V}CO_2$) observed during heavy exercise. At the onset of "fatigue" of the locomotor muscles, sensory cues would trigger the pre-programmed locomotor areas of the higher CNS to augment their output in order to maintain force output of the exercising locomotor muscles. Presumedly, the fall-out from this enhanced motor unit recruitment would include further stimulation of the medullary respiratory pattern generator and more ventilation.

An additional mechanical influence on the hyperventilation of heavy exercise may also be significant-especially as more and more of the tidal volume becomes flow limited. This occurs most often in those humans with greater than normal fitness who have increased $\dot{V}O_{2max}$ and higher than normal ventilatory demand, but without concomitant adaptation of the mechanical properties of the lung, airways or chest wall (4). Certain mechanical resistive and elastic "loads" are increased under these extraordinary conditions and reflexly-mediated feedback inhibition of ventilatory drive may occur *i.e.* so-called "central fatigue" of the inspiratory muscles. This phenomenon might explain in part why many of the highly trained show much less hyperventilation (than the untrained) during heavy exercise. Whether this mechanoreceptor-mediated inhibition does indeed occur and through what sensory nervous pathways have not been clearly delineated. Unfortunately, what is certain is that the combination of excitatory and inhibitory influences bearing on the control of breathing in heavy exercise make the understanding of this hyperventilatory response equally, if not more, complex than the already perplexing dilemma of the hyperpnea of moderate exercise.

ACKNOWLEDGMENTS

I am grateful to Ms. Gundula Birong for her preparation of the manuscript. Original research reported here was supported by NHLBI.

REFERENCES

1. Eldridge, F.L., D.E. Milhorn, J.P. Kiley and T.G. Waldrop. Stimulation by central command of locomotion, respiration and circulation during exercise. *Respir. Physiol.* 59:313-337, 1985.
2. Wasserman, K., B.J. Whipp, S.N. Koyal, and M.G. Cleary. Effect of Carotid Body Reserection on Ventilatory and Acid-Base Control during Exercise. *J. Appl. Physiol.* 39:354-358, 1975.
3. Pan, L.G., H.V. Forster, G.E. Forster, C.L. Murphy, and T.F. Lawry. Independence of exercise hyperpnea and acidosis during high intensity exercise in ponies. *J. Appl. Physiol.* 60:1016-1024, 1986.
4. Johnson, B.D., K.W. Saupe and J.A. Dempsey. Mechanical constraints on exercise hyperpnea in endurance athletes. *J. Appl. Physiol.* 73:874-886, 1992.

28

RESPIRATORY COMPENSATION, AS EVIDENCED BY A DECLINING ARTERIAL AND END-TIDAL PCO$_2$, IS ATTENUATED DURING FAST RAMP EXERCISE FUNCTIONS

B. W. Scheuermann and J. M. Kowalchuk

Faculty of Kinesiology and Department of Physiology
University of Western Ontario
London, Ontario, Canada N6A 3K7

INTRODUCTION

During low to moderate intensity exercise below the ventilatory threshold (T_{vent}), ventilation ($\dot{V}_E$) increases in proportion to CO$_2$ production ($\dot{V}CO_2$), resulting in arterial PCO$_2$ (P_aCO_2) remaining at, or increasing slightly above, resting levels. As the exercise intensity increases beyond the T_{vent}, $\dot{V}_E$ increases at a faster rate than $\dot{V}CO_2$ (i.e. hyperventilation with respect to $\dot{V}CO_2$) when the work rate (WR) is incremented slowly (i.e. a steady- or quasi-steady-state is reached during each step) (7,8). However, when the WR is increased rapidly (i.e. using step increments of $\leq$ 1 min or ramp exercise functions), $\dot{V}_E$ continues to increase in proportion to $\dot{V}CO_2$ as the end-tidal ($P_{ET}CO_2$) and arterial PCO$_2$ remain relatively constant (i.e. isocapnic buffering) (8,9). This isocapnic buffering phase reflects a specific ventilatory response to exercise above the T_{vent} where a combination of increased breathing frequency (f) and decreased time of expiration (T_E) effectively curtail the systematic increase in $P_{ET}CO_2$ and P_aCO_2 associated with exercise below T_{vent} (8). In this exercise paradigm respiratory compensation for the developing acidosis is delayed relative to the T_{vent} and follows the period of isocapnic buffering.

The extent of respiratory compensation may be dependent on the slope of the WR forcing function. Ward and Whipp (6) observed that the $P_{ET}CO_2$ was approximately 7 Torr lower at exhaustion following incremental exercise at 6 W·min^{-1} compared with 50 W·min^{-1}. Hughson and Green (5) observed that while P_aCO_2 was reduced below resting levels at exhaustion from slow ramp exercise (8 W·min^{-1}), P_aCO_2 was greater than resting levels following fast ramp exercise (65 W·min^{-1}).

Therefore, the purpose of this study was to examine the ventilatory response to slow and fast ramp exercise protocols and to determine the mechanism for the apparent lack of respiratory compensation for the metabolic acidosis during fast ramp exercise.

Modeling and Control of Ventilation, Edited by S. J. G. Semple, L. Adams, and B. J. Whipp
Plenum Press, New York, 1995

METHODS

Seven male subjects (age, 25.9 ± 4.8 yrs (mean ± SD); height, 1.78 ± 0.04 m; mass, 78.2 ± 6.3 kg) performed cycle ergometer exercise (Lode, H-300-R) to volitional fatigue on 2 separate occasions. Exercise consisted of 4 min 0 W pedalling, followed by progressive exercise whereby the WR increased as a ramp function at either 8 W·min^{-1} (Slow ramp, SR) or 65 W·min^{-1} (Fast ramp, FR).

Inspired and expired airflow were measured by a low resistance bi-directional turbine (VMM-110, Alpha Technologies). Fractional concentrations of O_2, CO_2, and N_2 were determined continuously (20 ml·s^{-1}) from the inspired and expired air sampled at the mouth by a mass spectrometer (Airspec MGA2000). Analog signals from the mass spectrometer

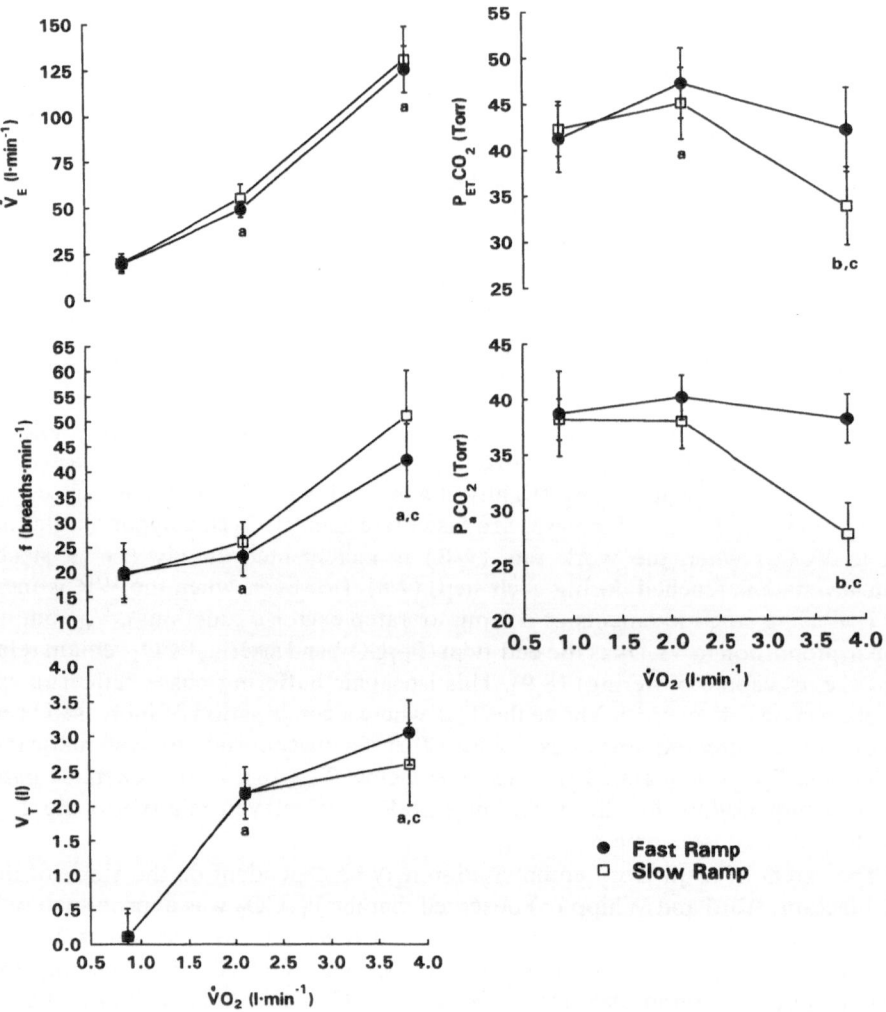

Figure 1. Ventilation ($\dot{V}_E$), respiratory pattern (tidal volume, V_T; breathing frequency, f), and end-tidal ($P_{ET}CO_2$) and arterial PCO_2 (P_aCO_2) at the $\dot{V}O_2$ corresponding to loadless cycling, the ventilatory threshold (T_{vent}) and exhaustion. Symbols: a, difference (p<0.05) from loadless cycling in FR and SR; b, difference (p<0.05) from loadless cycling in SR only; c, difference (p<0.05) between FR and SR.

and turbine transducer were sampled every 20 ms for the determination of ventilatory and gas exchange variables using breath-by-breath techniques, with transit delays in the analysis system and fluctuations in lung gas stores being accounted for in the computer algorithms (2). The T_{vent} was estimated using the following gas exchange criteria: *i*) an increase in $\dot{V}_E$ relative to $\dot{V}O_2$, *ii*) an increase in $\dot{V}CO_2$ relative to $\dot{V}O_2$, *iii*) an increase in $P_{ET}O_2$ without a decrease in $P_{ET}CO_2$, and *iv*) an increase in $\dot{V}_E/\dot{V}O_2$ without an increase in $\dot{V}_E/\dot{V}CO_2$ (3,9). The slope of the alveolar phase of CO_2 output was determined from off-line processing of the analog signal to reconstruct the intra-breath profile, after appropriate delay for wash-out of the airway deadspace (1). For each individual, a mean slope was determined on 5 breaths at each time period. Arterialized-venous blood was sampled from a dorsal hand vein into heparinized syringes (lithium heparin) and analyzed for blood gas and acid-base variables (Nova StatProfile 5, Nova Biomedical). Blood was sampled at rest, and at 2 min intervals during SR and 30 s intervals during FR to ensure a high density of samples during FR.

RESULTS

No differences were found for any of the variables during the loadless cycling that preceded the SR and FR functions, and thus, values were combined and reported as the mean ($\pm$ SD). Increases in $\dot{V}_E$ and $\dot{V}CO_2$ were similar (relative to $\dot{V}O_2$) in SR and FR during exercise below T_{vent} (Fig. 1); no differences were observed in the $\dot{V}O_2$ corresponding to T_{vent} between SR (2164 ± 403 ml·min^{-1}) and FR (2057 ± 318 ml·min^{-1}). Peak $\dot{V}_E$ was similar in SR and FR, but $\dot{V}CO_2$ was lower ($p<0.05$) in SR (Table 1), resulting in a higher ($p<0.05$) $\dot{V}_E/\dot{V}CO_2$ in SR (34.1 ± 4.3) compared to FR (26.7 ± 3.5). At exhaustion, the plasma $[H^+]$ was greater ($p<0.05$) than at rest, with no difference between conditions (SR, 50.0 ± 7.0 nmol·l^{-1}; FR, 46.7 ± 2.8 nmol·l^{-1}).

$P_{ET}CO_2$ was elevated ($p<0.05$) at the T_{vent} in both conditions (SR, 45 ± 4 Torr; FR, 47 ± 4 Torr) compared to loadless cycling (42 ± 3 Torr). At exhaustion, $P_{ET}CO_2$ was similar to loadless cycling in FR (42 ± 5 Torr) but was lower ($p<0.05$) in SR (34 ± 4 Torr) compared to loadless cycling and to FR at exhaustion. The P_aCO_2 responded similarly to that of $P_{ET}CO_2$ except that no increase was observed at the T_{vent} (Fig. 1).

Breathing frequency (*f*) and tidal volume (V_T) were greater ($p<0.05$) at the T_{vent} compared to loadless cycling (*f*, 19 ± 4 br·min^{-1}; V_T, 1145 ± 372 ml), with no differences between SR (*f*, 26 ± 4 br·min^{-1}; V_T, 2187 ± 374 ml) and FR (*f*, 23 ± 4 br·min^{-1}; V_T, 2176 ± 219 ml). At exhaustion, the *f* was higher in SR (51 ± 10 br·min^{-1}) than FR (42 ± 8 br·min^{-1}), and the V_T was lower in SR (2590 ± 590 ml) than FR (3050 ± 470 ml) (Fig. 1). The time of expiration (T_E) was 1.92 ± 0.47 s during loadless cycling, and decreased ($p<0.05$) similarly in both SR (0.56 ± 0.11 s) and FR (0.58 ± 0.15 s). The slope of the alveolar phase of CO_2

Table 1. Work rate (W), ventilation ($\dot{V}_E$) and gas exchange ($\dot{V}O_2$, $\dot{V}CO_2$) measured at loadless cycling and exhaustion following Fast and Slow ramp exercise

	0 Watts	Fast ramp	Slow ramp
Work Rate (W)	—	342 ± 30	$240 \pm 34^*$
$\dot{V}O_2$ (ml·min^{-1})	867 ± 133	3721 ± 718	3929 ± 628
$\dot{V}CO_2$ (ml·min^{-1})	757 ± 166	4731 ± 585	$3877 \pm 978^*$
$\dot{V}_E$ (l·min^{-1})	20 ± 4	125 ± 13	130 ± 28

*SR is significantly different from FR ($p < 0.05$).

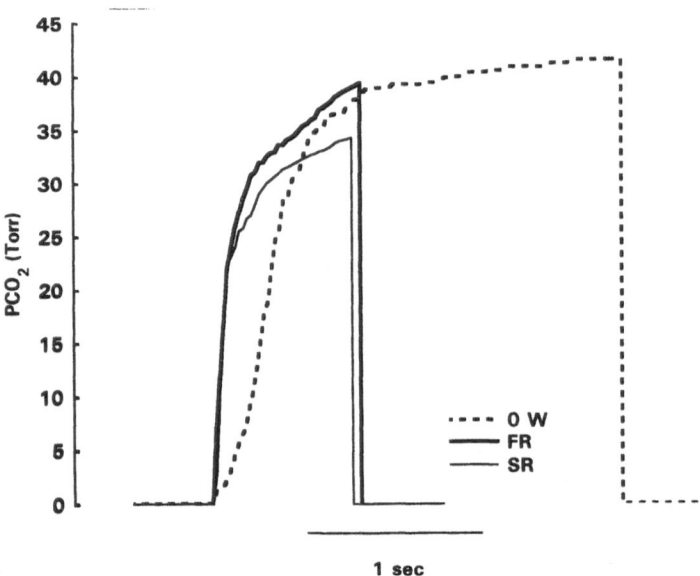

Figure 2. Three representative breath profiles of expired PCO_2 overlayed from a single subject at 0 W cycling and exhaustion during the FR and SR conditions. Note the similar T_E at exhaustion in FR and SR, and the higher alveolar slope of PCO_2 for FR compared to SR.

increased ($p<0.05$) from loadless cycling (2.6 ± 0.8 Torr·s^{-1}) to exhaustion; the alveolar slope of CO_2 was greater ($p<0.05$) in FR (19.0 ± 3.6 Torr·s^{-1}) than SR (13.0 ± 3.3 Torr·s^{-1}) at exhaustion (Fig. 2).

DISCUSSION

In the present study, respiratory compensation was demonstrated above the T_{vent} in SR by the decline in $P_{ET}CO_2$ and P_aCO_2 below values observed during loadless cycling. However, in spite of a similar acidosis in FR and SR, respiratory compensation was not apparent in FR as $P_{ET}CO_2$ and P_aCO_2 were similar to values observed during loadless pedalling.

The $P_{ET}CO_2$ is dependent upon the alveolar slope for CO_2, the time of expiration (T_E), and the mean alveolar (arterial) PCO_2 (4,8). Whipp and colleagues (8) demonstrated during a 1-min incremental exercise test, that there was an increase in f and a decrease in T_E during exercise above the T_{vent} which acted to offset the progressively increasing alveolar PCO_2 slope. Consequently, the rising alveolar slope of CO_2 was truncated, and $P_{ET}CO_2$ became constant (i.e. isocapnic buffering phase) (8). However, this mechanism cannot entirely explain the higher $P_{ET}CO_2$ at exhaustion during FR since the T_E was similar in FR and SR, in spite of a lower f during FR. As demonstrated in the present study, the greater $P_{ET}CO_2$ observed in FR compared to SR can be attributed, in part, to a higher alveolar PCO_2 slope during FR exercise (Fig. 2).

The slope of the alveolar PCO_2 is dependent upon the rate of venous return, and the difference between mixed venous PCO_2 and the instantaneous alveolar PCO_2 (1,4). Venous return and mixed venous PCO_2 were not measured, but there is reason to believe that a higher mixed venous PCO_2 was responsible, in part, for the higher alveolar PCO_2 slope found in

FR in this study. Peak $\dot{V}CO_2$ was greater in FR than SR, in agreement with Ward and Whipp (6), and could be attributed to either a higher cardiac output, a higher mixed venous - arterial CO_2 content difference, or both. There is no reason to believe that cardiac output (and venous return) were different at exhaustion in the two conditions. Arterial CO_2 content was higher in FR than SR, as demonstrated by both a higher P_aCO_2 (FR, 38 Torr; SR, 28 Torr) and arterial plasma $[HCO_3^-]$ (FR, 20 mmol·l^{-1}; SR, 14 mmol·l^{-1}). Thus, even if the mixed venous -arterial CO_2 content difference was similar between conditions, the higher arterial CO_2 content in FR would require that the mixed venous CO_2 content (and PCO_2) was also greater in FR than SR, thereby contributing to the higher alveolar slope for PCO_2.

CONCLUSION

The higher $P_{ET}CO_2$ and P_aCO_2 during FR compared to SR can be attributed to a higher alveolar PCO_2 slope as no difference was observed in T_E between ramp conditions. Although differences in the rate of venous return cannot be excluded, a higher mixed venous PCO_2 during FR may have contributed to the higher alveolar PCO_2 slope in FR. Thus, the apparent lack of respiratory compensation at exhaustion following the fast ramp protocol appears to reflect an inability of the respiratory system to respond adequately to a higher rate of CO_2 delivery to the lung associated with a higher alveolar PCO_2 slope, due in part to a higher mixed venous CO_2 content and PCO_2.

ACKNOWLEDGMENTS

This research was carried out at The Centre for Activity and Ageing (affiliated with the Faculties of Kinesiology and Medicine at the University of Western Ontario and The Lawson Research Institute at the St. Joseph's Health Centre). Financial support was provided by an operating grant to J.M.K. from the Natural Sciences and Engineering Research Council of Canada, and a travel grant from the Lawson Research Institute. B.W.S. is supported by a Natural Sciences and Engineering Research Council of Canada Student Fellowship.

REFERENCES

1. Allen, C.J. and N.L. Jones. Rate of change of alveolar carbon dioxide and the control of ventilation during exercise. *J. Physiol. (Lond)* 355: 1-9, 1984.
2. Beaver, W.L., N. Lamarra and K. Wasserman. Breath-by-breath measurement of true alveolar gas exchange. *J. Appl. Physiol. :Respirat. Environ. Exercise Physiol.* 51: 1662-1675, 1981.
3. Beaver, W.L., K. Wasserman and B.J. Whipp. A new method for detecting anaerobic threshold by gas exchange. *J. Appl. Physiol.* 60: 2020-2027, 1986.
4. Edwards, A.D., S.J. Jennings, C.G. Newstead and C.B. Wolff. The effect of increased lung volume on the expiratory rate of rise of alveolar carbon dioxide tension in normal man. *J. Physiol. (Lond)* 344: 81-88, 1983.
5. Hughson, R.L. and H.J. Green. Blood acid-base and lactate relationships studied by ramp work tests. *Med. Sci. Sports Exerc.* 14: 297-302, 1982.
6. Ward, S.A. and B.J. Whipp. Influence of body CO_2 stores on ventilatory-metabolic coupling during exercise. In: *Control of Breathing and its Modelling Perspective*, edited by Y. Honda, Y. Miyamoto, K. Konno and J.G. Widdicombe. New York: Plenum Press, 1992, p. 425-431.
7. Wasserman, K., A.L. Van Kessel and G.G. Burton. Interaction of physiological mechanisms during exercise. *J. Appl. Physiol.* 22: 71-85, 1967.
8. Whipp, B.J., J.A. Davis and K. Wasserman. Ventilatory control of the 'isocapnic buffering' region in rapidly-incremental exercise. *Respir. Physiol.* 76: 357-368, 1989.

9. Whipp, B.J. and S.A. Ward. Coupling of ventilation to pulmonary gas exchange during exercise. In: *Exercise: Pulmonary Physiology and Pathophysiology*, edited by B.J. Whipp and K. Wasserman. New York: Marcel Dekker, Inc., 1991, p. 271-307.

29

ACUTE VENTILATORY RESPONSE TO RAMP EXERCISE WHILE BREATHING HYPOXIC, NORMOXIC, OR HYPEROXIC AIR

M. L. Walsh and E. W. Banister

School of Kinesiology
Simon Fraser University
Burnaby, B.C., Canada V5A 1S6

INTRODUCTION

Ventilatory control during exercise is frequently investigated by examining the relation between ventilation and a potential stimulating biochemical, physiological, or neurological factor. Only rarely has ventilation been examined in relation to its primary functions: i.e., the rate of O_2 inspiration and the rate of CO_2 elimination. This study examined the primary functions of ventilation during ramp exercise while breathing either hypoxic, normoxic, or hyperoxic air under normobaric conditions.In addition, the ventilatory variables recorded were compared to absolute work rate and relative work rate since the anaerobic threshold is not difference between FIO_2 conditions when expressed as percent work rate relative to maximum work rate (6).

METHODS

Seven healthy male subjects gave their informed consent according to institution guidelines. Each subject performed 2 exercise bouts while breathing either 12 % O_2, 21 % O_2, or 40 % O_2, balance nitrogen. The order of the FIO_2 condition was randomized and the subjects were naive as to which gas they were inspiring. Each subject breathed the gas for 10 min at rest and 8 min at 30 W prior to progressive ramp cycle ergometry at 1 $W·3$ s^{-1} (20 $W·min^{-1}$) continuing until exhaustion. Each subject performed 6 tests, each separated by at least one day. $\dot{V}O_2$, $\dot{V}CO_2$, $\dot{V}E$, $\dot{V}I$, and $PETCO_2$ data were collected on a breath-by-breath basis. Appropriate compensations were made for gas transport time, analyzer response time, functional residual volume, and breathing valve dead space. The breath-by-breath data collection system was validated against the bag method and the correlations of $\dot{V}O_2$, $\dot{V}CO_2$, and $\dot{V}E$ between the two methods were all greater than 0.99 (4). The anaerobic threshold was estimated from the method of Beaver et al. (1) and referred to as the respiratory gas exchange threshold (RGET).

Modeling and Control of Ventilation, Edited by S. J. G. Semple, L. Adams, and B. J. Whipp
Plenum Press, New York, 1995

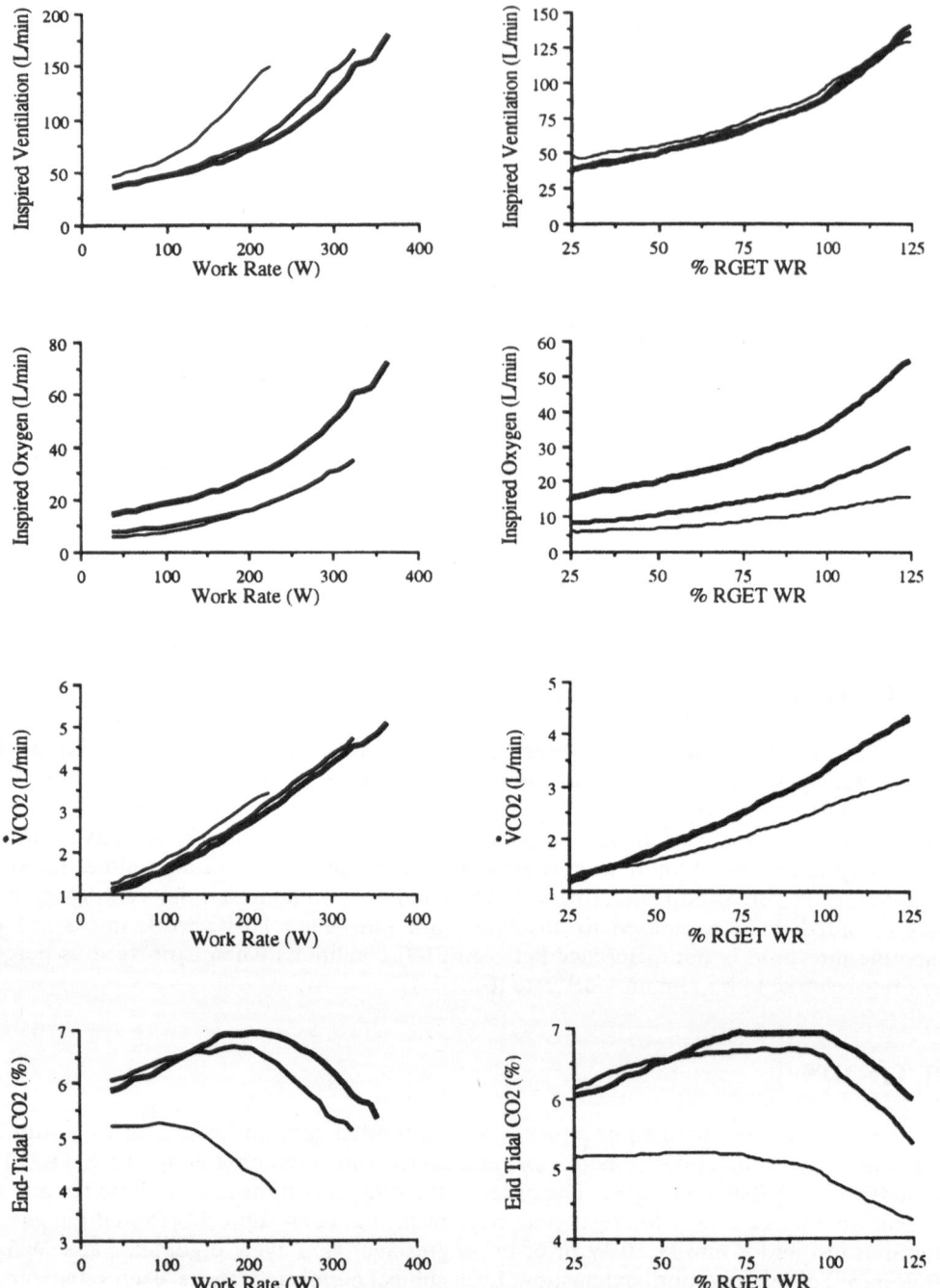

Figure 1. The relation between inspired ventilatory rate, inspired oxygen rate, V̇CO₂, and PETCO₂ (top to bottom respectively) with absolute work rate (left side) and relative RGET work rate (right side). Thin line is for hypoxia, medium line for normoxia, and thick line for hyperoxia.

RESULTS

The group mean maximum work rate was 238±19 (mean±S.D.), 332±18, and 352±20 W for hypoxia, normoxia, and hyperoxia respectively and were significantly different from each other. The maximum expired ventilatory rate was significantly lower in hypoxia (153±20 L/min) than in normoxia or hyperoxia (169±14 and 167±16 L/min, respectively).

The work rate equivalent to the RGET was 162±19, 237±19, 247±24 W for hypoxia, normoxia, and hyperoxia. When the RGET was expressed as a percentage of maximum work rate, there was no difference between the 3 inspired oxygen conditions (68±8, 72±5, and 70±5 % respectively).

The relations of inspired ventilatory rate, inspired oxygen rate, $\dot{V}CO_2$, and $PETCO_2$ with both work rate and % RGET work rate are presented in Figure 1.

DISCUSSION

A discussion of ventilatory control during exercise is dependent on how the results are presented. When ventilation and related variables are compared with absolute work rate it appears that ventilation may be regulated differently in hypoxia than in hyperoxia. Inspired ventilation was much greater and $PETCO_2$ much lower while breathing hypoxic air compared with breathing normoxic air. However, the rate of inspired oxygen was very similar between hypoxia and normoxia suggesting that oxygen delivery to the body is an important determinant of the ventilatory response. When comparing the hyperoxic ventilatory responses to those during normoxia it appears that the elimination of CO_2 from the body is more important than the delivery of oxygen into the body. It could be concluded that in normoxia, the ventilatory response is an exquisite balance between the need to delivery oxygen into the body and the need to expel CO_2 from the body.

When ventilatory rate is compared with % RGET WR rather than absolute work rate, inspired ventilation is not different between all 3 FIO_2 conditions. This remarkable similarity despite very different FIO_2 perturbations suggests ventilation is determined by a factor related to relative intensity of work rather than the absolute work rate for a given recruited muscle mass. The different inspired oxygen and $PETCO_2$ responses in all 3 FIO_2 conditions, when compared on a relative work rate basis, suggest that the primary functions of ventilation are not tightly coupled with the ventilatory response in the present study. The nearly identical inspired ventilation responses, related to relative work rate, suggests either a single undetermined ventilatory stimulation mechanism is predominating under all 3 FIO_2 conditions or that the net output of redundant ventilatory stimulation mechanisms is remarkably faithful in spite of different combinations of stimuli acting.

The elegant use of lower body pressure changes during exercise developed by Eiken and co-workers has demonstrated that, during exercise, ventilation decreases when lower body pressure is negative and ventilation increases when lower body pressure is positive (2, 3). These types of data have supported the suggestion that muscle perfusion or vasomotor tone is an important stimulant to ventilation (5). However, when lower body positive pressure is applied, the pattern of motor unit recruitment is altered based on electromyographic data and maximum aerobic power is reduced. When the ventilation data from lower body positive pressure tests are normalized to relative work rate rather than compared with absolute work rate, the ventilatory differences largely disappear.

The above contrast of ventilatory parameters with absolute and relative work indicate how an interpretation of data can change. If a perturbation changes the absolute power capability (e.g., $\dot{V}O_2max$), it will change the ventilatory response. However this does not preclude that the perturbation influences ventilation independently of relative work rate. Any mechanism that is claimed to regulate ventilation must be proven to do so independent of both absolute and relative work rate.

ACKNOWLEDGMENTS

This study was supported by a National Sciences and Engineering Research Council grant to E.W.B.

REFERENCES

1. Beaver, W.L., K. Wasserman, and B.J. Whipp. A new method for detecting anaerobic threshold by gas exchange. *J. Appl. Physiol.* 60: 2020-2027, 1986.
2. Eiken, O. and H. Bjurstedt. Dynamic exercise in man as influenced by experimental restriction of blood flow in the working muscles. *Acta. Physiol. Scand.* 131: 339-345, 1986.
3. Eiken, O., F. Lind, and H. Bjurstedt. Effects of blood volume distribution on ventilatory variables at rest and during exercise. *Acta. Physiol. Scand.* 127: 507-512, 1987.
4. Walsh, M.L. Oxygen uptake kinetics during exercise (Ph.D. thesis). Burnaby, Canada: Simon Fraser University, 1993.
5. Williamson, J.W., P.B. Raven, B.H. Foresman, and B.J. Whipp. Evidence for an intramuscular ventilatory stimulus during dynamic exercise man. *Respirat. Physiol.* 94: 121-135, 1993.
6. Yoshida T., M. Chida, M. Ichioka, K. Makiguchi, and Y. Suda. Effect of hypoxia on lactate variables during exercise. *J. Human. Ergol.* 16: 157-161, 1987.

30

VENTILATORY RESPONSES DURING RAMP EXERCISE IN HYPEROXIA

Y. Miyamoto and K. Niizeki

Department Electrical and Information Engineering
Faculty of Engineering
Yamagata University
Yonezawa, 992 Japan

INTRODUCTION

Hyperpnea during heavy exercise is generally thought to be closely coupled with the onset of anaerobic metabolism. Arterial acidosis stimulates the carotid chemoreceptors and the resultant hyperpnea produces arterial hypocapnia. This is supported by the finding that the asthmatic patients with carotid body resection do not hyperventilate during heavy exercise [14]. However, several investigators who have studied human subjects with dietary glycogen depletion have suggested a dissociation of hyperventilation and lactic acid production [7]. In ponies, significant arterial hypocapnia is always observed even during mild to moderate exercise and carotid body denervation accentuates rather than attenuates the hypocapnia during heavy exercise [11,12]. This group has claimed that hyperpnea during heavy exercise is not tightly dependent on arterial acidosis and that the carotid chemoreceptors contribute minimally to hyperventilation not only in ponies but also in humans [5]. The present study was undertaken to assess the role of the carotid chemoreceptors during incremental exercise in healthy human subjects. Carotid body sensitivity to humoral stimuli such as H^+, K^+, and plasma catecholamines is believed to be attenuated significantly by breathing an O_2-rich gas mixture [1,3]. Assuming that the nonlinear increase of the $\dot{V}E$ response during incremental ramp exercise is primarily dependent on the ventilatory drive mediated by the carotid bodies, hyperoxia would diminish hyperventilation during heavy exercise. Furthermore, as sustained hyperoxia has been shown to significantly reduce the blood lactate level during heavy exercise [13,15], bicarbonate buffering and the resultant production of extra CO_2 would disappear or would appear at a higher work rate when O_2 rich-gases are inhaled.

METHOD

Incremental ramp exercise tests were conducted employing five healthy young male subjects. No trained athlete was included in the subject group. Informed consent

Modeling and Control of Ventilation, Edited by S. J. G. Semple, L. Adams, and B. J. Whipp
Plenum Press, New York, 1995

was obtained after detailed explanation of the purpose and protocols of the experiment. The subject was seated on a bicycle ergometer with an electrically controlled braking system (Lode, model WLP-300) and breathed through a face mask with a respiratory valve (Tatebe, model DM-25p). A hot wire type pneumotachograph (Minato, model RF-2) fitted to the expiratory port of the valve continuously monitored expiratory airflow. We confirmed that the sensitivity of the pneumotachograph did not alter with changes in the gas composition over the range of physiological flow variations. Inspired and expired gases were directly sampled from the inside of the face mask and the composition was analyzed continuously with a medical mass spectrometer (Westron, model WSMR-1400). The subject pedaled at a constant rate of 60 rpm throughout the experiment. Ramp exercise was performed at an incremental rate of 15 w/min over the range from 0 (unloaded pedaling) to 300 w. Minute ventilation ($\dot{V}E$), the end-tidal pressures of O_2 and CO_2 ($PETO_2$ and $PETCO_2$), O_2 uptake($\dot{V}O_2$), and CO_2 output ($\dot{V}CO_2$) were determined breath by breath using a computer-aided monitoring system [9]. Heart rate was also monitored at the chest wall with bipolar electrodes. Experiments under hyperoxic conditions were performed in the same subject group by breathing a gas mixture consisting of 50 %O_2 and 50 % N_2. The test gas mixture was supplied to the subject via the inspiratory port of the respiratory valve from a large Douglas bag. The subject breathed the hyperoxic gas mixture for 15 min before the start of each experimental protocol to ensure complete washout of N_2 from the body. Experiments were performed under normoxic and hyperoxic conditions in a randomized order. Measurements were repeated three times for each individual. The breath by breath data from each experimental run were stored on a personal computer (NEC, model PC-98) together with timing signals for the exercise protocol. The data were then rearranged into a 5 -sec intervals for further analysis. Comparisons between normoxic and hyperoxic values were performed for a mean of three measurements in each subject by using Student's t-test for paired samples. A p value of <0.05 was regarded as significant.

RESULTS

Group mean profiles of the $\dot{V}E$, $\dot{V}CO_2$, $\dot{V}E/\dot{V}CO_2$ and $PETCO_2$ responses obtained during incremental exercise under normoxic and hyperoxic conditions are shown in Fig.1. The $\dot{V}E$ response in hyperoxia was similar to that in normoxia during mild to moderate exercise, but the nonlinear progressive rise in $\dot{V}E$ that was characteristic in normoxia became less significant with hyperoxia. No appreciable difference was observed between the $\dot{V}CO_2$ responses in normoxia and hyperoxia. $PETCO_2$ rose almost linearly with the work load during mild to moderate exercise regardless of the inspired O_2 level, but it leveled off at the transition from moderate to heavy exercise in normoxia. The isocapnic buffering point was extended to a very much higher range in hyperoxia. The ventilation equivalent for CO_2 ($\dot{V}E/\dot{V}CO_2$) determined in normoxia and hyperoxia fell almost linearly with incremental work load, but from a mid-load point both curves became dissociative. The normoxic curve reduced its falling rate with load significantly and finally turned upwardly at very heavy exercise while hyperoxic $\dot{V}E/\dot{V}CO_2$ fell consistently.

In individual $\dot{V}E$ response curves, a turning point was always detected from which the response increased with a steeper slope than previously, regardless of the inspired O_2 level ($\dot{V}E$-TP, Fig.2 left). The turning points were determined by visual inspection and the slopes were determined using the least square method on a computer screen. The lowermost inflection point of the curve was excluded from the determination of $\dot{V}E$-TP, because it can be regarded as an intrinsic delay appearing even in a linear system. The $\dot{V}E$-TP was 181±40 w (mean ± standard deviation) in normoxia, and 160±42 w in hyperoxia (no difference,

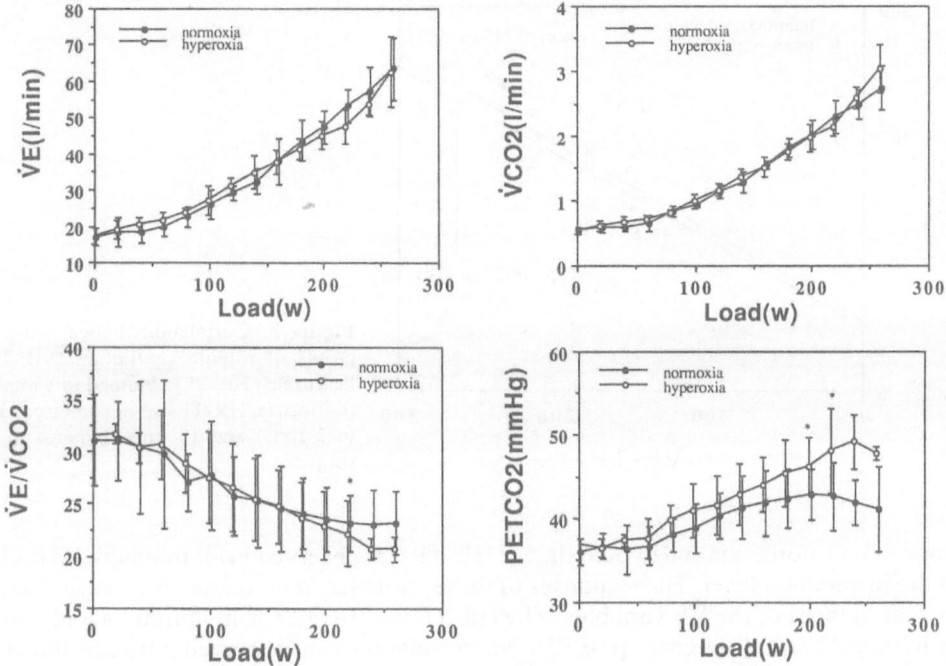

Figure 1. Response profiles of minute ventilation ($\dot{V}$ E), CO_2 output ($\dot{V}$ CO_2), ventilation equivalent for CO_2 ($\dot{V}$ E/$\dot{V}$ CO_2) and the end-tidal pressure of CO_2 (PETCO$_2$) during incremental ramp exercise (15 w/min) breathing air (normoxia) or 50% O_2 (hyperoxia). Points are the averages for five subjects and bars indicate the standard deviation. Asterisks indicate significant difference (P<0.05).

p<0.1). Before the $\dot{V}$E-TP, the increase of $\dot{V}$E with load showed an equal slope irrespective of the inspired O_2 level (0.153±0.042 l/min/w in normoxia and 0.153±0.035 l/min/w in hyperoxia). After the $\dot{V}$E-TP, the gradient of the slope increased significantly in both normoxia (0.247±0.037 l/min/w, p<0.03) and hyperoxia (0.219±0.041 l/min/w, p<0.01). There was no correlation between the $\dot{V}$E-TPs determined under normoxic and hyperoxic conditions.

Figure 2 also shows the HR responses during incremental exercise in one subject. HR increased from a specific work rate (HR-TP) close to the $\dot{V}$E-TP, with an accelerated

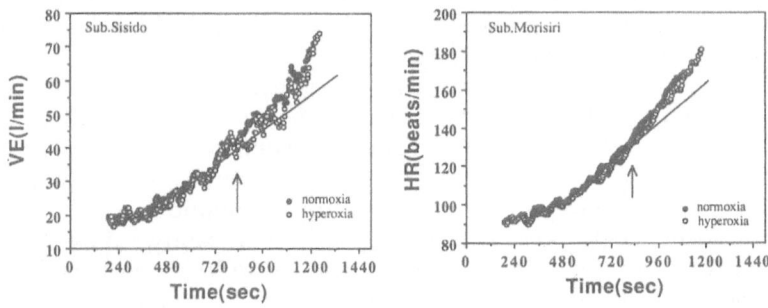

Figure 2. Minute ventilation ($\dot{V}$ E, left) and heart rate (HR, right) responses during incremental exercise (15 w/min, started from 180 sec) obtained in a subject breathing air (normoxia) and 50 % O_2 (hyperoxia).

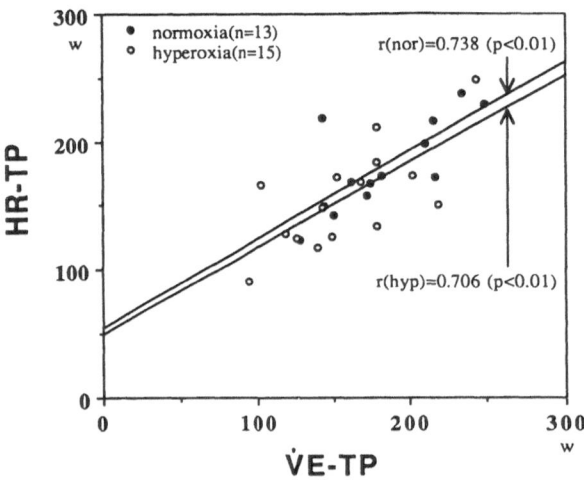

Figure 3. Correlations between the turning points of minute ventilation ($\dot{V}$ E-TP) and heart rate (HR-TP) obtained in a totals of 13 (normoxia, HR-TP was not precisely detected in 2 runs) and 15 runs (hyperoxia) in five subjects.

slope in both normoxia and hyperoxia. An HR-TP was observed in all five subjects regardless of the inspired O_2 level. The responses of these variables to incremental exercise were quite similar to those of the $\dot{V}E$ variables. The HR-TP was 183 ± 37 w in normoxia and 156 ± 40 w in hyperoxia (no difference, p=0.12). No correlation was observed between the HR-TPs determined under normoxic and hyperoxic conditions. Figure 3 shows the correlation between the $\dot{V}E$-TP and HR-TP values obtained in normoxia and hyperoxia. The correlation was significant irrespective of the inspired O_2 level (p<0.01).

DISCUSSION

Observation of the ventilatory and gas exchange responses during incremental ramp exercise in hyperoxia did not generally contradict the preceding hyperoxic studies employing step and other type exercise (2,6, 8,10). The isocapnic buffering observed in normoxia at the transition from moderate to heavy work did not appear in hyperoxia until a very heavy work load had been attained (see PETCO$_2$ tracings in Fig.1). This agrees with the observation that normoxic $\dot{V}E/\dot{V}CO_2$ that was consistently falling with load became flat near the isocapnic point and then turned upwardly by a further increase in work load, while $\dot{V}E/\dot{V}CO_2$ in hyperoxia decreased continuously during heavy exercise. These results would suggest the delayed onset of anaerobic metabolism under hyperoxic conditions and the depression of $\dot{V}E$ in the state of attenuated carotid body activity. Although the difference between the normoxic and hyperoxic $\dot{V}Es$ was not significant even during heavy exercise, the characteristic nonlinear progressive rise of $\dot{V}E$ during air breathing became less significant with hyperoxia. Hyperoxia would reduce the carotid chemoreceptor sensitivity to H^+, but at the same time the sustained administration of O_2-rich gas would intensify central chemoreceptor drive by an increase of arterial PCO$_2$, as predicted from the PETCO$_2$ response. Therefore, $\dot{V}E$ is not always suppressed in sustained hyperoxic conditions even during anaerobic exercise. As sustained hyperoxia has been shown to significantly reduce the blood lactate level during heavy exercise [13,15], the extra-production of $\dot{V}CO_2$ during heavy exercise by bicarbonate buffering should be suppressed when O_2 rich-gases are inhaled. However, this was not demonstrated in the present study.

We observed an upward turning point of the $\dot{V}E$ response with respect to work rate ($\dot{V}E$-TP), not only in normoxia but also in hyperoxia (Fig. 2). We also observed a clear upward turning point of HR (HR-TP) both in normoxia and in hyperoxia. The HR-TP correlated well with the $\dot{V}$E-TP irrespective of the inspired O_2 level (Fig. 3). This suggests that certain factors simultaneously regulating ventilation and circulation may be involved in producing the turning points of HR and $\dot{V}$E. Humoral factors such as H^+, K^+ or plasma catecholamines can largely be ruled out as responsible for these turning points since hyperoxia is believed to attenuate the sensitivity of the carotid bodies to these stimuli significantly [1,3]. Heart rate is primarily regulated by the autonomic nervous system. Recent studies on the noninvasive estimation of autonomic nervous activity utilizing heart rate variability during exercise have shown that heart rate is predominantly regulated by the withdrawal of parasympathetic tone during mild to moderate exercise, while it is accelerated by intensified sympathetic tone during heavy exercise [16]. It is conceivable that such changes in autonomic activity with exercise may play an important role in determining the turning points of $\dot{V}$E and HR. On the other hand, strong mental effort is required to continue severe exercise. In this context, the influence of central commands originating in the cortex and transmitted via the hypothalamus to the respiratory and circulatory centers in the medulla cannot be ruled out [4].

ACKNOWLEDGMENTS

This work was supported in part by Japanese Governmental Grant No. 5680750 in aid for developmental scientific research.

REFERENCES

1. Burger, R.E., Estavillo, J.A., Kumar, P, Nye, P.C.G., and D.J. Paterson. Effects of potassium, oxygen and carbon dioxide on the steady-state discharge of cat carotid body chemoreceptors, J. Physiol.Lond. 401:519-531, 1988.
2. Casaburi, R., Stremel, R.W., Whipp, B.J., Beaver, W.L., and K. Wsserman. Alteration by hyperoxia of ventilatory dynamics during sinusoidal work, J. Appl. Physiol. 48: 1083-1091, 1980.
3. Cunningham, D.J.C., Hey, E.N., Patrick, J.M., and B.B. Loyd. The effect of noradrenalin infusion on the relation between pulmonary ventilation and the alveolar PO_2 and PCO_2 in man, Ann. NY Acad. Sci. 109: 756-770, 1963.
4. Eldridge, F.L., Millhorn, D.E., Kiley, J.P., and T.G. Waldrop. Stimulation by central command of locomotion, respiration and circulation during exercise, Respir. Physiol. 59: 313-337, 1985.
5. Forster, H.V., Dunning, M.B., Lowry, T.F., Erickson, B.K., Forster, M.A., Pan, L.G., Brice, A.G., and R.M. Effros. Effect of asthma and ventilatory loading on arterial PCO_2 of humans during submaximal exercise, J. Appl. Physiol., 75: 1385-1394, 1993.
6. Griffiths, T.L., Henson, L.C.,and B.J. Whipp. Influence of inspired oxygen concentration on the dynamics of the exercise hyperpnea in man, J Physiol 380: 387-403, 1986.
7. Hughes, E.F., Turner, S.C., and G.A. Brooks. Effect of glycogen depletion and pedaling speed on "anaerobic threshold", J. Appl. Physiol., 52: 1598-1607, 1982.
8. Linnarsson, D. Dynamics of pulmonary gas exchange and heart rate changes at start and end of exercise, Acta Physiol. Scand. (suppl) 415: 1-68,1974.
9. Miyamoto, Y., Hiura, T., Tamura, T., Nakamura, T., Higuchi, J., and T. Mikami. Dynamics of cardiac, respiratory, and metabolic function in men in response to step work load. J Appl Physiol., 52:1198-1208, 1982.
10. Nakazono, Y., and Y. Miyamoto. Effect of hypoxia and hyperoxia on cardiorespiratory responses during exercise in man, Jpn J. Physiol., 37: 447-457, 1987.
11. Pan, L.G., Forster, H.V., Bisgard, G.E., Kaminiski, R.P., Dorsey, S.M., and M.A. Busch. Hyperventilation in ponies at the onset of and during steady-state exercise, J. Appl. Physiol., 54: 1394-1402, 1983.
12. Pan, L.G., Forster, H.V., Bisgard, G.E., Murphy, C.L., and T.F. Lowry. Independence of exercise hyperpnea and acidosis during hgh-intensity exercise in ponies, J. Appl. Physiol., 60: 1016-1024, 1986.

13. Wasserman, K., Whipp, B.J., and J.A. Davis. Respiratory physiology of exercise: metabolism, gas exchange and respiratory control. In: International review of Physiology, Respiratory Physiology III, Baltimore: Vol.23, University Park Press. p.150-211, 1981.

14. Wasserman, K., Whipp, B.J., Koyal, S.N., and M.G. Cleary. Effect of carotid body resection on ventilatory and acid-base control during exercise, J. Appl. Physiol., 39: 354-358, 1975.

15. Wilson, B.A., Welch, H.G., and J.N. Liles. Effects of hyperoxic gas mixtures on energy metabolism during prolonged work, J. Appl. Physiol. 39: 267-271, 1975.

16. Yamamoto, Y., Hughson, R.L., and J.C. Peterson. Autonomic control of heart rate during exercise studied by heart rate variability spectral analysis, J. Appl. Physiol., 71: 1136-1142, 1991.

RESPIRATORY COMPENSATION FOR THE METABOLIC ACIDOSIS OF SEVERE EXERCISE AS A MODULATOR OF MUSCULAR CAPILLARY O_2-UNLOADING

Yoshiyuki Fukuba[*] and Brian J. Whipp

Department of Physiology
St.George's Hospital Medical School
University of London
Cranmer Terrance, London SW17 0RE, United Kingdom

INTRODUCTION

Muscle O_2 uptake ($\dot{V}_TO_2$) is determined both by mass flow of O_2 and its subsequent diffusion into the muscle cells. Fick's equation and Law of diffusion are both representative of this process.

The convective flux may be expressed as:

$$\dot{V}_TO_2 = \dot{Q}_T \cdot (CaO_2 - CvO_2) = \dot{Q}_T \cdot [Hb] \cdot \beta \cdot (SaO_2 - SvO_2$$

$$= \dot{Q}_T \cdot [Hb] \cdot \beta \cdot (\lambda a \cdot PaO_2 - \lambda v \cdot PvO_2)$$

$$= \text{``}O_2\text{-delivery''} - \dot{Q}_T \cdot [Hb] \cdot \beta \cdot \lambda v \cdot PvO_2, \tag{1}$$

and the diffusive flux as:

$$\dot{V}_TO_2 = D_T \cdot (PmcO_2 - PmitO_2) = D_T \cdot (k \cdot PvO_2 - PmitO_2), \tag{2}$$

where Q_T is muscle blood flow; CaO_2 and CvO_2 are the O_2 content in arterial and venous blood; $[Hb]$ is haemoglobin concentration; β is the O_2-carrying capacity of Hb; SaO_2 and SvO_2 are the arterial and venous Hb saturations; D_T is a capillary-to-mitochondrial diffusion

[*] Present address; Department of Biometrics, Research Institute for Radiation Biology and Medicine, Hiroshima University, Hiroshima 734 Japan.

Modeling and Control of Ventilation, Edited by S. J. G. Semple, L. Adams, and B. J. Whipp
Plenum Press, New York, 1995

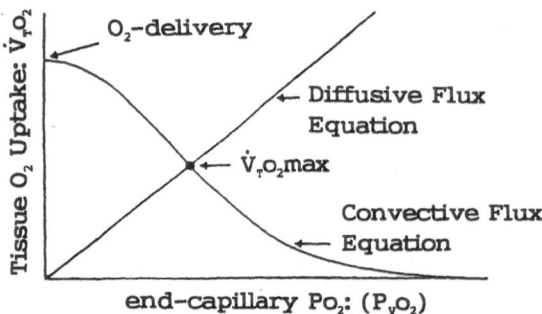

Figure 1. Graphical representation of tissue oxygen uptake with venous effluent PO_2 determined by both convective and diffusive O_2 flux (Eq.3).

'capacity'; $PmcO_2$, PvO_2, and $PmitO_2$ are mean-, and end-capillary, mitochondrial PO_2; k is a proportionality factor of $PvO_2 \propto PmcO_2$; and λa and λv are the conversion coefficients on the O_2-dissociation curve (O_2-DC) affected by pH, temperature, 2-3-DPG etc. As mitochondrial PO_2 can fall to near zero during severe intensity exercise (e.g., 3), the expression can therefore be simplified as:

$$\dot{V}_TO_2 = \text{``}O_2\text{-delivery''} - \dot{Q}_T \cdot [Hb] \cdot \beta \cdot \lambda v \cdot PvO_2 = D_T \cdot k \cdot PvO_2. \tag{3}$$

As seen in Figure 1, $\dot{V}_TO_2$ benefits from PvO_2 decreasing more from a convective viewpoint, but from a diffusive perspective it needs to decrease less. Thus the decrease in λv should be maximized: this is achieved by a rightward shift of the O_2-DC. The dominant factor shifting this curve to the right during severe exercise is the fall of the pH (i.e., Bohr

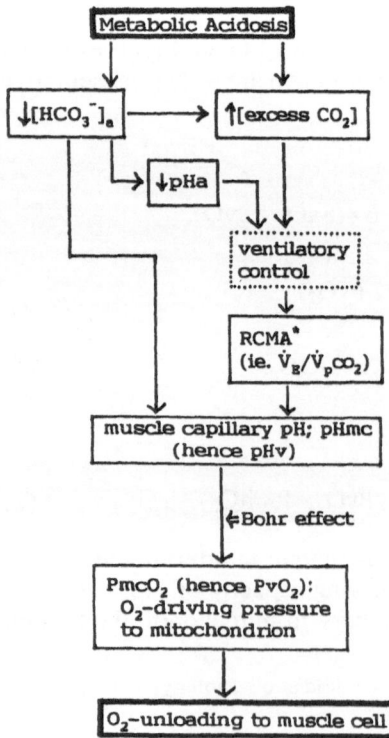

Figure 2. A schematic representation of the role of the respiratory compensation for the metabolic acidosis (*RCMA) on muscular O_2-unloading, via the Bohr effect.

effect). Wasserman and his associates, however, have recently demonstrated the important finding that this shift actually maintains PvO_2 at a relatively constant value of approximately 20 torr in humans (see Ref.4 for review).

However, in addition to the metabolic acidosis *per se*, ventilatory control also provided an important modulation of pH. That is, the reduction in arterial (and hence muscle capillary) pH will be attenuated by the degree of respiratory compensation for the metabolic acidosis (see, Figure 2). Therefore, we have modelled the influence on end-capillary pH (pHv) and PO_2 (PvO_2) of the Bohr shift interaction between the exercise-induced acidemia and the compensatory hyperventilation.

MODELLING

We used standard relationships for O_2-DC and the Bohr influences and also the blood CO_2 dissociation curve (CO_2-DC) and its Haldane effect. The details of the model is described below.

Assumptions

During severe intensity exercise, we assume that the entire system (lung, circulation, tissue) can be treated as though it is instantaneously in steady state.

1. The lung is functionally an 'ideal' gas exchanger, i.e., diffusion equilibrium between alveolar gas and pulmonary end-capillary blood, regional matching of ventilation to perfusion, and the absence of a right-to-left vascular shunt. Therefore, alveolar PCO_2 is equivalent to arterial PCO_2.
2. Pulmonary blood flow ($\dot{Q}_P$) is assumed to increase linearly with pulmonary oxygen uptake ($\dot{V}_PO_2$). The increase in $\dot{Q}_P$ and $\dot{V}_PO_2$ is assumed to be accounted for by the increase in blood flow to muscle ($\dot{Q}_T$) and muscle O_2 uptake ($\dot{V}_TO_2$). In other words, during exercise, the blood flow and O_2 uptake of non-exercising tissue ($\dot{Q}_N$ and $\dot{V}_NO_2$) remain at the same level, on average, as at rest.
3. Muscle end-capillary blood reflects mean muscle capillary blood for the modelling, while not being equal to it.
4. The gradient of $[HCO_3^-]$ between arterial and muscle end-capillary blood is assumed to be approximately 10 (mEq/l) from the recent experimental result (W.Stringer, personal communication).
5. V_D/V_T does not change further, within the severe intensity range.
6. Our model does not incorporate a temperature effect on the O_2-DC in addition to the pH effect.

Given Values

For the typical values:

1. $[Hb] = 15.0$ (g/dl).
2. $\beta = 1.34$ (ml/g).
At rest;
3. As blood flow (l/min), $\dot{Q}_P = 5$, $\dot{Q}_T = 1$, and $\dot{Q}_N = 4$.
4. As gas exchange parameters (l/min), $\dot{V}_PO_2 = 0.25$, $\dot{V}_TO_2 = 0.05$, $\dot{V}_NO_2 = 0.20$, $\dot{V}_PCO_2 = 0.20$, $\dot{V}_TCO_2 = 0.04$, and $\dot{V}_NCO_2 = 0.16$.
During severe intensity exercise;
5. Mitochondrial PO2 ($PmitO_2$) = 0.

6. $SaO_2 = 0.95$ (i.e., 95 %).

7. As blood flow, $\dot{Q}_P = 28$, $\dot{Q}_T = 24$, and $\dot{Q}_N = 4$.

8. As gas exchange parameters, $\dot{V}_PO_2 = 4.00$, $\dot{V}_TO_2 = 3.80$, $\dot{V}_NO_2 = 0.20$, $\dot{V}_PCO_2 = 4.80$, $\dot{V}_TCO_2 = 4.64$, and $\dot{V}_NCO_2 = 0.16$.

Independent variables

The main two independent variables considered here are end-capillary pH and PO_2 (i.e., pHv and PvO_2).

Controlled Variables

For the purpose of this study, we systematically vary two defining variables within a reasonable physiologically range; one is $[HCO_3^-]_a$ (mEq/l) - as an index of the degree of the metabolic acidosis, and the other is $\dot{V}_E/\dot{V}_PCO_2$ (l/l) - as an index of the compensatory ventilatory response to the metabolic acidosis.

By the integration of the Henderson-Hasselbalch and alveolar air equations (5),

$$\text{i.e.} \qquad pH = pK' + \log([HCO_3^-]/\alpha \cdot PCO_2), \qquad (4)$$

where $pK' = 6.1$ and $\alpha = 0.03$ (mM/torr at 37°C),

$$\text{and,} \qquad \dot{V}_A = \dot{V}_E \cdot (1 - V_D/V_T) = 863 \cdot \dot{V}_PCO_2/PaCO_2, \qquad (5)$$

therefore, the arterial pH can be expressed as,

$$pHa = pK' + \log\{([HCO_3^-]_a/25.8) \cdot (\dot{V}_E/\dot{V}_PCO_2) \cdot (1 - V_D/V_T)\}. \qquad (6)$$

This equation (5) shows three distinct components: a) the acid-base set-point component; $[HCO_3^-]_a$, b) the respiratory control component; $\dot{V}_E/\dot{V}_PCO_2$, and c) the ventilatory efficiency component; $(1 - V_D/V_T)$.

Dependent Variables

1. By giving some combination values of $[HCO_3^-]_a$ and $\dot{V}_E/\dot{V}_PCO_2$, pHa and $PaCO_2$ are computed from Eqs.6 and 4, respectively.

2. SvO_2 is computed from the O_2 convective equation of the tissue using the given values:

$$V_TO_2 = \dot{Q}_T \cdot [Hb] \cdot \beta \cdot (SaO_2 - SvO_2). \qquad (7)$$

3. Using $PaCO_2$ and SaO_2, $CaCO_2$ is computed by the following equation which was estimated from the standard CO_2-DC according to the Hb-O_2 saturation, letting $CCO_2 = CaCO_2$, $SO_2 = SaO_2$, and $PCO_2 = PaCO_2$:

$$CCO_2 = 5.5706 \cdot PCO_2^{0.36734} + \{ (-0.00544 \cdot SO2 + 0.00517) \cdot PCO_2 - 1.974 \cdot SO_2 + 1.8753 \}. \qquad (8)$$

4. $CvCO_2$ is computed from the CO_2 convective equation of the tissue using $CaCO_2$:

$$\dot{V}_T CO_2 = \dot{Q}_T \cdot (CvCO_2 - CaCO_2). \tag{9}$$

5. Using $CvCO_2$ and SvO_2, $PvCO_2$ is computed from the inverse function of Eq.8 on the CO_2-DC.
6. Under the assumption 4, we can get the pHv from Eq.4 using $PvCO_2$.
7. PvO_2 at pH=7.4 (PvO_2s) is computed from SvO_2 using Ellis' inverted dissociation curve equation (1), letting $PO_2s=PvO_2s$ and $SO_2=SvO_2$:

$$PO_2s \text{ (at pH=7.4)} = [A + B]^{1/3} - [B - A]^{1/3}, \tag{10}$$

where $A = 11700/[(1/SO_2) - 1]$, $B = [50^3 + A^2]^{1/2}$.
8. This standard PvO_2s is then corrected to pHv using the Bohr factor (2):

$$\Delta \ln(PvO_2)/\Delta pH = [PvO_2s/26.6]^{0.184} - 2.2 \tag{11}$$

where $\Delta pH = pHv - 7.4$.

RESULTS AND DISCUSSION

We computed pHv and PvO_2 according to the equations described in the above section. $[HCO_3^-]_a$ (as an index of metabolic acidosis) and $\dot{V}_E/\dot{V}_PCO_2$ (as an index of respiratory compensation) were varied systematically over a reasonable physiological range; 26 to 10 (mEq/l) and from 20 to 50 (l/l), respectively (Figure 3). The dots represent typical values at maximal exercise, i.e., $\dot{V}_E/\dot{V}_PCO_2$ and $[HCO_3^-]_a$ are 30 (l/l) and 14 (mEq/l), respectively. The model predicted the relationships among the variables within a plausible physiologically range, i.e., pHa=7.24, pHv=7.10, $PaCO_2$=33.7 (torr), $PvCO_2$=80.0 (torr), SvO_2=13.1 (%), and PvO_2=17.9 (torr).

As a result of the systematic computation, we confirmed that, in addition to the well recognized effect of the metabolic acidosis on pHv and PvO_2 during severe exercise, ventilatory control also had a significant modulating effect on pHv and PvO_2 (*Figure 3*). If the respiratory compensation for the metabolic acidosis ($\dot{V}_E/\dot{V}_PCO_2$) is below the normal ventilatory response range (< 30), e.g., as a result of insensitive ventilatory control mecha-

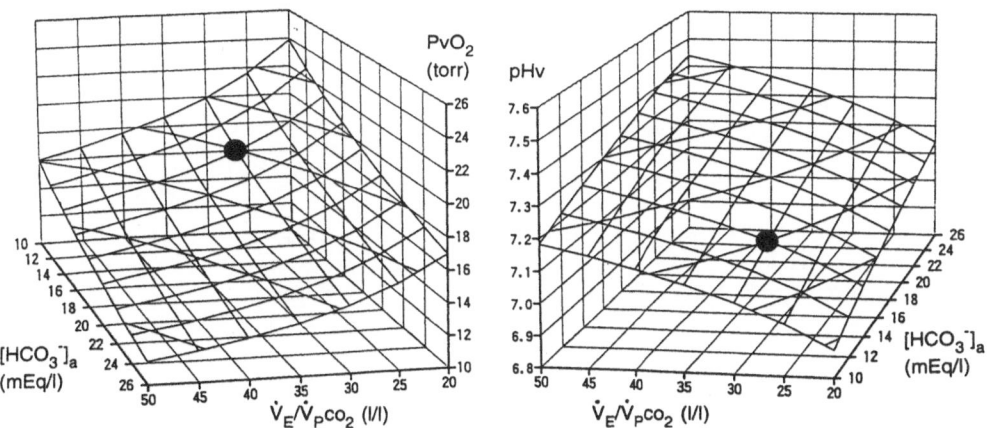

Figure 3. Computed values of pHv and PvO_2 within the specified physiological range of $[HCO_3^-]_a$ and $\dot{V}_E/\dot{V}_PCO_2$.

nisms, pHv decreased more severely - to values of 7.0 or less. On the other hand, if the compensatory hyperventilation results in values of $\dot{V}_E/\dot{V}_PCO_2$ above 30, the degree of the lowering of PvO_2 was significantly smaller than that of hypoventilatory range ($\dot{V}_E/\dot{V}_PCO_2 <$ 30). These results show that the compensatory hyperventilation during severe intensity exercise attenuates the acidaemia while also helping maintain the muscle capillary PO_2 level, i.e., in addition to the Bohr effect.

It should be noted, in the situation described above (i.e., $\dot{V}_E/\dot{V}_PCO_2=30$ and $[HCO_3^-]_a=14$), the estimated ($k \cdot D_TO_2$) value is approximately 200 ml $\cdot$ torr^{-1} $\cdot$ min^{-1}, this means that even 1 torr change of PvO_2 can correspond to 200 ml/min change of V_TO_2. In summary, the results of our model demonstrate that muscle capillary O_2-unloading, a major factor maintaining aerobic energy exchange during severe exercise, is dependent upon a balance between fixed acid production and the sensitivity of the ventilatory control mechanisms.

ACKNOWLEDGMENT

Y.Fukuba was supported by an overseas research fellowship from Uehara Memorial Foundation in Japan.

REFERENCES

1. Ellis, R.K. Determination of PO_2 from saturation. *J.Appl.Physiol.* 67:902, 1989.
2. Severinghaus, J.W. Simple accurate equations for human blood O_2 dissociation computations. *J.Appl.Physiol.* 46:599-602, 1979.
3. Severinghaus, J.W. Exercise O_2 transport model assuming zero cytochrome Po_2 at Vo_2max. *J.Appl.Physiol.* 77:671-678, 1994.
4. Wasserman, K. Coupling of external to cellular respiration during exercise: the wisdom of the body revisited. *Am.J.Physiol.* 266:E519-E539, 1994.
5. Whipp, B.J. and S.A.Ward. Respiratory responses of athletes to exercise. In: *Oxford Textbook of Sports Medicine. Eds.: M.Harries, L.J.Micheli, W.D.Stanish, and C.Williams. Oxford Univ.Press, 1994, pp.13-25.*

32

EFFECTS OF BASE LINE CHANGES IN WORK RATE ON CARDIORESPIRATORY DYNAMICS IN INCREMENTAL AND DECREMENTAL RAMP EXERCISE

Tatsuhisa Takahashi, Kyuichi Niizeki, and Yoshimi Miyamoto

Laboratory of Biological Cybernetics
Department of Electrical and Information Engineering
Yamagata University
4-3-16, Joh-Nan, Yonezawa 992, Japan

INTRODUCTION

The kinetics of ventilatory and gas exchange responses during the decremental (off) phase of trapezoidal and triangular ramp exercise in healthy subjects has been found to be significantly faster than those during the incremental (on) phase (1, 2). As an aid in exploring the mechanisms governing the on-off asymmetry of $\dot{V}O_2$ dynamics, a model analysis has been presented by Niizeki et al. (3). They suggested that the redistribution of blood flow during exercise would play a dominant role in producing the asymmetry of gas exchange response, assuming that the blood flow ratio of active muscles to total cardiac output increases linearly with the work intensity. Therefore, we tested the hypothesis that the asymmetrical response of gas exchange could be modulated by altering the absolute baseline of cardiac output. The dynamic properties of pulmonary gas exchange responses to trapezoidal ramp exercise in the upright position starting from two different baselines of 30- and 50-W were studied in five healthy men.

METHODS

Cardiac output ($\dot{Q}$) was determined by using thoracic impedance cardiography (model RGA-5, Nihon Koden). Heart rate (HR) was calculated from the R-R interval of the electrocardiogram (ECG). Minute ventilation ($\dot{V}E$), O_2 uptake ($\dot{V}O_2$), CO_2 output ($\dot{V}CO_2$), and gas exchange ratio (R) values were obtained breath by breath. Airflow was monitored with a hot-wire-type pneumotachograph (model RF-2, Minato). The composition of expired gas was continuously analyzed with a medical mass spectrometer (WSMR-1400, Westron). Five healthy young male subjects, aged 23.6±3.0 (mean±SD) yr, height 170.8±2.3 cm, and

Modeling and Control of Ventilation, Edited by S. J. G. Semple, L. Adams, and B. J. Whipp
Plenum Press, New York, 1995

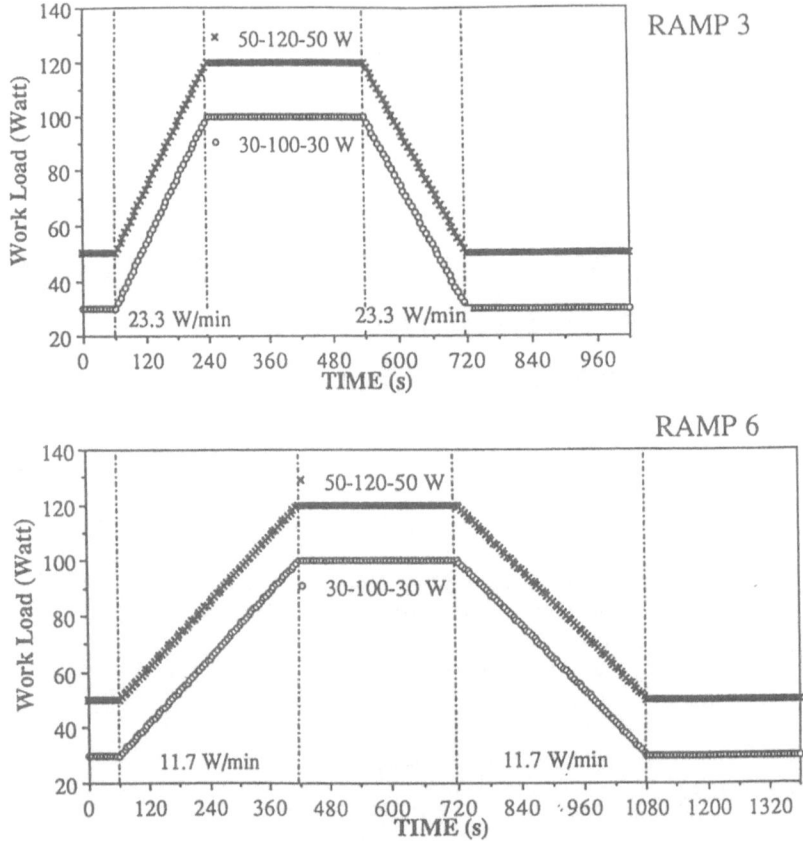

Figure 1. Schematic representation of the exercise protocols. Upper panel, RAMP 3 (slope= 23.3 W/min); lower panel, RAMP 6 (slope = 11.7 W/min).

weight 66.08.5 kg, volunteered for this study after informed consent was obtained. No trained athlete was included in the subject group. The subjects performed exercise using an electromagnetically braked cycle ergometer (Lode) in the upright position. During the exercise period, the subject pedaled at the constant rate of 60 rpm. The two ramp exercise protocols with different slopes of 23.3 and 11.7 W/min illustrated in Figure 1 (open circles); exercise was started from a base line of 30 W to the steady-state exercise of 100 W lasting for 5 min, which was followed by a recovery exercise with the same slopes. The other two ramp exercises with the same slopes were started from an increased base line of 50 W to a steady-state of 120 W (crosses). The periods at the base lines before the start of and after the recovery from the exercise lasted for 5 min each.

Breath-by-breath data from two repetitions of a given test condition for each subject were rearranged into a 5-s interval time scale, and then they were ensemble-averaged to yield a single data set per subject. Assuming a linear transfer function between the input exercise stimulus and the output response, we fitted the response data to a first-order exponential function with a time constant and a pure time delay. The output response, K(t), to a ramp forcing at any time may be expressed as

$$K(T)=Kb+K/T \{(t-td)-TC(1-exp[-(t-td)/TC])\}-K/T\{(t-td-T)-TC(1-exp[-(t-td-T)/TC])\}$$

for the incremental response and

$$K(T)=Ke-K/T \{(t-td)-TC(1-exp[-(t-td)/TC])\}+K/T\{(t-td-T)-TC(1-exp[-(t-td-T)/TC])\}$$

for the decremental response.
Kb: baseline steady-state values
Ke: exercise steady-state values
K: difference between Kb and Ke
T: ramp period (sec)
td: pure time delay (sec)
TC: time constant (sec)

  The sum of td and TC gave the mean response time (MRT). Kb and Ke were determined by taking the mean values during the last 1 min of each exercise level. Curve fittings were performed using a least-squares method.

  All values are expressed as mean$\pm$SD. Statistical analysis was performed using Student's paired t test. Statistical significant was accepted at P<0.05.

RESULT

  Normalized $\dot{V}O_2$ responses during trapezoidal ramp exercise are shown in Figure 2 together with the curves representing exponential models. Tables 1 and 2 summarize the steady-state values obtained before, during and after the stimulus exercise and the MRT determined from the ramp transients. The on-transient responses of $\dot{V}O_2$, $\dot{V}CO_2$, and $\dot{Q}$ during the incremental ramp exercise with a lower baseline started with a MRT that was longer compared with responses to higher baseline exercise, regardless of ramp slopes. However, the difference in MRT during the off-transients between lower and higher baseline exercise were scarcely found except for the $\dot{Q}$ response during 23.3 W/min exercise.

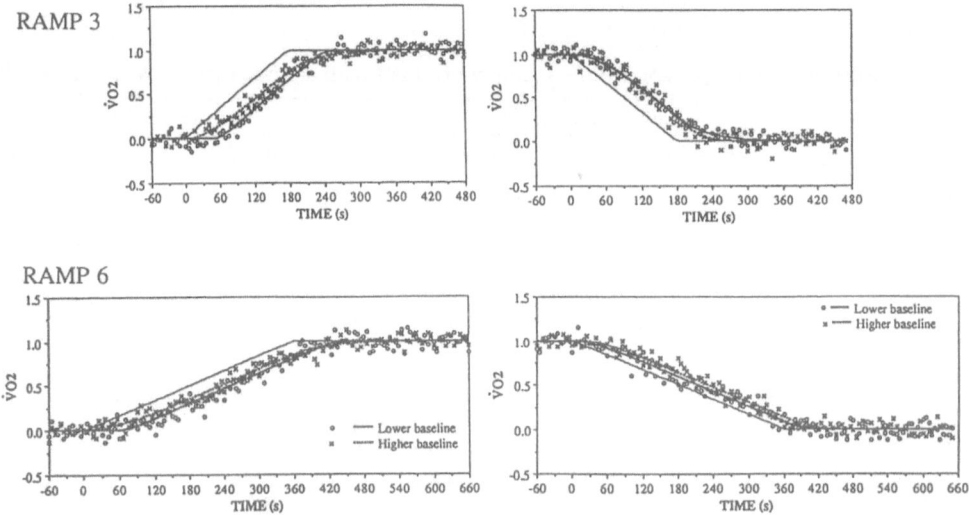

Figure 2. Open circles and crosses are normalized group mean responses of variables during ramp exercise with a baseline of 30- and 50-W, respectively. Each data points is average for 10 trials obtained in all 5 subjects. Thick solid lines and dotted lines represent the best-fit exponential functions to experimental data for lower and higher baseline ramp exercise, respectively. Thin solid lines show work load patterns.

Table 1. Steady-state values and MRT to ramp work load with a slope of 23.3 W/min (RAMP 3)

RAMP3 Variable	Lower baseline					Higher baseline				
	Steady states			MRT, s		Steady states			MRT, s	
	30 W	100 W	30 W	On	Off	50 W	120 W	50 W	On	Off
$\dot{V}E$, l/min	16.6 ± 2.6	24.4 ± 2.9	16.9 ± 1.7	90.8 ± 33.7	89.8 ± 32.9	18.3 ± 2.5	29.21* ± 3.2	20.2 ± 2.9	75.5 ± 9.7	77.4 ± 9.2
$\dot{V}O_2$, l/min	0.55 ± 0.11	0.96 ± 0.07	0.53 ± 0.08	63.0 ± 16.4	53.7† ± 9.9	0.62* ± 0.08	1.15* ± 0.07	0.57* ± 0.09	46.3 ± 9.1	40.4 ± 17.0
$\dot{V}CO_2$, l/min	0.45 ± 0.10	0.81 ± 0.06	0.46 ± 0.07	83.6 ± 18.7	71.0† ± 19.2	0.50* ± 0.07	0.98* ± 0.05	0.57* ± 0.06	69.2 ± 7.1	56.5‡ ± 4.6
R	0.81 ± 0.05	0.84 ± 0.03	0.85 ± 0.02			0.80 ± 0.05	0.84 ± 0.05	0.84 ± 0.06		
$\dot{Q}$, l/min	8.92 ± 2.2	10.64 ± 2.9	8.86 ± 1.8	57.2 ± 28.2	59.4 ± 33.7	9.47 ± 1.72	11.61 ± 1.94	9.77 ± 1.87	57.2 ± 32.5	43.1 ± 43.7
HR, beats/min	81.2 ± 5.5	96.0 ± 6.9	82.9 ± 6.8	45.3 ± 26.5	47.1 ± 17.0	81.4 ± 5.7	103.6 ± 5.8	87.2 ± 7.8	49.0 ± 31.9	52.8 ± 27.8

Values (mean ± SD) were obtained for 10 trials in 5 subjects.
*; significant difference (P<0.05) from the corresponding values for lower baseline exercise
† and ‡; significant difference (P<0.05 and <0.01, respectively) between the on- and off- responses.

The MRT of $\dot{V}O_2$ and $\dot{V}CO_2$ variables were significantly shorter during the off-transient response than that during the on-transient response when the baseline work rate was lower regardless of ramp slopes, leading a significant asymmetry (Fig.3). When the baseline work rate was higher, the MRT of these variables showed only a slight difference between the on- and off-transient responses, with the exception of $\dot{V}CO_2$ at RAMP 3 (P<0.01). The on-off asymmetry for the $\dot{Q}$ responses was, however, less significant for both the lower- and higher-baseline exercise.

Table 2. Steady-state values and MRT to ramp work load with a slope of 11.7 W/min (RAMP 6)

RAMP6 variable	Lower baseline					Higher baseline				
	Steady states			MRT, s		Steady states			MRT, s	
	30 W	100 W	30 W	On	Off	50 W	120 W	50 W	On	Off
$\dot{V}E$, l/min	16.3 ±3.5	23.9 ±2.7	17.2 ±3.1	105.6 ±53.1	91.4 ±24.0	17.7 ±2.8	28.7* ±3.1	19.4 ±2.5	83.7 ±17.3	103.0 ±67.1
$\dot{V}O2$, l/min	0.59 ±0.17	1.00 ±0.14	0.59 ±0.17	74.2 ±26.0	34.6† ±24.3	0.65 ±0.14	1.14* ±0.1	0.67 ±0.11	58.5 ±11.3	51.2 ±11.9
$\dot{V}CO2$, l/min	0.48 ±0.15	0.82 ±0.10	0.49 ±0.13	96.7 ±40.4	60.7 26.1	0.52 ±0.10	0.97* ±0.06	0.56* ±0.08	71.2 ±10.5	70.9 ±33.3
R	0.80 ±0.03	0.82 ±0.02	0.84 ±0.02			0.81 ±0.03	0.85 ±0.04	0.85 ±0.05		
$\dot{Q}$, l/min	9.62 ±2.2	11.07 ±2.1	9.50 ±1.7	79.4 ±23.9	64.2 ±28.6	9.47 ±1.70	11.46 ±2.55	9.82 ±1.62	51.0 ±26.6	±40.4
HR, beats/min	88.3 ±11.6	103.8 ±9.9	91.5 ±12.6	68.5 ±17.8	53.7 ±29.4	80.8 ±5.4	101.3 ±7.1	86.5 ±9.5	36.6 ±38.1	56.5 ±8.7

Values (mean ±SD) were obtained for 10 trials in 5 subjects.
*; significant difference (P<0.05) from the corresponding values for lower baseline exercise
†; significant difference (P<0.05) between the on- and off- responses.

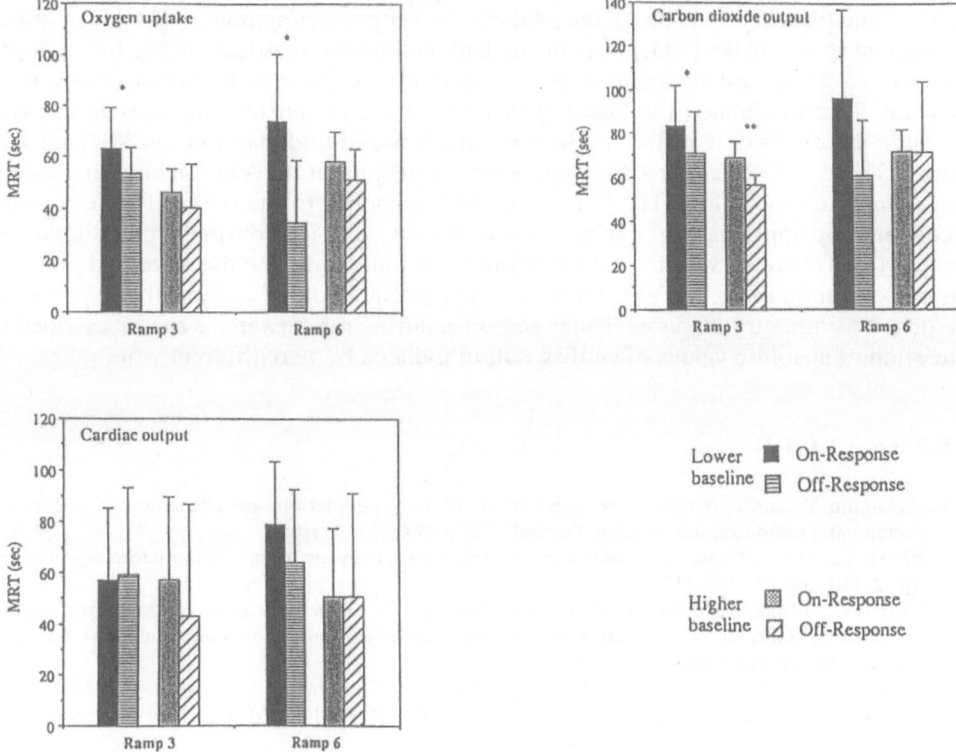

Figure 3. MRT during the on-transients and off-transients for V̇O₂, V̇CO₂, and Q̇. Black and gray bars show meanSD for MRT during the on-transient responses to lower and higher baseline ramp exercise, respectively, and horizontal- and oblique- striped bars show meanSD for MRT during the off-transient responses to lower and higher baseline ramp exercise, respectively. Asterisks indicate significant difference between the on- and off-responses (* $P<0.05$; ** $P<0.01$).

DISCUSSION

The present study demonstrated that the kinetics of ventilatory and gas exchange responses during the on-transient were slower than those during the off-transient when ramp exercise started from the lower baseline. These results are in good agreement with our previous observations (1). In the ramp exercise with the higher baseline, however, there was little difference between the MRT of ventilation and gas exchange for the on- and off-transients. The on-transient response kinetics became faster (a shorter MRT) when the baseline load was increased. However, the off-transient response kinetics for the higher baseline exercise were not different from those for the lower baseline. Therefore, the asymmetry of gas exchange responses between the incremental and decremental phases of the higher baseline exercise decreased by an amount corresponding to the shortening of the MRT. Recently, Niizeki et al. (3) presented a model simulation of the mechanism governing the on-off asymmetry of V̇O₂ kinetics during ramp exercise. They found that the redistribution of blood flow during exercise could play a dominant role in producing the asymmetry. Their simulation demonstrated significant on-off asymmetry of V̇O₂ kinetics in ramp exercise with a lower baseline and attenuation of the asymmetry during a higher baseline exercise. In the present study, we confirmed these phenomena experimentally. Their simulation also demonstrated that the muscle blood flow showed significant asymmetry of the incremental and decremental phases, even though the total cardiac output was symmetrical and the responses of pulmonary O_2 uptake ($\dot{V}O_2$) were dependent on the asymmetrical muscle blood flow. The time course of the cardiac output at the low work rate baseline differed from that at the high baseline. This modulated the blood flow ratio of active muscles to total cardiac output, although there was little difference between the low and high baselines in MRT for cardiac output. There is some evidence to support the concept that the $\dot{V}O_2$ kinetics are dependent on the muscle blood flow. The kinetics of $\dot{V}O_2$ response to the onset of step exercise are accelerated by applying negative pressure to the working legs during cycling in the supine posture (4). This observation can be interpreted as indicating that the increased perfusion of leg muscles modulates the $\dot{V}O_2$ kinetics. The present findings support the hypothesis that the on-off asymmetry of gas exchange response during ramp exercise can be ascribed to the alteration of absolute values of cardiac output induced by two different work rates.

REFERENCES

1. Miyamoto Y., and K. Niizeki. Dynamics of ventilation, circulation, gas exchange to incremental and decremental ramp exercise. J. Appl. Physiol. 72(6): 2244-2254, 1992.
2. Niizeki K., and Y. Miyamoto. Cardiorespiratory responses to cyclic triangular ramp forcings in work load. Jpn. J. Physiol. 41: 759-773, 1991.
3. Niizeki K., T. Takahashi, and Y. Miyamoto. Simulation of asymmetrical O_2 uptake during incremental and decremental ramp exercise. In: Modeling and Control of Ventilation, edited by S.J.G. Semle and L. Adams, Plenum Press, New York, 1995.
4. Hughson, R. L., J. E. Cochrane, and G. C. Butler. Faster O_2 uptake kinetics at onset of supine exercise with than without lower body negative pressure. J. Appl. Physiol. 75(5): 1962-1967, 1993.

33

SIMULATION OF ASYMMETRICAL O$_2$ UPTAKE KINETICS DURING INCREMENTAL AND DECREMENTAL RAMP EXERCISE

Kyuichi Niizeki, Tatsuhisa Takahashi, and Yoshimi Miyamoto

Laboratory of Biological Cybernetics
Department of Electrical and Information Engineering
Yamagata University
Yonezawa 992, Japan

INTRODUCTION

Pulmonary oxygen uptake ($\dot{V}_{O_2}$) increases exponentially after the onset of step exercise and decreases during recovery in a similar manner. During trapezoidal ramp work load exercise, however, the dynamic responses of $\dot{V}_{O_2}$ exhibits an asymmetry between the incremental and decremental phases (8). That is, during the incremental ramp loading, the mean response time (MRT) increases with decreasing ramp slope, whereas the MRT is shortened or unchanged during the decremental ramp loading.

$\dot{V}_{O_2}$ kinetics are influenced not only by those of muscle O$_2$ utilization but also by circulatory dynamics and O$_2$ store. There is some evidence which suggests that the pulmonary $\dot{V}_{O_2}$ kinetics can be influenced by the perfusion kinetics of the exercising legs (6,7). We hypothesized, therefore, that the redistribution of blood flow during exercise would be responsible for the asymmetry. The purpose of the present study was to develop a mathematical model accounting for the asymmetrical $\dot{V}_{O_2}$ response during ramp exercise.

METHODS

Model description

We assumed that the metabolic depiction of the body could be represented by the two compartments of the exercising muscles and the other organs and tissues of the body. The muscle compartment and the inactive tissue compartment were connected in parallel to the lung and perfused by separate regional circulations. Exercise was simulated by increasing the muscle cell $\dot{V}_{O_2}$ ($\dot{V}_{O_2}$mc) together with an appropriate increase in cardiac output ($\dot{Q}$). The $\dot{Q}$ was assumed to be controlled by an exponential equation with appropriate time constants ($\tau_{\dot{Q}} = 5{\sim}100$ s), but no pure time delay. We further assumed that the ratio of the

Modeling and Control of Ventilation, Edited by S. J. G. Semple, L. Adams, and B. J. Whipp
Plenum Press, New York, 1995

blood flow in the muscle compartment to total $\dot{Q}$ increased as a linear function of work rate during submaximal exercise (4). This relationship was described by the following equation:

$$Fm = K_f \cdot WR + 0.2 \qquad (1)$$

where Fm is the redistribution ratio for the muscle compartment, WR is the work rate, and K_f is the slope of the Fm-WR relationship. The value of K_f was initially set at 0.004, so that Fm was 0.32 and 0.6 at WR of 30 and 100 W, respectively. The kinetics of the adjustment of blood flow redistribution are unclear. We assumed that the kinetics of Fm increase as a first-order exponential function with a time constant, τ_f:

$$Fm(t) = Fm_b + (Fm_s - Fm_b) / RT \cdot \{t - \tau_f (1 - \exp(-t / \tau_f))\} \qquad (2)$$

where Fm_b and Fm_s are the redistribution ratios at baseline and during steady-state exercise, respectively. RT stands for the ramp period. We determined the τ_f value by comparing the model output with experimental $\dot{V}_{O_2}$ data, under the condition that the dynamic response characteristics of $\dot{Q}$ remained constant. The muscle blood flow ($\dot{Q}m$) is given by the following equation, employing a time constant for $\dot{Q}$ (τ_Q), for incremental ramp exercise:

$$\dot{Q}m(t) = Fm(t) \{\dot{Q}_b + (\dot{Q}_s - \dot{Q}_b)/RT \cdot [t - \tau_Q \cdot (1 - \exp(-t / \tau_Q))]\}) \qquad (3)$$

where $\dot{Q}_b$ and $\dot{Q}_s$ are the $\dot{Q}$ at baseline and during steady-state exercise, respectively. The blood flow for the inactive tissue compartment ($\dot{Q}t$) was then obtained as $\dot{Q} - \dot{Q}m$. The relationship between $\dot{V}_{O_2}$ mc and the O_2 content of blood leaving the muscle compartment ($C_{V_{O_2}m}$) can be expressed as follows using $\dot{Q}m$:

$$Vm \, dC\dot{V}O_2m(t) / dt = -\dot{V}_{O_2} \, mc(t) + \dot{Q}m(t) \{CaO_2 - C\dot{V}O_2m(t)\} \qquad (4)$$

where CaO_2 and Vm are the arterial O_2 content and the effective volume of the muscle tissue compartment, respectively. $\dot{V}_{O_2}$ mc(t) was also modeled so as to rise in a single exponential fashion with a time constant, τ_m. We assumed that oxidative chemical reactions provide all the energy for muscle contraction from the onset of moderate exercise (ignoring anaerobic ATP), and that the kinetics of muscle cell O_2 consumption were almost identical with the chemical energy transfer kinetics. An initial estimate of τ_m was chosen at 10 s, based on the result from the measurement of NADH fluorescence decay in the excised muscle preparation, that approximates the time course of mitochondrial O_2 consumption (3). The effect of altering the τ_m (1 ~ 60 s) on $\dot{V}_{O_2}$ kinetics was also explored.

Using the instantaneous values of $\dot{V}_{O_2}$ mc(t) and $\dot{Q}m(t)$ and the initial conditions, Eq.4 was numerically solved using a fourth order Runge-Kutta technique. The venous blood O_2 content in the other tissue compartment was also calculated using an equation similar to Eq.4. An initial estimate of 10 l for Vm was chosen by assuming that the active muscle mass was 15 kg with a water content of about 70 %. The effect of changes in Vm on $\dot{V}_{O_2}$ kinetics was also explored. $\dot{V}_{O_2}$ in the inactive tissue compartment was assumed to be constant regardless of the exercise work rate (0.3 l/min). Venous blood leaving the muscle compartment and blood leaving the inactive tissue compartment combine after a time delay, forming mixed venous blood. We calculated the transport delay by dividing the volume of the vessels extending between a given pair of compartments by the blood flow as follows:

$$t_{dm} = V_{ML} / \dot{Q}m(t), \; t_{dt} = V_{TL} / \dot{Q}t(t) \qquad (5)$$

where t_{dm} is the time delay between the muscle and lung compartments, t_{dt} is the delay between the inactive tissue and lung compartments, and V_{ML} (1.3 1) and V_{TL} (0.5 1) are the volumes of the vessels respectively connecting the muscle and inactive tissue compartments to the lungs. The mixed venous O_2 content (CvO_2) was therefore calculated as a blood flow-weighted average of O_2 in regional venous blood returning from the muscle and inactive tissue compartments as follows:

$$Cvo_2 = \{\dot{Q}m \cdot CVo_2m(t) \cdot \delta(t-t_{dm}) + \dot{Q}t(t) \cdot Cvo_2t(t) \cdot \delta(t-t_{dt})\} / \dot{Q}(t) \tag{6}$$

where $\delta(t)$ is a delta function.

The rate of O_2 uptake in the muscle ($\dot{V}_{O_2}$ m) and lung compartments was obtained by the Fick equations using the instantaneous blood flow and venous O_2 content in each compartment. The CaO_2 was modeled to remain constant at 20 vol% throughout exercise. The fractions of metabolism/work rate ($\Delta\dot{V}_{O_2}/\Delta WR$) and blood flow rate/work rate ($\Delta\dot{Q}/\Delta WR$) were set at 0.01 l/min/W and 0.06 l/min/W, respectively.

Comparative Data

Experimental data were collected from 7 healthy young male subjects who exercised on a computer controlled cycle ergometer (Lode 300) in the upright position. The experimental procedure has been described elsewhere in detail (8,10). The dynamic responses of $\dot{V}_{O_2}$ to moderate exercise with step and trapezoidal ramp work load profiles were measured. Work rate at the baseline level was selected at either 30 W or 50 W, and the increment of work rate from the baseline was set to 70 W. Two different ramp slopes (R3=23.3 W/min and R6=11.7 W/min) were used for trapezoidal ramp exercise. The R3 ramp protocol consisted of an incremental exercise period of 3 min, a plateau period of 5 min, and a decremental exercise period of 3 min. The incremental and decremental exercise periods of the R6 ramp protocol were each 6 min, and the plateau period was the same as that in the R3 protocol.

RESULTS

The parameter values (τ_Q, τ_f, and Vm) were first selected so as to simulate the experimental $\dot{V}_{O_2}$ response to stepwise changes of the work rate. The effect of altering the kinetics of $\dot{Q}$ (τ_Q) and blood flow redistribution (τ_f), and that of different Vm size on the $\dot{V}_{O_2}$ response to the on-transients of step exercise are shown in Fig.1, together with experimental group mean values obtained from 14 trials in 7 subjects. In this figure, the $\dot{V}_{O_2}$ responses are represented in a normalized form. Under the condition of a constant τ_f value, alteration of τ_Q produced a significant difference in the phase 1 response (Fig.1B). It can be seen that the faster τ_Q led to a quick rise in the phase 1 response but a slower rise in the phase 2 response. When a τ_Q value of 40 s was chosen, the simulated phase 1 response closely matched the experimental data. The size of Vm (Fig.1A) and the redistribution kinetics (Fig. 1 C) predominantly affected the phase 2 response profile, but alteration of these parameters had relatively little effect on the phase 1 response. When a Vm value of 7.5 l and a τ_f value of 10 s were chosen, the simulated $\dot{V}_{O_2}$ closely matched the experimental data judging from the residual sum of squares (Fig.1D).

Using the same model structure, we simulated the $\dot{V}_{O_2}$ response during trapezoidal ramp exercise to confirm whether there was any asymmetry of the responses between the incremental and decremental phases. Figures 2A and 2B show the effects of τ_Q and τ_f on

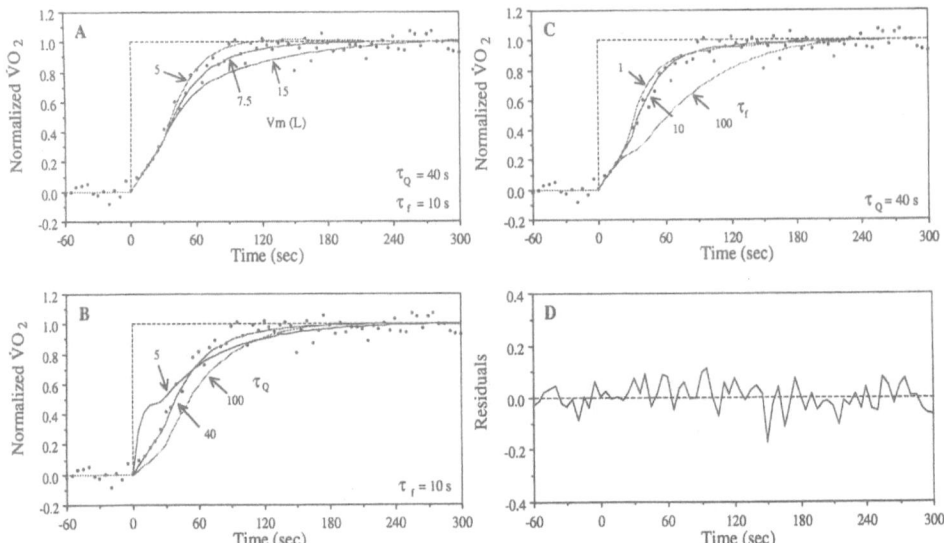

Figure 1. Influence of changes in Vm (A), changes in τ_Q (B), and changes in τ_f (C) on the simulated oxygen uptake ($\dot{V}_{O_2}$) response for on-transitions of step exercise from 30 to 100 W, together with the residuals profile (D) when τ_Q and τ_f were set at 40 and 10 s, respectively. Data points are experimental data in a normalized form.

normalized simulated $\dot{V}_{O_2}$ responses during the incremental phase of R6 trapezoidal ramp exercise. The model output follow the experimental data when the selected parameter values were τ_Q =40 s and τ_f = 10 s, respectively, which were the same values as those used for the step on-transition of work load. As is seen, the effect of τ_Q and τ_f on the $\dot{V}_{O_2}$ profile was

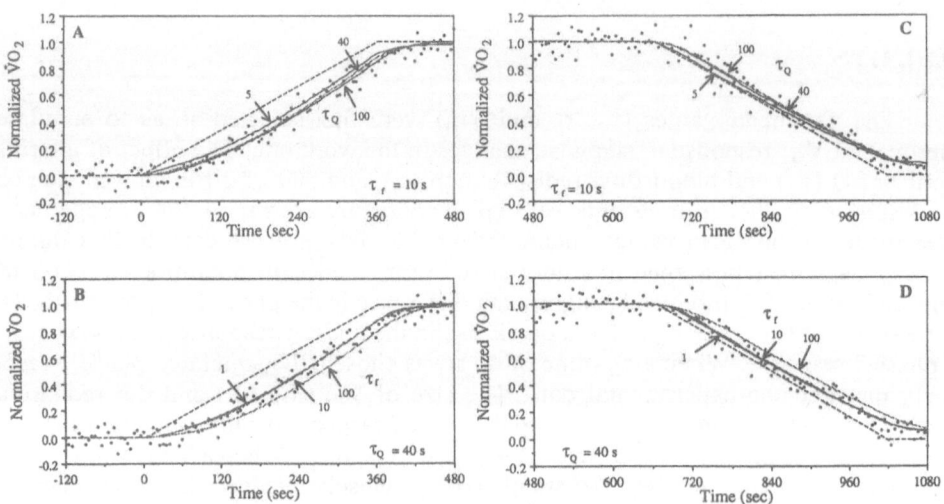

Figure 2. Influence of changes in τ_Q (A) and changes in τ_f (B) on the $\dot{V}_{O_2}$ response during the incremental phase of trapezoidal ramp exercise with a slope of 11.7 W/min (R6) and the corresponding experimental data in a normalized form. Comparable responses during the decremental phase are shown in C and D.

relatively small. The mean response time (MRT), assessed in terms of the algebraic sum of a pure time delay and a time constant determined by fitting the transient $\dot{V}_{O_2}$ response to a monoexponential function, was close to the experimental value (60 s for simulated response vs. 66 s for experimental response). Figures 2C and 2D show comparable simulation results for the decremental phase of R6 ramp exercise. The kinetics of $\dot{V}_{O_2}$ remained almost unchanged even when the values of τ_Q and τ_f were varied together over a wide range, as in the simulation of the incremental response. Note that both the simulated and experimental responses are significantly faster than those during the incremental phase of exercise. The MRT for the decremental phase of model output was 30 s, which was apparently smaller than that for the incremental phase, even though identical values were chosen for τ_Q and τ_f for the incremental and decremental phases.

Figure 3 shows the simulated MRT values for the $\dot{V}_{O_2}$, $\dot{V}_{O_2}$ m, and $\dot{Q}$m kinetics in response to trapezoidal ramp exercise with various ramp slopes. Note that the predicted MRT of V_{O_2} were prolonged during the incremental phase of exercise and shortened during the decremental phase along with a decreasing ramp slope, i.e., there was a slope dependency of the MRT. Similar characteristics was seen for the kinetics of $\dot{V}_{O_2}$ m and $\dot{Q}$m.

DISCUSSION

In the present simulation, we attempted to deduce the transient responses of V_{O_2}m by assuming first-order kinetics of muscle cell metabolism ($\dot{V}_{O_2}$ mc) and $\dot{Q}$, as well as a linear increase in the blood redistribution ratio with work rate (Fm). Without incorporating any asymmetrical kinetics of $\dot{V}_{O_2}$ mc, we showed that the model can demonstrate the asymmetry of the $\dot{V}_{O_2}$ response between incremental and decremental ramp work loads. The model also showed that the response of $\dot{V}_{O_2}$ to incremental ramp exercise became longer with a decreasing ramp slope, as reported in previous studies (9,11).

Recently, intramuscular phosphocreatine (PCr) has come to be used as an indirect measure of oxidative metabolism, since a close relationship between $\dot{V}_{O_2}$ and PCr has been demonstrated (2,12). In human muscles, the PCr change measured by ^{31}P-NMR has been reported to fall almost exponentially with a time constant of about 20~40 s (2,12). The relatively small estimate used for the τ_m value in the present model might be subject to debate. In order to examine the effect of τ_m on the asymmetry, we computed the $\dot{V}_{O_2}$ kinetics using various values for τ_ms. Table 1 shows the effect of different τ_m values on $\dot{V}_{O_2}$ kinetics for the

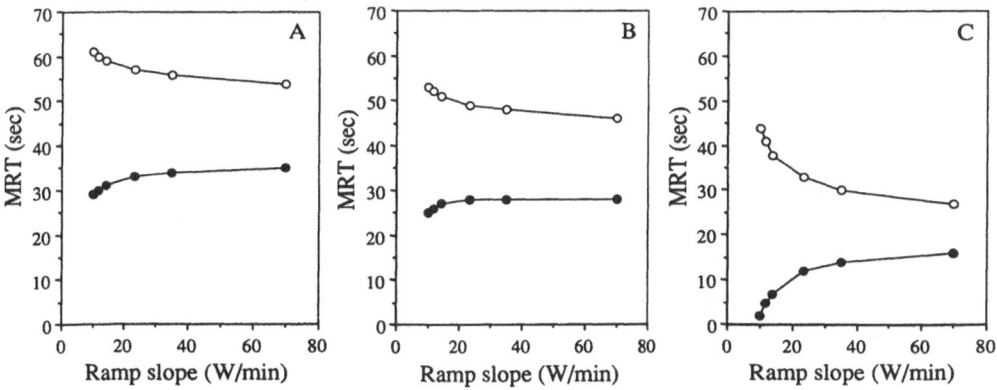

Figure 3. Simulated MRT for $\dot{V}_{O_2}$ (A), $\dot{V}_{O_2}$ m (B) and $\dot{Q}$m (C) responses to incremental (open circles) and decremental (closed circles) ramp exercise with various ramp slopes (30 W ~ 100W exercise).

Table 1. Effect of different time constants for muscle cell O_2 consumption (τ_m) on Vm size and mean response time (MRT) for $\dot{V}_{O_2}$ to incremental and decremental phases of ramp exercise (R6:11.7W/min)

	τ_m, s						
	1	10	20	30	40	50	60
Vm, l	12	7.5	4.5	2	0.1	0	0
MRT (incremental), s	63	60	46	45	48	58	67
MRT (decremental), s	29	30	37	43	47	57	68

R6 ramp exercise protocol. If the simulated response is to fit the experimental data, Vm should be decreased as τ_m increases. Vm value was determined so that the simulated $\dot{V}_{O_2}$ response fit the experimental response to the on-transient of step exercise. Throughout the simulation, τ_Q and τ_f were set at 40 and 10 s for the incremental and decremental phases, respectively. When τ_m was greater than 40 s, the decremental $\dot{V}_{O_2}$ response was retarded, resulting in no asymmetry between the incremental and decremental MRT. This implies that, if the muscles are assumed to be a single homogeneous compartment (i.e., Vm = 0), the model output never leads to asymmetry of the $\dot{V}_{O_2}$ response even when $\dot{Q}m$ is changed asymmetrically. It was necessary to manipulate the value of τ_m to less than that for the incremental phase to follow the actual experimental data for the decremental phase. However, there is no reason to assume that muscle cell metabolism is faster during the decremental phase.

Recently, Barstow et al. (1) and Cochrane and Hughson (5) have proposed a model to infer the control of muscle $\dot{V}_{O_2}$ dynamics, in which the entire enhancement in $\dot{Q}$ was directed to the muscle compartment, with the flow for the other tissue compartment remaining constant. Selection of a constant $\dot{Q}t$ resulted in a significantly longer MRT during the decremental phase compared with the original model output.

To predict the effect of $\dot{Q}m$, we further simulated the kinetics $\dot{V}_{O_2}$ response when the base work rate was increased to 50 W and the stimulus work rate to 120W, where Fm was supposed to increase from 0.4 (50 W) to 0.68 (120 W). The same model parameter values except for the baseline values of $\dot{Q}$ and $\dot{V}_{O_2}$ mc were used for the simulation. The MRT values of $\dot{V}_{O_2}$ responses in the incremental phases tended to decrease compared with those of lower baseline exercise, reflecting the decrease in the transport delay by the elevated $\dot{Q}$ (Fig.4). In contrast, the decremental phase MRT values were less influenced by the baseline work rate. Therefore, the model predicted that the asymmetric feature becomes less pronounced when

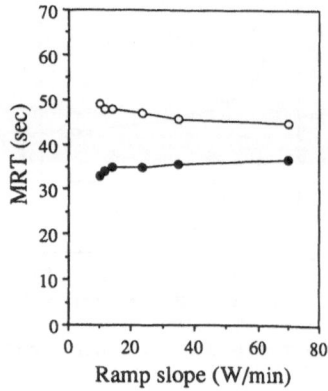

Figure 4. Simulated MRT for $\dot{V}_{O_2}$ responses to trapezoidal ramp exercise from a baseline of 50 W to a stimulus level of 120 W.

the baseline work rate is increased. The simulation accorded well with the experimental observation.

Hughson et al. (6) reported that the $\dot{V}_{O_2}$ response in the supine position adapted more slowly compared to the response in the upright posture at the onset of step exercise. Application of negative pressure to the exercising muscles of the lower body results in faster O_2 uptake kinetics (7). These results were interpreted as indicating that the $\dot{V}_{O_2}$ kinetics might have been accelerated by an increase in perfusion of the exercising muscles. All of the findings cited above and the results of the present model analysis support the concept that the muscle $\dot{V}_{O_2}$ kinetics ($\dot{V}_{O_2}$ m) would be affected by perfusion of the exercising muscles and thus they are not equivalent to the intracellular $\dot{V}_{O_2}$ kinetics ($\dot{V}_{O_2}$ mc) during the transient state of exercise. If the present model is valid, the conclusion can be derived that the redistribution of blood flow to the working muscles is a predominant factor in producing the asymmetry of $\dot{V}_{O_2}$ responses observed between the incremental and decremental ramp exercise.

ACKNOWLEDGMENT

This study was partly supported by a grant to Dr. Y. Miyamoto from the Ministry of Education and Culture of Japan (05680750).

REFERENCES

1. Barstow, T.J., N. Lamarra, and B.J. Whipp. Modulation of muscle and pulmonary O_2 uptakes by circulatory dynamics during exercise. *J. Appl. Physiol.* 68:979-989, 1990.
2. Binzoni, T., G. Ferretti, K. Schenker, and P. Cerretelli. Phosphocreatine hydrolysis by ^{31}P-NMR at the onset of constant-load exercise in humans. *J. Appl. Physiol.* 73:1644-1649, 1992.
3. Chapman, J.B. Fluorometric studies of oxidative metabolism in isolated papillary muscle of the rabbit. *J. Gen. Physiol.* 59:135-154, 1972.
4. Clausen, J.P. and N.A. Lassen. Muscle blood flow during exercise in normal man studied by the [133]Xenon clearance method. *Cardiovasc. Res.* 5: 245-254, 1971.
5. Cochrane, J. E. and R.L. Hughson. Computer simulation of O_2 transport and utilization mechanisms at the onset of exercise. *J. Appl. Physiol.* 73: 2382-2388, 1992.
6. Hughson, R.L., Xing, H.C., Borkhoff, C., and Butler, G.C. Kinetics of ventilation and gas exchange during supine and upright cycle exercise. *Eur. J. Appl. Physiol.* 63:300-307, 1991.
7. Hughson, R.L., Cochrane, J.E., and Butler, G.C. Faster O_2 uptake kinetics at onset of supine exercise with than without lower body negative pressure. *J. Appl. Physiol.* 75:1962-1967, 1993.
8. Miyamoto, M. and K. Niizeki Dynamics of ventilation, circulation and gas exchange to incremental and decremental ramp exercise. *J. Appl. Physiol.* 72: 2244-2254, 1992.
9. Swanson, G.D. and R.L. Hughson. On the modeling and interpretation of oxygen uptake kinetics from ramp work tests. *J. Appl. Physiol.* 65:2453-2458, 1988.
10. Takahashi, T., K. Niizeki, and Y. Miyamoto. Effects of base line changes in work rate on cardiorespiratory dynamics in incremental and decremental ramp exercise. In: Modeling and Control of Ventilation, edited by S.J.G. Semple and L. Adams. New York : Plenum Press, 1995.
11. Yoshida, T. Gas exchange responses to ramp exercise. *Ann. Physiol. Anthrop.* 9(2):167-173, 1990.
12. Yoshida, T. and H. Watari. ^{31}P-Nuclear magnetic resonance spectroscopy study of the time course of energy metabolism during exercise and recovery. *Eur. J. Appl. Physiol.* 66:494-499, 1993.

REFERENCES

34

CORE TEMPERATURE THRESHOLDS FOR VENTILATION DURING EXERCISE

Temperature and Ventilation

M. D. White and M. Cabanac

Department of Physiology
Laval University
Ste-Foy, Quebec, Canada

INTRODUCTION

An increase of body temperatures by about 1.0 °C in humans at rest increases the ventilation rate (8). Recently in humans we reported that passive hyperthermia, above thresholds of tympanic (T_{ty}) and esophageal temperatures (T_{es}), induced an increase in ventilation rate that was proportional to the increase in core temperatures (4). This increased ventilation during hyperthermia leads to a small but significant increase respiratory heat loss (10) and this thermally induced hyperventilation in humans (3), as it does in other species (1), appears to be a thermolytic mechanism that could be contributing to selective brain cooling (SBC). SBC in hyperthermic humans occurs when the T_{ty} is lower than T_{es}. During exercise, however, the relationship between ventilation and core temperatures is not so clear.

In normo- and hypothermic exercising subjects there was no association between the ventilatory equivalent for carbon dioxide ($\dot{V}I \cdot \dot{V}CO_2^{-1}$) and rectal temperature (16). However, for hyperthermic exercising subjects, at increased rectal temperatures, greater $\dot{V}_1 \cdot \dot{V}CO_2^{-1}$ and ventilatory equivalents for oxygen ($\dot{V}_1 \cdot \dot{V}O_2^{-1}$) were evident (14). This result suggests as core temperature increases during exercise that there could be core temperature thresholds for ventilation. The goal of the present study was to examine whether such temperature thresholds for ventilation exist in subjects rendered hyperthermic by exercise. Unlike previous studies (14, 16), T_{ty} and T_{es} rather than rectal temperature were used as estimates of core temperature because the latter is not accepted to accurately reflect thermal transients of the core (7, 11). Recently T_{ty} was shown to be a good estimate of the brain temperature and T_{es} of the trunk temperature (12).

METHODS

Five endurance trained male subjects (1.7 ± 0.03 m height, 69.2 ± 3.5 kg weight), volunteered to participate in the study. A medical examination was a prerequisite for

Modeling and Control of Ventilation, Edited by S. J. G. Semple, L. Adams, and B. J. Whipp
Plenum Press, New York, 1995

173

participation, and subjects were informed of potential risks associated with the experimental protocol. Subjects were also tested using a submaximal exercise protocol to determine their level of fitness and ensure that they would be able to undergo a maximal exercise test. Subjects wore only shorts and a T-shirt. They had not eaten nor exercised for 3 h prior to the experiments. A medical emergency kit including a defibrillator was available at all times.

Core temperature was measured in the esophagus (T_{es}) with a thermocouple placed 0.385 ± 0.004 m past the nares, a position corresponding to the left ventricle (13). Tympanic temperature (T_{ty}) was measured with a thermocouple placed on the tympanic membrane according to the criteria previously described (2). Skin temperature (T_{sk}) was measured at 4 sites (forehead, right arm, chest, and thigh), and values were expressed as the unweighted mean.

The subject breathed from a low resistance mouth-piece connected to a 2-way valve. Collins tubing connected the 2-way valve to a fluted mixing box, and a rotometer was placed on the inspiratory side of the breathing circuit (KL Engineering, model KLE 511). Expiratory gas was drawn by a pump at a rate of $0.4 \, l \cdot min^{-1}$ from the mixing box for analysis of carbon dioxide (Beckman CO_2 analyzer, model LB2) and oxygen contents (Appl. Electroch. oxygen analyzer, model S-3A). The inspired minute ventilation ($\dot{V}_I$) was determined with the rotometer, and values were totaled on a modified spirogram (Pneumoscan, model S301C). Instruments were controlled by a Macintosh computer using an Omega data acquisition system. Data were sampled and recorded at 30-s intervals.

All sessions were conducted between 13:00 and 15:00. The subject rested in the climatic chamber conditions of 25°C and approximately 35% RH for 30 min prior to the session. The wind speed in the chamber, determined with a constant temperature hot wire anemometer (Thermonetics, model HWA-103), was negligible. Each subject participated in one session and the session began with a 5-min rest with the subject seated on an electrically braked cycle ergometer. Then the exercise started at a prescribed pedaling rate of 60 rpm. The starting workrate was 40 W and the workrate was increased at increments of 40 W each 2 min until the subjects exhaustion.

Calibrations of thermocouples were made in regulated temperature hot water baths. Gas analyzers were calibrated against gases of known values of O_2 and CO_2 prior to each experiment. The flowmeter was calibrated by the manufacturer and verified with a Tissot Spirometer. Thresholds for ventilation were determined following the cumulative sum technique of Ellaway (6). These values were verified with values determined from plots of ventilatory equivalents against core temperatures. A two tailed dependent t-test was employed for means comparisons. The level of significance was set at 5%.

RESULTS

T_{ty} increased from an initial 36.82 ± 0.10°C to a maximum of 37.72 ± 0.17°C. Likewise, T_{es} increased from an initial 36.72 ± 0.08°C to a maximum of 38.00 ± 0.27°C. Mean skin temperature was either stable or increasing from an initial value of 33.07 ± 0.07°C to a final values of 34.96 ± 0.23°C. The same was apparent for forehead skin temperature that increased from 35.15 ± 0.27°C to 36.86 ± 0.22°C at the end of exercise. Typical results from one subject are given for $\dot{V}_I \cdot \dot{V}CO_2^{-1}$ and $\dot{V}_I \cdot \dot{V}O_2^{-1}$ plotted against T_{ty} and T_{es} in Figure 1.

Figure 1 shows three distinct sections. Initially there was a decrease in $\dot{V}_I \cdot \dot{V}CO_2^{-1}$ and $\dot{V}_I \cdot \dot{V}O_2^{-1}$ while temperatures decreased slightly or remained stable. Next, ventilation increased independently of temperatures until core temperature thresholds were reached. Following this the ventilation increased in a manner approximately proportionate to either core temperatures. Above the core temperature thresholds the two core temperatures tended to diverge. In all five subjects core thresholds for ventilation were evident at both sites of

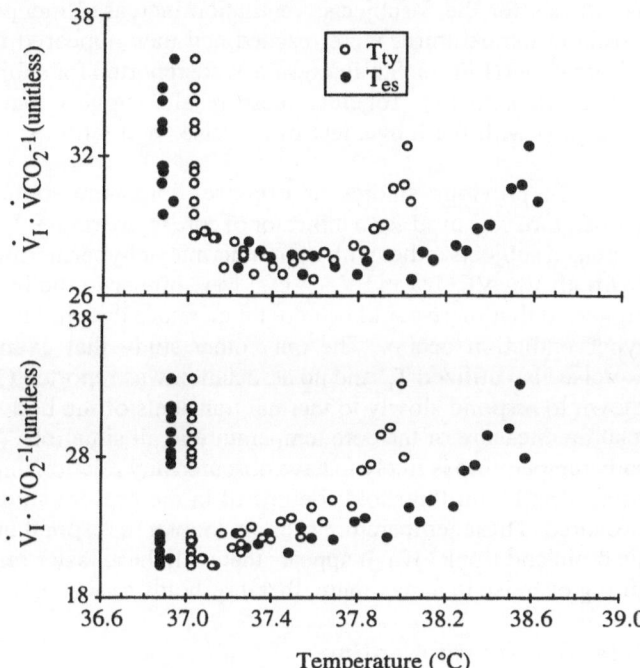

Figure 1. Typical core temperature thresholds for $\dot{V}_I \cdot VCO_2^{-1}$ and $\dot{V}_I \cdot VO_2^{-1}$ from one subject.

temperature measurement. Figure 2 gives the average core thresholds for the group measured at either the tympanic or esophageal site. For both $\dot{V}_I \cdot \dot{V}CO_2^{-1}$ (p <0.005) and $\dot{V}_I \cdot \dot{V}O_2^{-1}$ (p <0.005) the mean T_{ty} threshold was significantly less than the mean T_{es} threshold.

DISCUSSION

In this study we asked whether the excess ventilation seen at higher intensities of exercise could be related to increases in core temperature. The results support the working

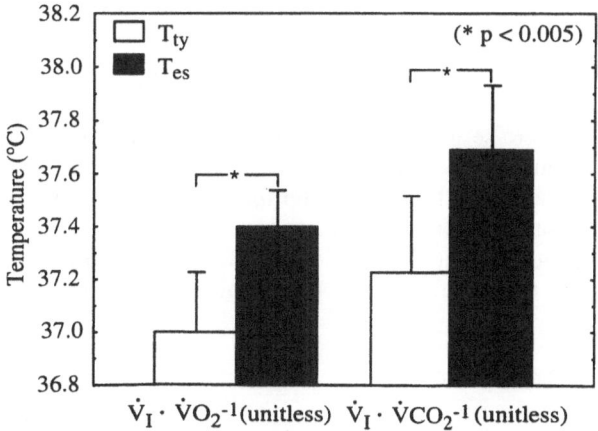

Figure 2. Mean core temperature thresholds for $\dot{V}_I \cdot \dot{V}CO_2^{-1}$ and $\dot{V}_I \cdot \dot{V}O_2^{-1}$.

hypothesis for the 5 subjects: ventilation increased independently of temperature until core threshold temperatures were reached and then appeared to increase proportionately to core temperatures (Fig. 1). Similar results were reported for subjects rendered hyperthermic in a hot bath immersion (4). Together these results suggest that, with or without the stimulation associated with the movement of exercise, ventilation is stimulated by higher core temperatures.

In previous studies of exercise, core temperature and ventilation (14, 16) rectal temperature was used as an indicator of core temperature. Peterson and Vjeby-Christensen (14) exercised subjects either with normothermic or hyperthermic rectal temperatures, and they saw that both $\dot{V}_I \cdot \dot{V}CO_2^{-1}$ and $\dot{V}_I \cdot \dot{V}O_2^{-1}$ were higher in the hyperthermic condition. Their results suggested that there could be core temperature threshold(s) above which a thermally induced hyperventilation occurs. The only other study that examined core and ventilation during exercise also utilized T_{re} and no association was reported (16). However, rectal temperature is known to respond slowly to thermal transients of the body core and is thought not to give an accurate measure of the core temperature in all situations (7, 11). The slow response of T_{re} to body temperature is likely to have obscured any relationship between ventilation and temperatures. Ventilation thresholds described in the present study appeared when T_{ty} and T_{es} were measured. These temperature sites are known to respond quickly to thermal transients both in the brain and trunk (12). It appears that with these faster responding sites of core temperatures during exercise to a maximum, that thresholds for ventilation were evident.

Selective Brain Cooling

A possible physiological rationale for an increased ventilation during hyperthermia is that the hyperventilation is in part related to the SBC of hyperthermic humans. In panting animals ventilation increases as a function of core temperature (9) and the subsequent respiratory heat loss (RHL) is a primary heat loss mechanism that results in a selective cooling of their brains (1, 3, 5). Although sweating is the main heat loss effector in hyperthermic humans, studies that have manipulated avenues of RHL have shown that T_{ty}, as an index of brain temperature (12), is sensitive to such changes (15, 17, 18). Increased respiratory heat loss from the nose (18) together with nasal mucosal vasodilatation during hyperthermia (17) are thought to participate in a coordinated cooling mechanisms of the human brain. The increased ventilation seen during the hyperthermia of exercise (Fig. 1) or during passively induced hyperthermia (4) would appear to be complementary to such a cooling mechanism. The lower thresholds for T_{ty} (Fig. 2) and the divergence of T_{ty} and T_{ty} in the higher intensity exercise are seen as evidence of such a SBC.

CONCLUSION

During graded exercise to maximal workrates indexes of minute ventilation when expressed against tympanic and esophageal increased independently of temperature until core thresholds were reached. Above these temperature thresholds ventilation increased proportionately to tympanic and esophageal temperatures as the two temperatures tended to diverge. The results suggest that the hyperventilation seen during higher intensities of exercise could be in part a thermolytic response involved in human selective brain cooling.

ACKNOWLEDGMENTS

This work was supported by grants from the Natural Sciences and Engineering and Research Council of Canada and the Defense and Civil Institute of Environmental Medicine.

A word of thanks to Mr. A. Samson for his attentive help during the experiments. Additionally thanks go to Dr. M. Walsh for his advice on the calculations employed for threshold detections.

REFERENCES

1. Baker, M. A. Brain cooling in endotherms in heat and exercise. *Ann. Rev. Physiol.* 44: 85-96, 1982.
2. Brinnel, H. and M. Cabanac. Tympanic temperature is a core temperature in humans. *J. Thermal Biol.* 14: 47-53, 1989.
3. Cabanac, M. Selective brain cooling in humans: "fancy" or fact. *FASEB* 7: 1143-1147, 1993.
4. Cabanac, M. and M. D. White. Core temperature thresholds for hyperpnea during passive hyperthermia in humans. *Eur. J. Appl. Physiol.* 71: 71-76, 1995.
5. Caputa, M. Selective brain cooling an important component of thermal physiology. In: *Contributions to Thermal Physiology,* edited by Z. Szelényi and M. Székely. Pecs, Hungary: Permagon Press, Budapest, 1980, p. 183-192.
6. Ellaway, P. H. Cumulative sum technique and its application to the analysis of peristimulus time histograms. *Electroencephalogr. Clin. Neurophysiol.* 45: 302-304, 1978.
7. Gerbrandy, J., E. S. Snell and W. I. Cranston. Oral, rectal and esophageal temperatures in relation to central temperature control in man. *Clinical Science* 13f: 615-623, 1954.
8. Haldane, J. S. The influence of high air temperatures. *J. Hygiene* 55: 497-513, 1905.
9. Hammel, H. T., D. C. Jackson, J. A. J. Stolwijk, J. D. Hardy and S. W. B. Stromme. Temperature regulation by hypothalamic proportional control with an adjustable set point. *J. Appl. Physiol.* 18(6): 1146-1154, 1963.
10. Hanson, R. de. G. Respiratory heat loss at increased core temperature. *J. Appl. Physiol.* 37(1): 103-107, 1974.
11. Livingstone, S. D., J. Grayson, J. Frim, C. L. Allen and R. E. Limmer. Effect of cold exposure on various sites of core temperature measurements. *J. Appl. Physiol.* 54(4): 1025-1031, 1983.
12. Mariak, Z., J. Lewko, J. Luczaj, B. Polocki and M. D. White. The relationship between directly measured human cerebral and tympanic temperatures during changes in brain temperatures. *Eur. J. Appl. Physiol.* 69: 545-49, 1994.
13. Mekjavic, I. B. and M. E. Rempel. Determination of esophageal probe insertion length based on standing and sitting height. *J. Appl. Physiol.* 69: 376-379, 1990.
14. Petersen, E. S. and H. Vejby-Christensen. Effects of body temperature on steady state ventilation in exercise. *Acta Physiol. Scand.* 89: 342-351, 1973.
15. Rasch, W., P. Samson, J. Côté and M. Cabanac. Heat loss from the human head during exercise. *J. Appl. Physiol.* 71: 590-595, 1991.
16. Whipp, B. J. and K. Wasserman. Effect of body temperature on the ventilatory response to exercise. *Resp. Physiol.* 8: 354-360, 1970.
17. White, M. D. and M. Cabanac. Nasal mucosal vasodilatation in response to passive hyperthermia in humans. *Eur. J. Appl. Physiol.* 70: 207-212, 1995.
18. White, M. D. and M. Cabanac. Physical dilatation of the nares lowers the thermal strain in exercising subjects. *Eur. J. Appl. Physiol. 70: 200-206, 1995.*

35

MODELLING THE EFFECT OF TAPER ON PERFORMANCE, MAXIMAL OXYGEN UPTAKE, AND THE ANAEROBIC THRESHOLD IN ENDURANCE TRIATHLETES

P. C. Zarkadas, J. B. Carter, and E. W. Banister

School of Kinesiology
Simon Fraser University
Burnaby B.C. Canada V5A 1S6

ABSTRACT

The purpose of this study was to determine the nature of taper required to optimize performance in Ironman triathletes. Eleven triathletes (26±4 yrs, 77.0± 6.5 kg) took part in 3 months of training interspersed with two taper periods, one of 10 days (Taper 1) and another six weeks later for 13 days (Taper 2). Reducing training volume by 50 % in an exponential fashion ($\tau \leq 5$ days) in one group of triathletes during Taper 1 resulted in a 46 second (4 %) improvement in their 5 km criterion run time and a 23 W (5 %) increase in maximal ramp power output above the same measurement at the beginning of taper. A 30 % step reduction in training volume in the second group did not result in any significant improvement in physical performance on the same measures. Training volume was reduced exponentially from the end of training in both a high volume group ($\tau \geq 8$ days) and a low volume group ($\tau \leq 4$ days) during Taper 2. Criterion run time improved significantly by 74 seconds (6 %) and 28 seconds (2%) in the high and low volume groups respectively, while maximal ramp power increased significantly by 34 W (8 %) only in the low volume taper group. Maximal oxygen uptake increased progressively from 62.9 ± 5.8 ml·kg^{-1}·min^{-1} two weeks prior to taper, to a significantly higher level 68.9 ± 4.2 ml·kg^{-1}·min^{-1} during the final week of Taper 2 ($p \leq 0.05$). The anaerobic threshold determined by a non-invasive method was also observed to increase from 70.9 % to 74.9 % of a subject's maximal oxygen uptake during Taper 2. These results demonstrate that proper placement of training volume during taper is a key factor in optimizing performance for a specific competition and a high volume of training in the immediate days preceding an event may be detrimental to physical performance.

Modeling and Control of Ventilation, Edited by S. J. G. Semple, L. Adams, and B. J. Whipp
Plenum Press, New York, 1995

INTRODUCTION

Taper is a highly specialized form of reduced training that decays in a systematic non-linear fashion (11). Recent evidence has demonstrated that much of the performance decrement and loss of training adaptation that occurs with detraining (10) may be minimized if training is simply maintained at a reduced level (4,6,7,12,14,17,18) or tapered (8,9,13,19). There appears to be advantage to the use of a taper protocol compared with a standard step reduction in training volume. In one study, a 70 % reduction of normal training volume for 3 weeks did not significantly improve a 5 km run performance or muscular power in distance runners (12). In contrast, a 7 day taper with an 85 % reduction in weekly training volume improved a 5 km race time and increased muscular power (9,19). These results indicate that when training volume is reduced progressively to an extremely low amount physical performance is improved more than by a standard single step reduction in training volume up to 2/3 volume, which only appears to maintain performance (12).

Although it is well known $\dot{V}O_2$max increases with training (18) current studies of taper have demonstrated an improvement in physical performance without an increase in $\dot{V}O_2$max (9,19). Improvement in physical performance has been attributed to adaptation at the muscle level where recovery from the fatigue of heavy training rather than improvement in $\dot{V}O_2$max is considered the determining factor. However, heavy training imposes a fatiguing stress on both cardiorespiratory and peripheral muscular systems. If training of sufficient intensity and duration to stimulate adaptation is performed both muscular and aerobic power should improve. The latter of these two effects has yet to be experimentally demonstrated, however.

METHODS

Subjects

Eleven male triathletes volunteered to take part in this study. They gave their written consent after being medically approved for participation by a physician and after having been fully informed of the nature, risks and benefits of their participation. All tests performed on these subjects received approval of Simon Fraser University's Human Subjects Ethics Approval Committee.

Quantifying Training Dose

A systems model of training proposed by Calvert et al. (2), described by Morton et al. (16), and elaborated to explain the structural features of a training program by Fitz - Clarke et al. (5) was used to quantify and describe the pattern of training and peaking. In this system the training stimulus **w(t)** is calculated and presented as a training impulse (**TRIMP**) expressed in arbitrary training units (**ATU**). A single training bout is described by the product of the duration of the session (**D**) in minutes and the ratio of exercise heart rate (**HRex**) to maximum heart rate (**HRmax**), both above a resting value (**HRrest**). The latter measure is termed the delta heart rate ratio

(Δ **HR Ratio**). Thus training undertaken at any time **t** may be expressed as an area under the curve represented by the pseudointegral:

w(t) = Duration of Training x **Hrex - HRrest / HRmax - HRrest**
      = D x Δ **HR Ratio**

An intensity weighting factor **Y** is used to correct the bias introduced into **w(t)** from long non-strenuous training. This metabolic intensity factor reflects the exponential rise of blood lactate as the fractional elevation of exercise heart rate above rest and gives more credit to short intense sessions which provide a greater training stimulus. Thus overall:

$$w(t) = D \times \Delta \text{ HR Ratio} \times Y$$

Study Design

The entire experiment lasted 98 days and required each athlete to follow two separate periods of heavy training of approximately 40 days (d) each followed by a peaking or taper period prior to a designated real competition. Prior to Taper 1 subjects were randomly assigned to one of two taper groups, matched for age, weight, and initial criterion 5 km run time. Group 1 (n = 6) tapered their training volume by 50 % in an exponential decay while Group 2 (n = 3) implemented a single step reduction in training by 30 % of their initial volume.

Following the mid season competition heavy training was resumed and subsequent to this period all subjects were again randomly assigned to 2 different exponential taper groups each lasting 13 days with respectively a high volume (τ 3 8 d) in Group A (n = 7) or a low volume taper (τ 2 4 d) in Group B (n = 4). All subjects competed in the 1993 Canadian Ironman Championship in Penticton, B.C. (3.9 km swim, 180.2 km bike, 42.2 km run) following Taper 2.

Training Heart Rate

Average training heart rate for each workout was recorded at one minute intervals via telemetry (Polar Vantage XL) in serial daily files for weekly downloading to computer.

Criterion Physical Performance

In order to monitor any change in physical performance at least one and sometimes two 5 km criterion runs were performed each week to the best of the subject's current ability. Subjects ran all out on their own 5 km course which was chosen to be as flat as possible, and each subject ran this course at a time when there would be a minimal possible interruption to their effort (traffic, obstacles etc.).

Ergometry

A ramp test to exhaustion was also performed weekly by each subject on a cycle ergometer in order to assess the serial change in maximal power output and in respiratory gas exchange, both of which were measured on selected dates during training and taper. The cycle ergometer (Lode: Groningen, Holland) was electromagnetically braked and externally controlled by a computer. The ramp test protocol consisted of a 4 minute warm-up period at a work rate of 30 watts followed by a 30 watts / min ramp until exhaustion or until a subject could not maintain pedal rate above 80 rpm. On three ramp tests chosen two weeks prior to taper and during each week of Taper 2 a real time breath-by-breath $\dot{V}O_2$max acquisition was made using a Macintosh IIfx computer equipped with a National Instruments software package (LabView II).

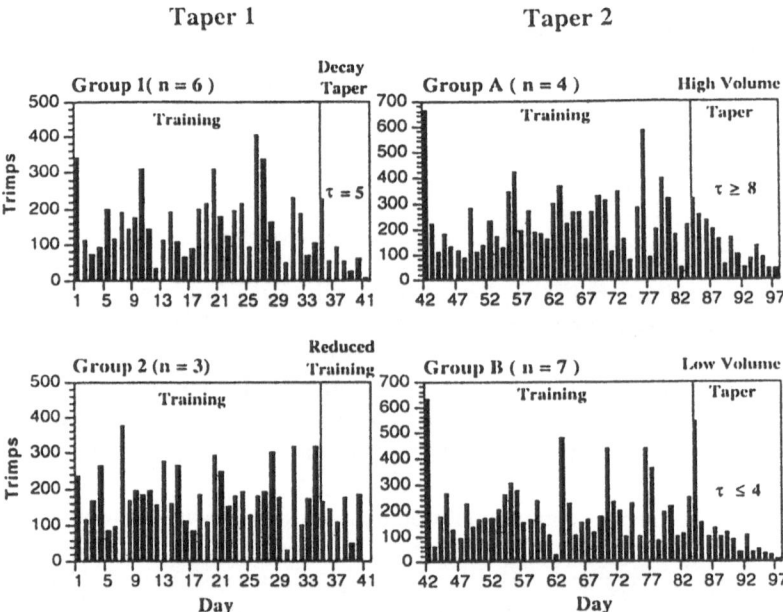

Figure 1. Comparison of the profile of training and taper for Taper 1 (decay taper vs reduced training) and Taper 2 (high and low volume exponential taper).

Statistics

The improvement in physical performance, $\dot{V}O_2$max, and the anaerobic threshold with taper was statistically analyzed using a paired t-test for significance. The difference between the two groups during training and taper periods, for all variables which include performance on a 5 km criterion run, and a 30 watt ramp maximal ergometer test was evaluated using a Generalized Linear Model to account for the unbalanced number of subjects in each group.

RESULTS

The effect of a 10 day taper (Taper 1) and a 13 day taper (Taper 2) on physical performance and physiological parameters are summarized in Tables 1-3 while the profile of each taper is shown graphically in Figure 1 .

Table 1. Performance measures for both a criterion 5 km run and a maximal ramp on a cycle ergometer immediately preceding and at one or two times during Taper 1

		Exponential decaytaper (n =		Step reduction intraining volume		
			First 5 days of	Last 5 days of		First 5 days of	Last 5 days of
		Pre taper	taper	taper	Pre taper	taper	taper
Run	Time (sec)	1149±109	1136 ± 86	1103 ±87 *	1121 ± 56	1106 ± 67	1108 ± 60
Cycle	Power (W)	423 ± 25	—	446 ± 32 *	412 ± 9	—	418 ± 15 *

Values are means±S.D. * Significant improvement compared with Pre Taper.† Significant group difference.

Table 2. Performance measures for both criterion 5 km run and maximal cycle ergometry immediately preceding and at two times during Taper 2

		Low volume taper			High volume taper		
		Pre taper	1st week of taper	2nd week of taper	Pre taper	1st week of taper	2nd week of taper
Run	Time (sec)	1167±80	1126 ±90 *	1093 ±90 *	1159 ± 60	1142 ±62 *	1131 ±53 *
Cycle	Power (W)	433 ± 36	440 ± 26	467 ±28 *†	394 ± 45	405 ± 42 *	409 ± 33

Values are means±S.D.* Significant improvement compared with Pre taper. † Significant group difference.

Table 3. Relative $\dot{V}O_2$max and anaerobic threshold (θ_{an}) measured two weeks prior to taper and at two times during Taper 2 (n = 8)

	Pre-taper	First week of taper	Last week of taper
$\dot{V}O_2$max (ml $\cdot$ kg $^{-1} \cdot$ min $^{-1}$)	62.9 ± 5.8	67.4 ± 5.6 *	68.6 ± 4.2*
θ_{an} (% of $\dot{V}O_2$max)	71 ± 8	73 ± 6	75 ± 6 *

Values are means±S.D.* Significantly larger than pre-taper, paired t-test.

DISCUSSION

Taper 1

The effect of a ten day exponential taper (Taper 1) on two physical performance measures resulted in a mean 23 W (5 %) increase in maximal ramp power on a cycle ergometer, and a mean 46 second (4 %) decrease in performance time on a 5 km criterion run compared with performance in the week prior to taper. These results are similar to those reported by Houmard et al. (9) in runners who improved their performance on a 5 km race by 3 % with a 7 day decay taper.

A step reduction in training volume during the same peaking period resulted in a mean 6 Watt (2 %) increase and mean 13 second (1 %) decrease during the ramp and criterion run respectively, compared with the pre-taper value. These results were not significantly improved from pre-taper. Houmard and Johns (8) recently suggested that tapered training is more effective in producing optimal performance than simply reducing training in the days and/or weeks preceding an event, however, no study has previously characterized the various types of taper possible or compared their various effectiveness quantitatively. This study compared the effectiveness of a step decrease and exponential decay protocols. The group comparison made above showed an exponential decay protocol is superior to a step decrement protocol on both the criterion run time and maximal ramp power output during the last five days of Taper 1, yet a firm conclusion cannot be reached due to the slight difference in training volume between the step and exponential taper groups.

Taper 2

The effect of a high volume ($\tau \geq 8$ d) versus a low volume exponential decay taper ($\tau \leq 4$ d) on physical performance was compared in Taper 2. Results of this 13 day taper produced a significant improvement in almost all criterion measures in both groups. The criterion run time decreased in the final week of Taper 2 compared with pre-taper values by 28 seconds (2 %) and 74 seconds (6 %) in the high and low volume taper groups respectively. Maximal ramp power was significantly higher in the final week of Taper 2 only in the low volume taper group by 34 Watts (8 %). There was also a significant interaction in maximal ramp power between groups during the last week of Taper 2 suggesting that the low volume taper group was the optimal taper. This conclusion is also supported by the large improvement in the low volume group which was on average 46 seconds faster on a 5 km criterion run compared with the higher volume group in the final week of Taper 2. These results suggest that a longer taper (13 d versus 10 d) is superior preparation for competition. This observation is supported by a study by Costill *et al.* (3) who tapered swimmers for 21 days and observed performance improvement right until the end of taper.

Taper Volume

The results of both Taper 1 and Taper 2 suggest that additional training volume in the immediate days preceding competition whether it is derived from a reduced training protocol or by an exponential taper with a slower decay time constant is detrimental to physical performance. It is apparent that training volume may be reduced by as much or more than 50 % from its previous average with optimal results during taper of up to 13 days. Frequency of training in both Taper 1 and Taper 2 decreased significantly by 1 day / week in the final half of each taper in both groups. Since no concomitant impairment in performance was observed, it may be beneficial for an athlete to take one or two days of complete rest in the week prior to competition. This may allow maximal recovery from training fatigue and optimize preparation for competition.

Human training studies have investigated separately the effect of step reduced training (6,7,12,14,17,18) and taper training (9,19) on physical performance although taper has not been characterized as exponential decay. However, previous studies are difficult to interpret precisely or compare since they define training volume simply by the duration of training. This is inappropriate since volume depends on both duration and intensity. The latter of these is itself a complex of heart rate intensity and a true metabolic intensity factor defined from heart rate (1).

Taper Intensity

The average intensity of training during both Taper 1 and Taper 2, measured quantitatively via telemetry and heart rate monitor did not change compared with the average intensity during training. Exercise intensity is a key factor in either maintaining (6,7), or improving physical performance (9,19), and much of the performance improvement in the present study may be attributed to the maintenance of intense training during both Taper 1 and Taper 2. Serial criterion performance measurement repeated 2-3 times / week during taper besides providing good characterization of the change in serial performance was also a key factor in providing the necessary intensity of training during taper and a means by which to assess performance in this study. Results from this study suggest that an optimal taper would require sufficient decay in training volume ($\leq$ 50 % reduction) while intensity of training is maintained or increased above approximately 70 % Δ HR Ratio to observe improvement in physical performance. Thus for a given volume of training if intensity was increased the duration of a session must decrease.

Cardiorespiratory Results

Relative $\dot{V}O_2$max measured on a maximal ramp test to exhaustion in 8 subjects significantly increased during the final 13 day taper (Taper 2). $\dot{V}O_2$max increased from 62.9 $\pm$ 5.8 ml·kg^{-1}·min^{-1} measured two weeks prior to taper to a significantly higher value of 67.4 $\pm$ 5.6 ml·kg^{-1}·min^{-1} during the first week of taper to 68.6 $\pm$ 4.2 ml·kg^{-1}·min^{-1} during the second week of taper. An increase in $\dot{V}O_2$max with taper has not been previously reported in the literature although $\dot{V}O_2$max has been reported to increase with training (17) and decrease with a prolonged reduction in training volume (7). Recent taper studies have reported improvement in physical performance with no increase in $\dot{V}O_2$max (9,19). These latter treadmill studies, however, have either constrained the subject to run at a fixed speed uphill or themselves control the speed. The tasks are distracting and cast doubt on the ability of the subject to run maximally under such conditions. It also appears that training intensity is a key factor in maintenance of aerobic power (7). The 9 % increase in $\dot{V}O_2$max observed in this study between the pre-taper and final week of taper is very similar to the 8 % increase in maximal power output on the cycle ergometer in the low volume taper group.

The anaerobic threshold determined by a non invasive technique described by Whipp et al. (20) improved from 71 % of $\dot{V}O_2$max to a significantly higher value of 75 % of $\dot{V}O_2$max. An improvement in the anaerobic threshold has not been previously reported with taper. These values are slightly less than 83 % of $\dot{V}O_2$max reported previously by Medelli (15) in highly trained triathletes due in part perhaps to the different method of determining the anaerobic threshold or the difference in fitness level of these subjects.

ACKNOWLEDGMENT

This study was supported by funds from the Natural Sciences and Engineering Research Council of Canada to E.W.B..

REFERENCES

1. Banister, E.W. Modeling elite athletic performance. IN: Physiological testing of elite athletes. Eds. H. Green, McDougal, J., Wenger, H., Champaign. Human Kinetics Publishers, 103-121, 1991.
2. Calvert, T.W., E.W. Banister, M.V. Savage, and T. Bach. A systems model of the effects of training on physical performance. *IEEE Trans. Syst. Man Cybernet.* 6:94-102, 1976.
3. Costill, D.L., R. Thomas, R.A. Robergs, D. Pascoe, C. Lambert, S. Barr, and W.J. Fink. Adaptations to swimming training: influence of training volume. *Med. Sci. Sports Exerc.*, 23 (3): 371-377, 1991.
4. Cullinane, E.M., S.P. Sady, L. Vadeboncoeur, M. Burke, and P.D. Thompson. Cardiac size and VO2max do not decrease after short term exercise cessation. *Med. Sci. Sports Exerc.* 18(4):420-424, 1986.
5. Fitz-Clarke, J.R., R.H. Morton, and E.W. Banister. Optimizing athletic performance by influence curves. *J. Appl. Physiol.* 71(3):1151-1158, 1991.
6. Hickson, R.C., C. Foster,M.L. Pollock, T.M. Galassi, and S. Rich. Reduced training intensities and loss of aerobic power, endurance, and cardiac growth. *J. Appl. Physiol.* 58(2): 492-499, 1985.
7. Hickson, R.C., C. Kanakis,J.R. Davis,M. Moore,and S. Rich Reduced training duration effects on aerobic power, endurance, and cardiac growth. *J. Appl. Physiol.*, 53(1):225-229, 1982.
8. Houmard, J.A., R.A. Johns. Effects of taper on swim performance. *Sports Med.* 17(4):224-232, 1994.
9. Houmard, J.A., B.K. Scott, C.L. Justice, and T.C. Chenier. The effects of taper on performance in distance runners. *Med. Sci. Sports Exerc.*, 26(5): 624-631, 1994.
10. Houmard, J.A.,T. Hortobagyi, R.A.Johns, N.J. Bruno, C.C. Nute, M.H. Shinebarger, and J.W. Welborn. Effect of short-term training cessation on performance measures in distance runners. *Int. J. Sports Med.* 13(8):572-576, 1992.
11. Houmard, J.A. Impact of reduced training on performance in endurance athletes. *Sports Med.* 12(6):380-393, 1991.
12. Houmard, J.A., D.L. Costill, J.B. Mitchell, S.H. Park, R.C. Hickner, and J.N. Roemmich. Reduced training maintains performance in distance runners. *Int. J. Sports Med.* 11:46-52, 1990.
13. Johns, R.A.,J.A. Houmard, R.W. Kobe, T.Hortobagyi,N.J. Bruno, J.M. Wells, and M.H. Shinebarger. Effects of taper on swim power, stroke distance, and performance. *Med. Sci. Sports Exerc.* 24(10):1141-1146, 1992.

14. McConnell, G.K., D.L.Costill, J.J.Widrick, M.S.Hickey, H. Tanaka, and P.B. Gastin. Reduced training volume and intensity maintain aerobic capacity but not performance in distance runners. *Int.J.Sports Med.* 14(1):33-37, 1993.

15. Medelli, J., Y.Maigourd, B. Bouferrache, V.Bach, M. Freville, and J.P.Libert. Maximal oxygen uptake and aerobic-anaerobic transition on treadmill and bicycle in triathletes. *Jap.J.Physiol.*, 43:347-360, 1993.

16. Morton, R.H., J. Fitz-Clarke, and E.W. Banister. Modeling human performance in running. *J. Appl. Physiol.* 69: 1171-1177, 1990.

17. Neufer, P.D. The effect of detraining and reduced training on physiological adaptations to aerobic exercise. *Sports. Med.* 8:302-321,1989.

18. Neufer, P.D., D.L. Costill, R.A. Fielding, M.G. Flynn, and J.P. Kirwan. Effect of reduced training on muscular strength and endurance in competitive swimmers.*Med. Sci. Sports Exerc.* 19(5): 486-490, 1987.

19. Shepley, B., J.D. MacDougall, N. Cipriano, J.R. Sutton, M.A. Tarnopolsky, and G. Coates. Physiological effects of tapering in highly trained athletes. *J. Appl. Physiol.* 72(2):706-711, 1992.

20. Whipp, B.J., J.A. Davis, and K.Wasserman. Ventilatory control of the isocapnic buffering region in rapidly-incremental exercise. *Resp. Physiol.*, 76:357-368, 1989.

36

IS THE SLOW COMPONENT OF EXERCISE $\dot{V}_{O_2}$ A RESPIRATORY ADAPTATION TO ANAEROBIOSIS?

Karlman Wasserman, William W. Stringer, and Richard Casaburi

Division of Respiratory and Critical Care Physiology and Medicine
Harbor-UCLA Medical Center
1000 W. Carson St., Torrance, California 90509

INTRODUCTION

The increased rate of O_2 consumption ($\dot{V}_{O_2}$) during exercise reflects the rate of aerobic regeneration of adenosine triphosphate (ATP). In 1972, Whipp and Wasserman (9) reported that a steady-state for aerobic regeneration occurred by 3 minutes only for work rates below the lactic acidosis threshold (*LAT*). For work rates above the *LAT*, $\dot{V}_{O_2}$ continued to increase past 3 minutes, the rate quantified as the $\dot{V}_{O_2}$ increase between 3 and 6 minutes ($\Delta\dot{V}_{O_2}(6\text{-}3)$). $\Delta\dot{V}_{O_2}(6\text{-}3)$ has a high linear correlation with blood lactate increase with the regression passing through the origin (8). Others (summarized in ref 8) have confirmed an association between blood lactate increase and the slow component, but a causal mechanism remains unclear. The slow component increase in $\dot{V}_{O_2}$, as illustrated in figure 1, could reflect either 1) a developing inefficiency in the aerobic regeneration of ATP, 2) an adaptation to an anaerobic state in which the O_2 supply to the muscles was improving, thereby facilitating the aerobic regeneration of ATP, or 3) a combination of both processes.

THEORY

Bioenergetic Efficiency

The mechanism by which ATP regeneration might require more than the normal amount of O_2 can be envisioned by referring to the biochemical scheme for ATP production (Fig 2). During aerobic glycolysis, the cytosolic redox state ($NADH+H^+/NAD^+$) and therefore lactate/pyruvate are kept constant (see insert in Fig 2). The cytosolic $NADH+H^+$ is reoxidized by pathway A, with mitochondrial flavine adenine dinucleotide (FAD) being the predominant coenzyme in this reoxidation. This coenzyme enters the electron transport chain at a lower energy level than mitochondrial $NADH+H^+$ and only 2 ATP are generated for each oxygen atom consumed instead of the 3 ATP when mitochondrial $NADH+H^+$ is

Modeling and Control of Ventilation, Edited by S. J. G. Semple, L. Adams, and B. J. Whipp
Plenum Press, New York, 1995

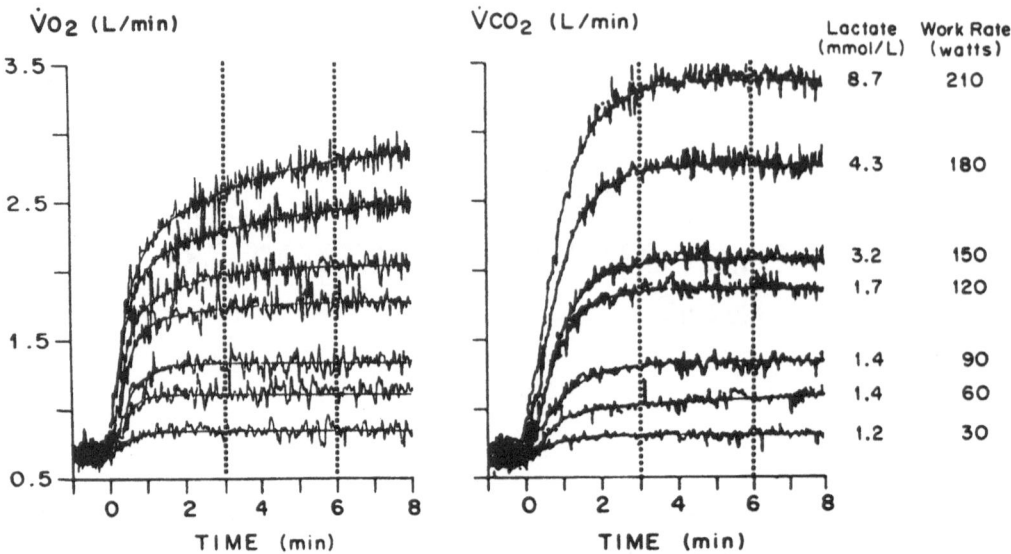

Figure 1. Time course of $\dot{V}_{O_2}$ and $\dot{V}_{CO_2}$ for 7 leg cycling work rates performed on different days in a single subject. Each curve is second-by-second average of 4-8 replicate studies. 1 min post-exercise antecubital venous lactate concentration and work rates are shown to right of respective curves. (Modified from reference 2).

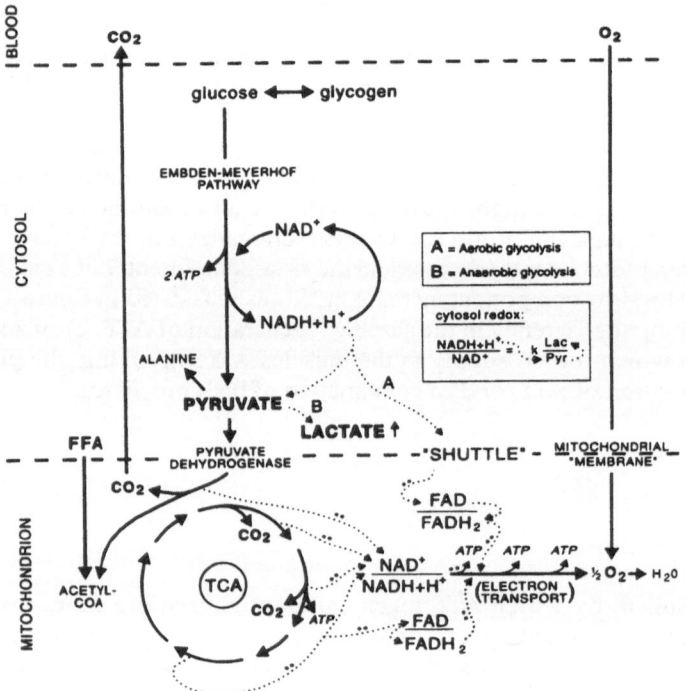

Figure 2. Pathways of aerobic and anaerobic glycolysis. During aerobic glycolysis, cytosolic NADH+H$^+$ is reoxidized by the mitochondrial membrane shuttle (pathway A), mitochondrial coenzymes, the electron transport chain and O$_2$. During anaerobic glycolysis, pyruvate reoxidizes cytosolic NADH+H$^+$ to NAD$^+$ with increase in lactate production and lactate/pyruvate ratio (pathway B). See text for effect of changing pathways for electrons to the electron transport chain on bioenergetic efficiency. (From reference 8).

the source of the electrons for the electron transport chain. Under normal cytosolic redox states only 1/6 of the electron transport is through FAD, the remaining being through the more efficient NAD^+. If, however, cytosolic $NADH+H^+/NAD^+$ increased, (evidenced by the increased conversion of glycogen to lactate and the increase in lactate/pyruvate ratio, pathway B of Fig 2), the lowered cytosolic redox state would raise the potential for electron transport through the mitochondrial membrane shuttle to FAD, making $FADH_2$ more competitive with $NADH+H^+$ for sites on the electron transport chain. This would change the balance between the two pathways into the electron transport chain. With the lower ATP yield of the former ($P:O_2=4$) than the latter ($P:O_2=6$), this change in pathways for electron entry into the electron transport chain provides a potential mechanism to explain a reduction in bioenergetic efficiency found during exercise as judged from the higher than predicted $\dot{V}o_2$ measured past 3 minutes in time (10) when the exercise intensity is above the *LAT*.

Critical Capillary PO₂

While a mechanism requiring increased O_2 can be hypothesized when the lactate/pyruvate (cytosolic $NADH+H^+/NAD^+$) increases, how is this extra O_2 made available to the cell? An O_2 driving pressure must exist in the capillary to overcome the diffusive resistance between the capillary red cell and the myocyte mitochondria. While the Po_2 is very high on the arterial side of the capillary bed (Fig 3), it is much lower toward the venous end of the capillary bed. The change in capillary Po_2 as it transits the capillary bed must depend on the arterial Po_2, muscle blood flow/O_2 consumption ($Qm/\dot{V}o_2m$) ratio, haemoglobin concentration and the Bohr effect. All of the O_2 cannot be extracted from the blood because there must continue to be a Po_2 gradient for diffusion. Wittenberg and Wittenberg (11) have estimated the critical capillary Po_2 to be between 15 to 20 Torr. Mathematical modelling of capillary Po_2 shows that the driving pressure toward the venous end of the capillary bed will reach a critical level when the $Qm/\dot{V}o_2m$ ratio falls below 6 at a haemoglobin concentration of 15 gm/dl. At a $Qm/\dot{V}o_2m$ <6, the exercise will require anaerobic ATP production and induce a lactic acidosis.

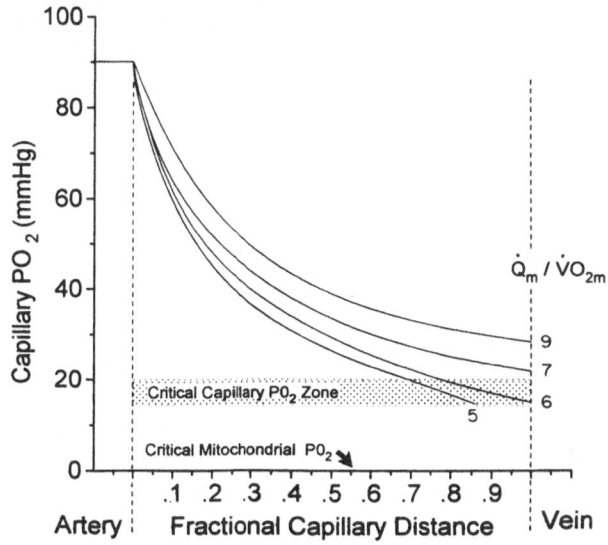

Figure 3. Model of muscle capillary bed O_2 partial pressure (Po_2) as blood travels from artery to vein. The model assumes haemoglobin concentration of 15 g/dl, arterial Po_2 of 90 mmHg and a linear O_2 consumption along the capillary. The rate of fall of capillary Po_2 depends on the muscle blood flow/muscle $\dot{V}o_2$ ratio ($Qm/\dot{V}o_2$). The curves include a Bohr effect due to CO_2 production. (From reference 8).

HYPOTHESIS: ROLE OF LACTIC ACID PRODUCTION IN MUSCLE O₂ SUPPLY

As classically viewed, blood lactate increases during exercise when ATP is anaerobically regenerated. In this presentation, we describe an additional, and perhaps a more important, physiological role for the lactic acidosis of exercise. This role has come to light because of the observation that 1) blood lactate does not increase until the capillary P_{O_2} has reached its critical value (4,7), 2) the mechanism for the O_2 extraction from exercising muscle capillary blood for work above the *LAT* is the Bohr effect rather than a falling capillary P_{O_2} (7) and 3) patients with muscle enzyme disorders associated with an inability to make lactate, e.g., myophosphorylase deficiency, fail to extract O_2 from the capillary blood normally at maximal exercise (5). The purpose of this analysis is to present a hypothesis to explain the slow component of $\dot{V}_{O_2}$ increase and why it is linked to the lactate increase induced by heavy exercise. The hypothesis states that the lactic acidosis is essential in order to perform heavy intensity exercise because it provides a mechanism by which O_2 can be maximally extracted from the capillary blood yet maintain an adequate capillary diffusion pressure. This mechanism can account for the observed lactic acidosis-linked slow component in the $\dot{V}_{O_2}$ kinetics without a simultaneous slowing in $\dot{V}_{CO_2}$ kinetics (Fig.1).

CRITICAL CAPILLARY P_{O_2}, MUSCLE LACTATE RELEASE, ACID-BASE CHANGE AND O_2 EXTRACTION

Critical Capillary P_{O_2}

To determine the end capillary P_{O_2}, oxyhemoglobin (O_2Hb sat'n), lactate concentration and acid-base changes during leg cycling exercise as related to exercise $\dot{V}_{O_2}$, we sampled femoral vein and arterial blood at rest and during progressively increasing work rate and constant work rate exercise in 10 normal subjects (7). Blood was analyzed for lactate by a lactate electrode (Yellow Springs), pH, P_{CO_2} and P_{O_2} by a blood gas analyzer (Instrumentation Laboratories) and O_2Hb sat'n with an IL Co-Oximeter. Gas exchange was measured, breath-by-breath (7). A "floor" in femoral vein blood P_{O_2} was found to occur in the mid work rate range at a value which was virtually the same as the previously postulated critical capillary P_{O_2} estimated from anatomical considerations (11).

In both the study on normal subjects (7) and heart disease subjects (4), the lactate concentration did not increase in femoral vein blood during leg cycling exercise until the critical capillary P_{O_2} was reached (Fig. 4). The critical capillary P_{O_2} was the same for a given subject whether measured during constant work rate or progressively increasing work rate exercise (4,7).

Mechanism of O_2 Extraction

The end capillary P_{O_2} was the same for constant work rate exercise performed just below and considerably above the *LAT* (Fig. 5). The entire decrease in O_2Hb sat'n above the *LAT* was accounted for by the Bohr effect. This is illustrated in a more conventional way by plotting femoral vein O_2Hb sat'n vs P_{O_2} on the lower part of the O_2Hb dissociation curve (Fig. 6). It can be easily appreciated that the decrease in end capillary blood O_2Hb sat'n is completely accounted for by the pH decrease. The unloading of O_2 from Hb when performing heavy work depends on the development of a lactic acidosis. It is probably also of physiological importance that the critical capillary P_{O_2} is at a value above which the Bohr

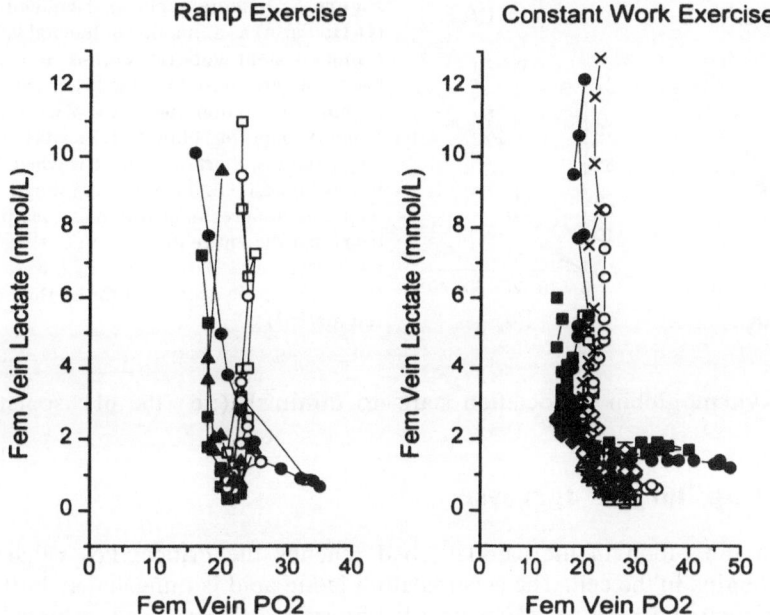

Figure 4. Femoral vein lactate as function of femoral vein Po_2 for incremental (ramp) exercise in 5 subjects (left panel) and 10 constant work rate exercise tests (five below and five above the LAT) in 5 subjects (right panel). The highest Po_2 values are at the start of exercise. Different symbols represent different subjects. (From reference 7).

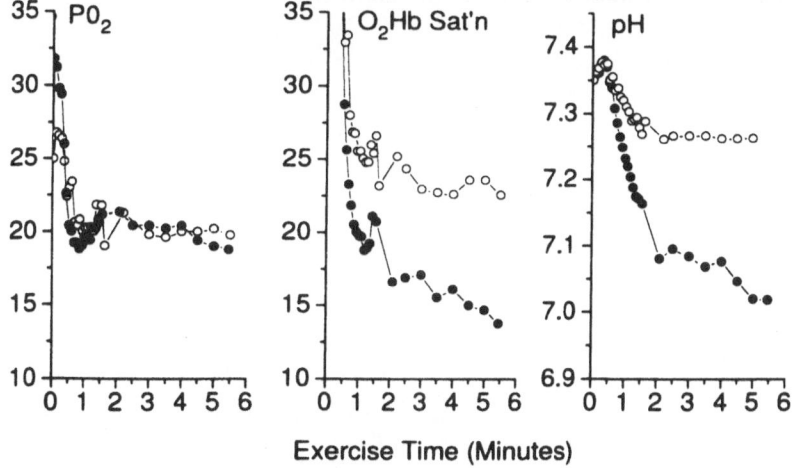

Figure 5. Femoral venous Po_2, oxyhemoglobin saturation (O_2Hb Sat'n), and pH as related to time of exercise for 2 constant work rate tests, 1 below (O) and 1 above (●) *LAT*. Each curve is the average of 5 subjects. Below- and above-*LAT* work rates averaged 113 and 265 W, respectively. (From reference 7).

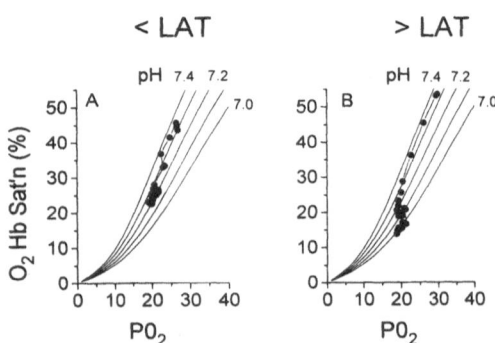

Figure 6. Femoral vein oxyhemoglobin saturation (O_2Hb Sat'n) as a function of femoral vein P_{O_2} for the 6-min constant work rate exercise tests shown in figure 5. Superimposed are the lower part of oxyhemoglobin dissociation curves for pH values of 7.0–7.4. Panel A: data for below *LAT*. Panel B: data for above *LAT* exercise. Start of exercise is where O_2Hb saturation is highest. Oxyhemoglobin saturations fall on pH isopleths in agreement with measured pH. This indicates that the entire decrease in O_2Hb saturation that takes place after P_{O_2} reaches its lowest value (above the *LAT*) can be accounted for by Bohr effect. (From reference 7).

effect on oxyhemoglobin dissociation starts to diminish (i.e., the pH isopleths start to converge).

Source of Capillary H⁺ Increase

When the blood in the capillary bed reaches the critical P_{O_2} value, anaerobic metabolism begins in the cell. The accumulating lactic acid is immediately buffered **in the cell** (8) resulting in the increase in $\dot{V}_{CO_2}$ noted in figure 1. At the same time, the cell consumes plasma HCO_3^- as it exchanges with cellular lactate (8). Thus as exercise proceeds above the *LAT*, the end-capillary P_{CO_2} markedly increases and HCO_3^- decreases (Fig 7), causing an accelerating decrease in pH.

MECHANISM OF THE 'SLOW COMPONENT'

The possibility that lactate, in itself, (6), catecholamines (1,3) or increased body temperature of exercise (1) stimulate aerobic metabolism to account for the slow component of $\dot{V}_{O_2}$ increase has not been supported by investigations which have examined it. Theoretically, mechanisms which increase aerobic metabolism independent of acid-base changes

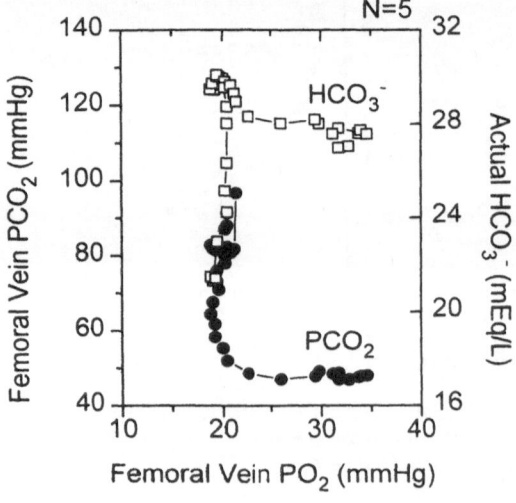

Figure 7. Femoral vein (approximating end-capillary) P_{CO_2} and HCO_3^- as a function of femoral vein P_{O_2} during constant work rate heavy exercise. Values are the average for the 5 normal subjects whose data are shown in figure 6B. Values at the start of exercise are at the right and move leftward as $\dot{V}_{O_2}$ increases (data taken from subjects reported in ref 7).

caused by lactate accumulation should increase $\dot{V}co_2$ in a parallel fashion to the increase in $\dot{V}o_2$. This is not the case, as shown in figure 1. The increase in $\dot{V}co_2$ is greater than $\dot{V}o_2$ and $\dot{V}co_2$ kinetics become dissociated from $\dot{V}o_2$ kinetics (2).

The Growing Bohr Effect and the Slow Component

The slow component in $\dot{V}o_2$ which becomes evident only when performing heavy intensity exercise, may be accounted for by the facilitation of O_2Hb dissociation by the lactic acidosis and explain the persistent correlation between increasing $\dot{V}o_2$ and increasing blood lactate (8). Aerobic and anaerobic metabolism take place in the same capillary bed during heavy exercise as illustrated in figure 8. Toward the arterial end of the capillary bed, O_2 dissociates from haemoglobin primarily due to a falling Po_2 with a small Bohr effect due to

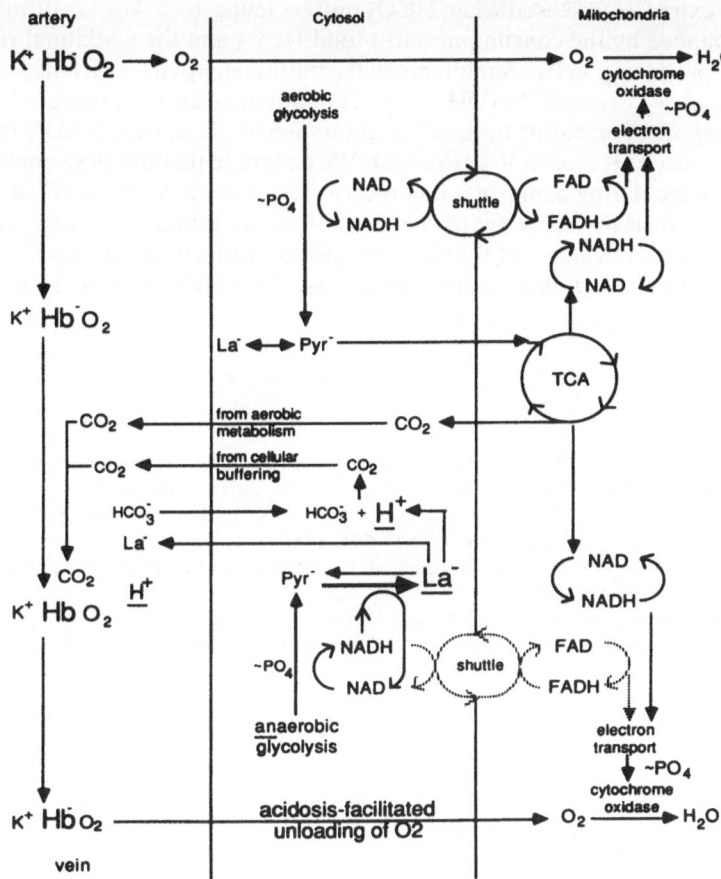

Figure 8. Scheme of changing capillary oxyhemoglobin (O_2Hb) saturation during blood transit from artery to vein during heavy intensity exercise. At arterial end of capillary, O_2Hb dissociates primarily due to decrease in Po_2 with energy generation primarily by aerobic glycolysis. As blood approaches the venous end of capillary where Po_2 becomes critically low, the mitochondrial membrane redox shuttle fails to reoxidize cytosolic $NADH+H^+$ to NAD^+ at a rate sufficient to keep the cytosolic $NADH+H^+/NAD^+$ from increasing. Accordingly, pyruvate is converted to lactate in proportion to the change in cell redox state (see Fig. 2 and associated text). The effect is an increase in cell lactate with a stoichiometric increase in H^+. The sum of aerobically produced CO_2 and anaerobically produced CO_2 (from buffering), along with decreasing blood HCO_3^- further acidifies the capillary blood resulting in acidosis-facilitated dissociation of O_2Hb. (From reference 8).

an increase in P_{CO_2} resulting from aerobic metabolism. When P_{O_2} reaches its critical value toward the venous end of the capillary bed, anaerobic glycolysis begins causing accumulation of lactic acid. The acid is buffered by HCO_3^- in the cell with release of additional CO_2 and decrease in plasma HCO_3^- as the cell consumes HCO_3^- in exchange for the lactate efflux from the cell (8). The resulting local blood acidosis facilitates further oxyhemoglobin dissociation without reducing the P_{O_2} driving pressure.

Lactic Acidosis as a Facilitator of the Slow Component

This analysis and the accompanying data support the hypothesis that when the critical capillary P_{O_2} is reached, lactic acid starts to accumulate at a rate depending on the level of exercise above the *LAT*. This affects both $\dot{V}_{O_2}$ and $\dot{V}_{CO_2}$. $\dot{V}_{O_2}$ slowly increases at a rate proportional to the exercise lactic acidosis. $\dot{V}_{CO_2}$ increases in a different pattern from $\dot{V}_{O_2}$ because of the extra CO_2 released when HCO_3^- buffers lactic acid. The resulting extracellular acidosis is enhanced by the consumption of blood HCO_3^- and the additional release of CO_2 by lactic acid-producing cells. Simultaneously, the bioenergetic efficiency may decrease because the increased cytosolic $NADH+H^+/NAD^+$ ratio increases the potential for mitochondrial FAD transport of electrons to the electron transport chain over NAD^+, with a reduced yield of ATP for each O_2 molecule consumed. We conclude that the slow component in $\dot{V}_{O_2}$ kinetics takes place during exercise when the work rate is above the *LAT*, a level of work which may be intrinsically less efficient because of the simultaneous change in the cytosolic redox state. We also conclude that the slow component reflects an *adaptative* mechanism to meet the O_2 requirement of heavy intensity exercise for which the increasing acidosis plays an essential role.

REFERENCES

1. Casaburi, R., Storer, T.W., Ben-Dov, I. and Wasserman, K. Effect of endurance training on possible determinants of $\dot{V}_{O_2}$ during heavy exercise. J.Appl.Physiol. 62:199-207, 1987.
2. Casaburi, R., Barstow, T.J., Robinson, T. and Wasserman, K. Influence of work rate on ventilatory and gas exchange kinetics. J. Appl. Physiol. 67:547-555, 1989.
3. Gaesser, G.A. Influence of endurance training and catecholamines on exercise $\dot{V}_{O_2}$ response. Med. Sci. Sports Exerc. 26:1341-1346, 1994.
4. Koike, A., Wasserman, K., Taniguchi, K. and Hiroe, M. Critical capillary oxygen partial pressure and lactate threshold in patients with cardiovascular disease. J. Am. Coll. Cardiol. 23:1644-1650, 1994.
5. Lewis, S.F. and Haller, R.G. The pathophysiology of McArdle's disease: clues to regulation in exercise and fatigue. J. Appl. Physiol. 61:391-401, 1986.
6. Poole, D.C., Gladden, L.B., Kurdak, S. and Hogan, M.C. L(+)-Lactate infusion into working dog gastrocnemius: no evidence lactate *per se* mediates $\dot{V}_{O_2}$ slow component. J. Appl. Physiol. 76:787-792, 1994.
7. Stringer, W.W., Wasserman, K., Casaburi, R., Porszasz, J., Maehara, K. and French, W. Lactic acidosis as a facilitator of oxyhemoglobin dissociation during exercise. J. Appl. Physiol. 76:1562-1567, 1994.
8. Wasserman, K. Coupling of external to cellular respiration during exercise: the wisdom of the body revisited. Am. J. Physiol. 266:E519-E539, 1994.
9. Whipp, B.J. and Wasserman, K. Oxygen uptake kinetics for various intensities of constant-load work. J. Appl. Physiol. 33:351-356, 1972.
10. Whipp, B.J. and Wasserman, K. Effect of anaerobiosis on the kinetics of O_2 uptake during exercise. Federation Proc. 45:2942-2947, 1986
11. Wittenberg, B.A. and Wittenberg, J.B. Transport of oxygen in muscle. Ann. Rev. Physiol. 51:857-878, 1989.

37

EFFECTS OF AGE ON $\dot{V}O_2$ KINETICS DURING CALF AND CYCLING EXERCISE

P. D. Chilibeck, D. H. Paterson, and D. A. Cunningham

The Centre for Activity and Ageing
Faculties of Kinesiology and Medicine
The University of Western Ontario
London, Ontario, Canada

INTRODUCTION

The kinetics of oxygen uptake ($\dot{V}O_2$) adjustment to moderate intensity cycling exercise are slowed as a function of age (1,5). This implies that older individuals must rely on anaerobic systems, to a greater extent, to meet energy requirements during transitions to exercise, increasing the possibility of early fatigue. The purpose of this study was to compare $\dot{V}O_2$ kinetics between old and young subjects during an exercise of a smaller muscle mass (ankle plantar flexion). Ankle plantar flexion was performed using an ergometer, developed to measure kinetics with ^{31}P-nuclear magnetic resonance spectroscopy (NMRS), where phosphocreatine kinetics correspond to muscle respiratory kinetics (3,10,15). Since plantar flexor muscles of older individuals have decreased capillarization and oxidative enzyme activity (4,7) and a greater amount of "non-muscle" tissue than young (14), we expected to find a slow oxygen delivery, or oxygen utilization at the start of exercise. This would result in an impairment of $\dot{V}O_2$ kinetics with age.

METHODS

Six older subjects (5 females, 1 male; aged 66.0±7.7y) were compared to 11 younger subjects (4 females, 7 males; aged 26.9±2.5y). Subjects initially performed a ramp exercise test on a cycle ergometer for determination of ventilatory threshold (V_ET), and a ramp exercise test on the plantar flexion ergometer for determination of ^{31}P-NMRS measured intracellular threshold (11). On subsequent laboratory sessions, $\dot{V}O_2$ kinetics were determined by having subjects perform 10 to 18 square wave transitions from unloaded to loaded ankle plantar flexion at a workrate corresponding to 90% of intracellular threshold (5.1±2.0 W for older subjects and 10.9±3.5 W for younger subjects), and three transitions from unloaded to loaded pedalling on a cycle ergometer at 90% of $\dot{V}_ET$ (37.2±16.6 W for older subjects and 114.5±34.9 W for younger subjects). Pulmonary $\dot{V}O_2$ was measured breath-by-

Modeling and Control of Ventilation, Edited by S. J. G. Semple, L. Adams, and B. J. Whipp
Plenum Press, New York, 1995

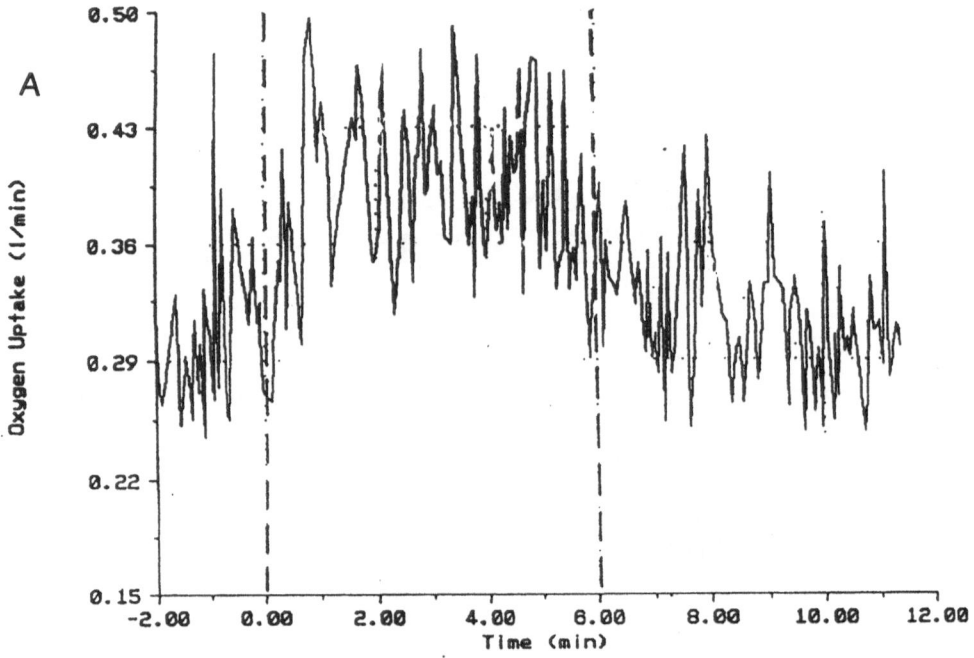

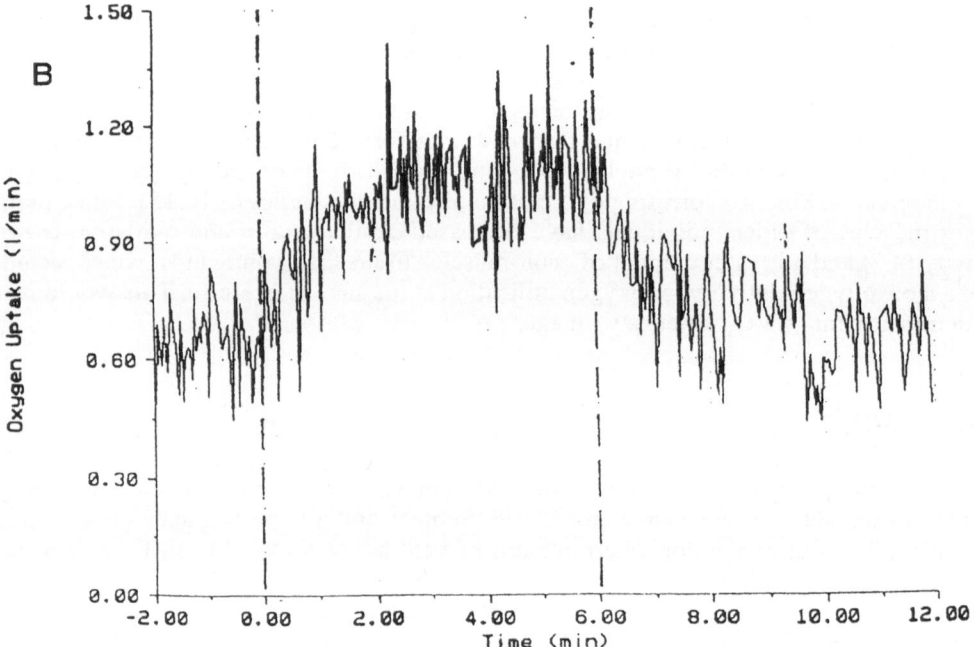

Figure 1. Breath-by-breath alveolar $\dot{V}O_2$ during on- and off- transitions to a) plantar flexion and b) cycling exercise (single transition for older subject).

breath using a turbine and mass spectrometer (Airspec 2000) [1]. $\dot{V}O_2$ transients were interpolated to 1 second, time aligned and averaged. The $\dot{V}O_2$ signal obtained during plantar flexion was of smaller magnitude than during cycle ergometry (Figure 1).

Thus, a greater number of plantar flexion transitions were averaged to obtain an adequate signal to noise ratio (9). $\dot{V}O_2$ on- and off- transients were fit with a mono-exponential equation of the form:

$$\dot{V}O_2(t) = a\{1 - e^{-[(t-\delta)/\tau]}\},$$

where $\dot{V}O_2(t)$ is $\dot{V}O_2$ at any time (t); a is the amplitude of the response (difference between steady state and baseline $\dot{V}O_2$); τ is the time constant and δ the time delay of the response. Curves were fit starting at phase 2 (20 s) of the transient (16), for determination of $\tau\dot{V}O_2$.

RESULTS

Maximal $\dot{V}O_2$ values, from the cycle ergometer ramp tests were 43.7±7.6 and 20.5±4.1 ml·kg⁻¹·min⁻¹ for young and old subjects, respectively. This indicates that subjects in both groups were of "average" fitness for their age (13). On- and off- transients of averaged

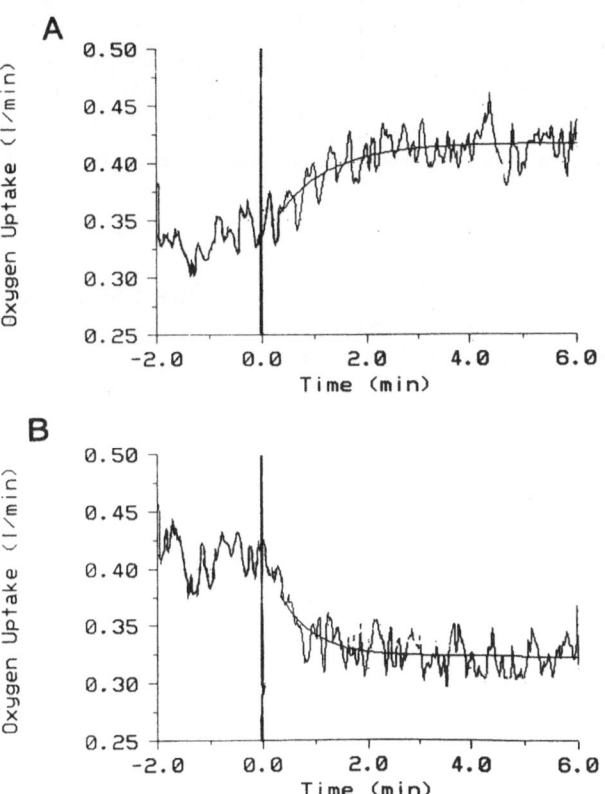

Figure 2. Ankle plantar flexion. a. On-transient of interpolated and averaged repeats (n = 12) with mono-exponential fit . b. Off-transient

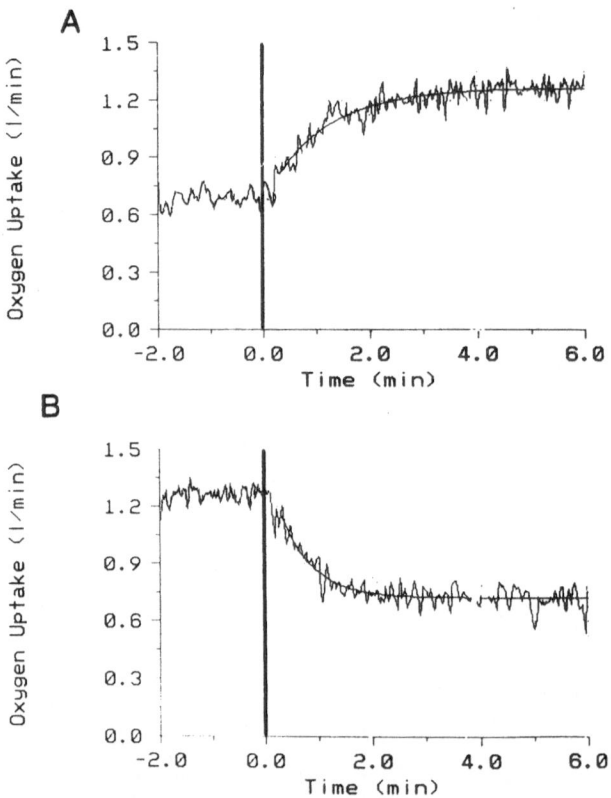

Figure 3. Cycle exercise. a. On-transient of interpolated and averaged repeats (n = 3) with mono-exponential fit. b. Off-transient.

repeats and the mono-exponential fits are shown in Figure 2 for ankle plantar flexion and Figure 3 for cycle ergometry, for an older subject.

Average $\tau\dot{V}O_2$ results for old and young groups are shown in Table 1. On- and off- transients were significantly slower during cycling in old compared to young, but significantly faster in old during the plantar flexion exercise, and not different from the young.

Table 1. $\tau\dot{V}O_2$ for ankle plantar flexion and cycle exercise in young and old subjects

| | $\tau\dot{V}O_2$ (s) | | | |
| | Cycle transients | | Plantar flexion transients | |
	On	Off	On	Off
Young	36.9 ± 11.3	34.9 ± 5.9	46.0 ± 9.6	32.4 ± 13.6
Old	$53.9 \pm 13.5^*$	$49.7 \pm 14.1^*$	$39.3 \pm 15.7^\P$	$35.2 \pm 12.1^\P$

All values are means±SD.
*Significantly different from young (p<0.05).
¶Significantly different between modes (p<0.05).

DISCUSSION

$\dot{V}O_2$ kinetics during cycle ergometry were slower in the old as previously demonstrated (1,5). Unexpectedly, $\dot{V}O_2$ kinetics of the old group were not slowed during the plantar flexion exercise. We hypothesized that plantar flexion $\dot{V}O_2$ kinetics would be slower with age, for the following reasons: Levels of oxidative enzymes and capillarization of gastrocnemius muscle are decreased with age (4,7) and older individuals have a greater amount of "non-muscle" tissue (fat and connective tissue) within the plantar flexor compartment (14). One would therefore expect oxygen transport to muscle or oxygen utilization within muscle to be slower with age, resulting in slowed $\dot{V}O_2$ kinetics. McCully et al. (12) and Keller et al. (8) found time constants of approximately 55 seconds for phosphocreatine (PCr) resynthesis during recovery from plantar flexor exercise in older individuals. PCr kinetics are thought to correspond to muscle respiratory kinetics (10,15). The PCr kinetics reported by McCully et al. (12) and Keller et al. (8) appear similar to the cycling $\dot{V}O_2$ kinetics in older subjects reported by Babcock et al. (1), but differ from the plantar flexion $\dot{V}O_2$-off kinetics of the old subjects in the present study. Our unexpected results may be due to one of the following factors: 1) The circulatory demands during exercise of the small muscle mass of the plantar flexors may not challenge the oxygen transport system (6). Thus, if cardiac output kinetics are slowed with ageing, one would expect that $\dot{V}O_2$ kinetics would be slowed during exercise of a large muscle mass (cycling), but not during exercise of a small muscle mass (plantar flexion). 2) Slowed $\dot{V}O_2$ kinetics with age may only occur in an exercise mode (cycling) in which muscle groups are not accustomed to the activity. Babcock et al. (2) have demonstrated that cycling $\dot{V}O_2$ kinetics in old subjects were faster following specific training and approached values reported in young subjects. For a muscle group used in walking and accustomed to daily activity (ankle plantar flexors), $\dot{V}O_2$ kinetics are not slowed with age, however.

CONCLUSIONS

We found that $\dot{V}O_2$ kinetics are slowed with age during cycling, but not ankle plantar flexion exercise. The differences with ageing between the two types of exercise may be related to the muscle mass involved, and the circulatory demands, or slowed $\dot{V}O_2$ kinetics with age may only occur in a mode (cycling) in which the muscle groups are not accustomed to the activity. For a muscle group used in walking (calf plantar flexion), $\dot{V}O_2$ kinetics are not slowed.

ACKNOWLEDGMENT

Supported by the Natural Science and Engineering Research Council, Canada.

REFERENCES

1. Babcock, M.A., D.H. Paterson, D.A. Cunningham, and J.R. Dickinson. Exercise on-transient gas exchange kinetics are slowed as a function of age. *Med. Sci. Sports Exerc.* 26: 440-446, 1994.
2. Babcock, M.A., D.H. Paterson, and D.A. Cunningham. Effects of aerobic endurance training on gas exchange kinetics of older men. *Med. Sci. Sports Exerc.* 26: 447-452, 1994.
3. Barstow, T.J., S. Buchthal, S. Zanconato, and D.M. Cooper. Muscle energetics and pulmonary oxygen uptake kinetics during moderate exercise. *J. Appl. Physiol.* 77: 1742-1749, 1994.

4. Coggan, A.R., R.J. Spina, M.A. Rogers, D.S. King, M. Brown, P.M. Nemeth, and J.O. Holloszy. Histochemical and enzymatic comparison of the gastrocnemius muscle of young and elderly men and women. *J. Gerontol.* 47: B71-B76, 1992.

5. Cunningham, D.A., J.E. Himann, D.H. Paterson, and J.R. Dickinson. Gas exchange dynamics with sinusoidal work in young and elderly women. *Respiratory Physiology* 91: 43-56, 1993.

6. Hughson, R.L. Exploring cardiorespiratory control mechanisms through gas exchange dynamics. *Med. Sci. Sports Exerc.* 22: 72-79, 1990.

7. Keh-Evans, L., C.L. Rice, E.G. Noble, D.H. Paterson, D.A. Cunningham, and A.W. Taylor. Comparison of histochemical, biochemical and contractile properties of triceps surae of trained aged subjects. *Canadian Journal on Aging* 11: 412-425, 1992.

8. Keller, U., R. Oberhansli, P. Huber, L.K. Widmer, W.P. Aue, R.I. Hassink, S. Muller, and J. Seelig. Phosphocreatine content and intracellular pH of calf muscle measured by phosphorus NMR spectroscopy in occlusive arterial disease of the legs. *Eur. J. Clin. Invest.* 15: 382-388, 1985.

9. Lamarra, N., B.J. Whipp, S.A. Ward, and K. Wasserman. Effect of interbreath fluctuations on characterizing exercise gas exchange kinetics. *J. Appl. Physiol.* 62: 2003-2012, 1987.

10. Marsh, G.D., D.H. Paterson, J.J. Potwarka, and R.T. Thompson. Transient changes in high energy phosphates during moderate exercise. *J. Appl. Physiol.* 75: 648-656, 1993.

11. Marsh, G.D., D.H. Paterson, R.T. Thompson, and A.A. Driedger. Coincident thresholds in intracellular phosphorylation potential and pH during progressive exercise. *J. Appl. Physiol.* 71: 1076-1081, 1991.

12. McCully, K.K., R.S. Fielding, W.J. Evans, J.S. Leigh, and J.D. Posner. Relationships between in vivo and in vitro measurements of metabolism in young and old human calf muscles. *J. Appl. Physiol.* 75: 813-819, 1993.

13. Paterson, D.H. Effects of ageing on the cardiorespiratory system. *Can. J. Spt. Sci.* 17: 171-177, 1992.

14. Rice, C.L., D.A. Cunningham, D.H. Paterson, and M.S. Lefcoe. Arm and leg composition determined by computed tomography in young and elderly men. *Clin. Physiol.* 9: 207-220, 1989.

15. Whipp, B.J., and M. Mahler. Dynamics of gas exchange during exercise, In: *Pulmonary Gas Exchange*, Vol II, J.B. West (Ed.). New York: Academic Press, pp 33-96, 1980.

16. Whipp, B.J., S.A. Ward, N. Lamarra, J.A. Davis, and K. Wasserman. Parameters of ventilatory and gas exchange dynamics during exercise. *J. Appl. Physiol.* 52: 1506-1513, 1982.

$\dot{V}O_2$ ON-TRANSIENT KINETICS WITH A CENTRALLY ACTING CALCIUM CHANNEL BLOCKER

R. J. Petrella, D. A. Cunningham, and D. H. Paterson

Faculties of Medicine and Kinesiology and
The Centre for Activity and Ageing
The University of Western Ontario
London, Ontario, Canada

INTRODUCTION

We (1) have recently shown that maximal aerobic capacity ($\dot{V}O_{2max}$) is increased significantly in normal elderly subjects coincident with improved left ventricular diastolic function following ingestion of the centrally acting Ca^{2+} channel blocker, verapamil. In addition, we have shown that exercise training has resulted in increased $\dot{V}O_{2max}$ and faster gas exchange kinetics among the elderly (2). Hughson and Smyth (3) reported a slower increase of $\dot{V}O_2$ to steady state with submaximal exercise and Tesch and Kaiser (4) reported a lower $\dot{V}O_{2max}$ under the influence of β-adrenergic blockade in normal young subjects. The slowed responses in elderly subjects (2) and younger subjects (3) during β-adrenergic blockade may reflect impaired central or peripheral components of the $\dot{V}O_2$ response (5,6). Whipp et al. (7) have described the kinetics of the gas transport system by a first order model and termed the initial 20s of the $\dot{V}O_2$ response "cardiodynamic". This period reflects the central or cardiac component of the $\dot{V}O_2$ on-transient (Phase 1). The effects of a centrally acting Ca^{2+} channel blocker which increased $\dot{V}O_{2max}$ and improved left ventricular function in the elderly (1), may also be reflected in improved kinetics in the non-steady state response to an exercise perturbation. This may result from improvement in the central component of the $\dot{V}O_2$ on-transient response. In order to examine this, we have analyzed the gas exchange responses in elderly men to a step increase in power output to a sub-anaerobic threshold level (40 W) before and 4 hours after ingestion of 240mg of verapamil SR.

METHODS

Six older men (aged 67 ± 2 y) pedalled at 0 W on a cycle ergometer (Lode) for six min and then the work rate was initiated instantaneously (square wave) to 40 W under computer control and continued for six min at this rate. The load was then abruptly turned

Modeling and Control of Ventilation, Edited by S. J. G. Semple, L. Adams, and B. J. Whipp
Plenum Press, New York, 1995

off (back to 0 W) and the subject pedalled for six min. Each subject underwent four repeats of the protocol. Gas exchange was measured breath-by-breath using a mass spectrometer (Airspec 2000) and the data interpolated to 1 s, and ensemble averaged. The kinetics were modelled with a first order exponential equation with three components (amplitude, time constant and delay) from 20 s to six min of exercise (phase 2) of the form:

$$\dot{V}O_2(t) = a\{1 - e^{-[(t-\delta)/\tau]}\},$$

where $\dot{V}O_2(t)$ is the $\dot{V}O_2$ at any time (t); a is the amplitude of the response; τ is the time constant and δ the time delay of the response.

Subjects had previously performed a ramp exercise test on the cycle ergometer for determination of maximal aerobic capacity ($\dot{V}O_{2max}$). Continuous variables are expressed as mean ± SD. Student's paired t-test was used to analyze differences between the pre and post verapamil values. Differences were considered to be significant at $p < 0.05$.

RESULTS

Clinical characteristics of the subjects are shown in Table 1. A significant increase in $\dot{V}O_{2max}$ (22.1 ±2.9 versus 24.8 ±2.6) was observed after verapamil ingestion ($p < 0.05$). No difference was observed in resting heart rate or blood pressure following verapamil ingestion.

There were no significant differences in the steady state values for $\dot{V}O_2$ ($\dot{V}O_2SS$) after verapamil ingestion (Table 2). Of the total amplitude of the $\dot{V}O_2$ response, 64% (±8%) of the response occured at 20 s of the on-transient ($\dot{V}O_220s$) before verapamil compared to 83% (±7%) after verapamil ingestion. This resulted in a significant increase in $\dot{V}O_220s$ after verapamil ingestion ($p < 0.05$). Figure 1 depicts the $\dot{V}O_2$ on-transient before and after verapamil ingestion for subject #5. No difference was observed in the phase 2 on-transient ($\tau\dot{V}O_2$) after verapamil ingestion.

Figure 2 depicts the pre and post verapamil results for $\dot{V}O_{2max}$, $\dot{V}O_220s$, $\dot{V}O_2SS$ and $\tau\dot{V}O_2$.

Table 1. Summary of individual clinical characteristics before (pre) and four hours after (post) verapamil ingestion

Subject	Age	BSA	$\dot{V}O_2max_{Pre}$	$\dot{V}O_2max_{Post}$	HR_{Pre}	HR_{Post}	BP_{Pre}	BP_{Post}
1	65	2.2	23.9	26.0	76	74	148/82	144/80
2	68	2.2	25.4	28.2	70	70	126/70	128/72
3	71	1.7	20.4	22.0	72	71	136/80	134/80
4	69	2.2	21.0	23.6	62	64	122/70	122/70
5	65	2.1	24.4	26.8	70	72	150/84	146/80
6	68	1.9	17.5	22.3	48	46	118/80	118/76
×	67	2.0	22.1	24.8*	66	66	133/78	132/76
(±SD)	(2)	(0.2)	(2.9)	(2.6)	(10)	(10)	(13)/(6)	(11)/(5)

*Denotes a significant difference from the pre value ($p < 0.05$).

BSA = body surface area in m^2; $\dot{V}O_2max$ = maximal oxygen uptake in $ml \cdot kg^{-1} \cdot min^{-1}$; HR = resting heart rate in $beats \cdot min^{-1}$; BP = resting systolic/diastolic blood pressure in mmHg.

Table 2. Summary of individual $\dot{V}O_2$ on-transient responses before (pre) and four hours after (post) verapamil ingestion

Subject	τ		VO_2 20s		VO_2SS		TD	
	Pre	Post	Pre	Post	Pre	Post	Pre	Post
1	38.2	43.1	0.80	0.93	1.21	1.23	2.2	0.47
2	54.9	56.5	0.66	0.75	0.89	0.91	2.41	1.55
3	54.6	51.9	0.95	1.03	1.24	1.16	0.17	2.33
4	45.9	45.0	0.88	0.95	1.16	1.21	8.28	1.18
5	40.0	41.9	1.05	1.21	1.84	1.81	3.77	0.73
6	55.4	54.9	0.96	1.00	1.15	1.14	1.75	4.26
$\times$	48.2	48.9	0.88	0.97*	1.24	1.24	3.10	1.75
($\pm$SD)	(7.8)	(6.3)	(0.13)	(0.14)	(0.31)	(0.30)	(2.79)	(1.39)

*Denotes a significant difference from the pre value ($p < 0.05$).
τ = the time constant of the monoexponential fit; VO_2 20s = the oxygen consumption 20s into the on-transient response; $\dot{V}O_2$ SS = the oxygen consumption at steady state; TD = the time delay of the monoexponential fit.

DISCUSSION

With verapamil intake the increment of $\dot{V}O_2$ in the initial 20 s was increased, however, no change in the phase 2 response of the $\dot{V}O_2$ on-transient was observed in the present study

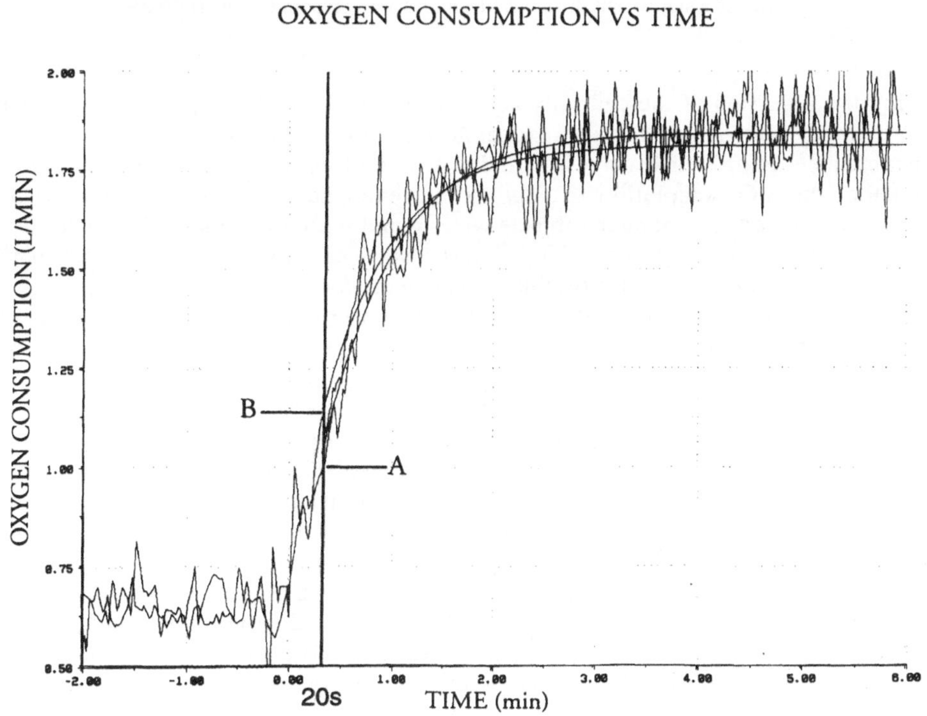

OXYGEN CONSUMPTION VS TIME

Figure 1. $\dot{V}O_2$ on-transient response curve for subject #5 before (pre) and four hours after (post) the ingestion of verapamil.

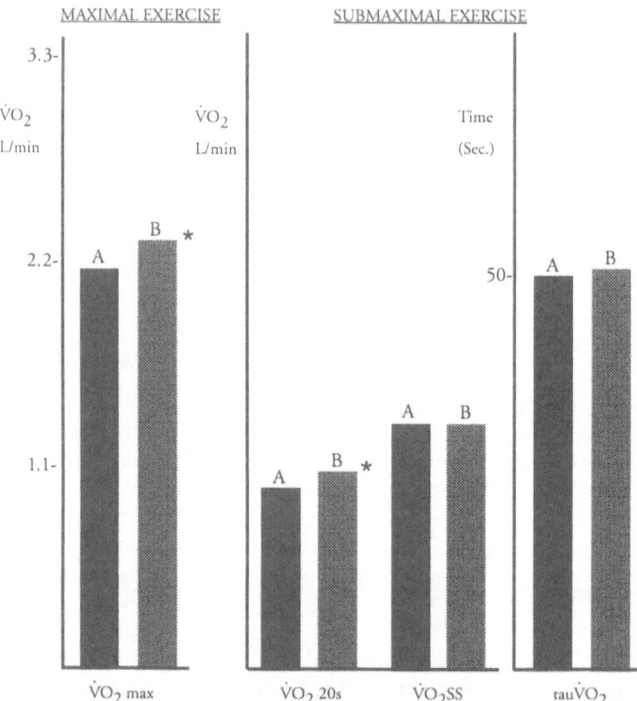

Figure 2. Summary of key selected variables before (pre) and 4 hours after (post) verapamil ingestion.
* denotes a significant difference from the pre verapamil value (p<0.05).

following the ingestion of the centrally acting calcium channel antagonist, verapamil. We have previously shown that the gas exchange kinetics are slowed in older compared to younger subjects using square waves (2) and sinusiodal forcing (8). Hughson and Smyth (3) showed that the $\dot{V}O_2$ adaptation to submaximal exercise ($\dot{V}O_2$ uptake kinetics) during β-adrenergic blockade in younger subjects was slowed without a change in the steady state $\dot{V}O_2$. Maximal exercise capacity ($\dot{V}O_{2max}$) has also been shown to be reduced in young subjects after beta blockade (4). Possible mechanisms for the slowed $\dot{V}O_2$ response at the onset of exercise include reduced delivery of oxygen to the exercising tissues by the heart (central component) (5), or reduced oxidative processes in the muscle (peripheral component) (6). We have recently shown that the centrally acting calcium channel blocker verapamil improved cardiac filling and $\dot{V}O_{2max}$ in normotensive and hypertensive elderly subjects (1), but had no effect in young or physically fit elderly subjects (unpublished data). The predominantly central activity of this drug suggests that the increased $\dot{V}O_{2max}$ (1) and the increase in phase 1 or the $\dot{V}O_2$ at 20 s of the on-transient in this study may be due to improved cardiac filling. There is no change in Phase 2 of the on-transient for the subjects in this study. If however, these subjects are divided into two groups according to their $\dot{V}O_{2max}$, all three subjects with a lower $\dot{V}O_{2max}$ (19.6 ±1.9) showed a reduction in τ after verapamil ingestion while those with a higher $\dot{V}O_{2max}$ (24.6 ±0.8) had little or no change in τ after verapamil. This raises the possibility that the level of cardiorespiratory function may affect not only the initial cardiodynamic phase but also the phase 2 kinetics as shown with verapamil ingestion. Indeed, we have shown that physically fit elderly had little change in left ventricular filling (unpublished data) while sedentary elderly improved cardiac filling (1). The age-related decline in cardiac function and in particular, diastolic filling at rest and during exercise, has been ascribed to altered intracellular calcium handling (9,10). Thus, improved calcium handling with verapamil ingestion (11) may contribute to the improved central or cardiac contribution to the $\dot{V}O_2$ response to a steady-state submaximal exercise perturbation in elderly men.

CONCLUSIONS

Thus, given the previously described increase in $\dot{V}O_{2max}$ following verapamil intake in these subjects, and the present observations of: *1.* an increase in the increment of $\dot{V}O_2$ in the initial 20 s of exercise and *2.* no change in phase 2 of the on-transient, it would suggest that verapamil improved left ventricular pump function and the central component of the $\dot{V}O_2$ response.

ACKNOWLEDGMENTS

The Centre for Activity and Ageing is affiliated with The Lawson Research Institute of St. Joseph's Health Center.

Research approved by the University Health Sciences Review Committee on Human Research, and supported by grants from Natural Science and Engineering Research Council (Canada) and Searle Canada.

REFERENCES

1. Petrella,R.J., P.M.Nichol, D.H.Paterson, and D.A.Cunningham. Verapamil improves left venricular filling and exercise performance in elderly normotensive and hypertensive individuals.Can.J.Cardiol. 10:973-981,1994.
2. Babcock,M.A., D.H.Paterson, D.A.Cunningham, and J.R.Dickinson. Exercise on-transient gas exchange kinetics are slowed as a function of age. Med.Sci.Sports Exerc.26:440-446,1994.
3. Hughson,R.L., and G.A.Smyth. Slower adaptation of VO_2 to steady state of submaximal exercise with β-blockade. Eur.J.Appl.Physiol. 52:107-110,1983.
4. Tesch,P.A., and P.Kaiser. Effect of β-adrenergic blockade on maximal oxygen uptake in trained males. Acta Physiol.Scand. 112:351-352,1981.
5. Hughson,R.L., and M.Morrissey. Delayed kinetics of respiratory gas exchange in the transition from prior exercise. J.Appl.Physiol. 52:921-929,1982.
6. Wasserman,K. New concepts in assessing cardiovascular function. Circulation, 78:1060-1079,1988.
7. Whipp,B.J., S.A.Ward, N.Lamarra, J.A.Davis, and K.Wasserman. Parameters of ventilatory and gas exchange dynamics during exercise. J.Appl.Physiol. 52:1506-1523,1982.
8. Cunningham,D.A., J.E.Himan, D.H.Paterson, and J.R.Dickinson. Gas exchange dynamics with sinusiodal work in young and elderly women. Respiratory Physiology 91:43-56,1993.
9. Wei,J.Y. Age and the cardiovascular system. N.Engl.J.Med. 327:1735-1739,1992.
10. Gaasch,W.H. Diagnosis and treatment of heart failure based on left ventricular systolic or diastolic dysfunction. JAMA 271:1276-1280,1994.
11. Arrighi,J.A., V.Dilsizian, P.Perrone-Filardi, J.G.Diodati, S.L.Bacharach, and R.O.Bonow. Improvement of the age-related impairment in left ventricular diastolic filling with verapamil in the normal human heart. Circulation 90:213-219,1994.

39

DYNAMICS OF THE PULMONARY O$_2$ UPTAKE TO BLOOD FLOW RATIO ($\dot{V}O_2/\dot{Q}$) DURING AND FOLLOWING CONSTANT-LOAD EXERCISE

Takayoshi Yoshida[1] and Brian J. Whipp[2]

[1] Exercise Physiology Laboratory
Faculty of Health and Sports Sciences
Osaka University
Toyonaka, 560 Osaka, Japan
[2] Department of Physiology
St George's Hospital Medical School
Tooting, United Kingdom

INTRODUCTION

For skeletal muscle during exercise, the difference between the content of its arterial inflow (CaO$_2$) and the ratio of its metabolic rate ($\dot{V}O_2$) to blood flow ($\dot{Q}$) determines the O$_2$ content of its venous effluent (CvO$_2$), ie:

$$CvO_2 = CaO_2 - (\dot{V}O_2/\dot{Q}) \tag{1}$$

Consequently, as CvO$_2$, and hence PvO$_2$, is an important determinant of both mean capillary PO$_2$ and the driving pressure for O$_2$ transfer to mitochondria, muscle tissue oxygenation is crucially dependent upon the ratio $\dot{V}O_2/\dot{Q}$ (6, 9). And although both total cardiac output and skeletal muscle blood flow have been demonstrated to increase as linear function of $\dot{V}O_2$ during steady-state exercise, the relationship has an intercept on the $\dot{Q}$-axis with the result that both muscle and mixed venous O$_2$ content decrease hyperbolically as work rate increases (11).

During the non-steady-state phase of the exercise CvO$_2$ will be determined by the relative time constant (τ) for $\dot{V}O_2$ and $\dot{Q}$ in addition to the final steady-state values or 'gains', i.e.

$$\text{on: } CvO_2 = CaO_2 - \left\{ \frac{VO_{(rest)} + \Delta\dot{V}O_2(1 - e^{-t/\tau\dot{V}O_2})}{Q(rest) + \Delta Q(1 - e^{-t/\tau Q})} \right\}$$

Modeling and Control of Ventilation, Edited by S. J. G. Semple, L. Adams, and B. J. Whipp
Plenum Press, New York, 1995

(2)

$$\text{off: } CvO_2 = CaO_2 - \left\{ \frac{\dot{V}O_2 + \Delta\dot{V}O_2 \cdot e^{-t/\tau\dot{V}O_2}}{\dot{Q}(\text{rest}) + \Delta\dot{Q} \cdot e^{-t/\tau\dot{Q}}} \right\}$$

(3)

At work rates where the $\dot{V}O_2$ and $\dot{Q}$ transients evidence first order kinetics, $\tau\dot{Q}$ has been reported to be smaller or similar to $\tau\dot{V}O_2$ (3, 4, 7). It is important, with respect to tissue oxygenation, to know whether there is on-off symmetry for the $\dot{Q}$ kinetics, as has been demonstrated for $\dot{V}O_2$ (5, 8, 10). Were $\tau\dot{Q}$ also to be appreciably shorter than $\tau\dot{V}O_2$ in the recovery phase then, as evidenced from equation 1, tissue oxygenation could be stressed more after the exercise than during its performance.

It was therefore the purpose of these experiments to determine the kinetics of $\dot{Q}$ with respect to $\dot{V}O_2$ at the on and off transients of constant load exercise in normal subjects.

METHODS AND MATERIALS

Subjects: Six healthy male volunteered as subjects for one-legged cycle ergometry. Each subject was fully informed of the purpose of this study and possible risks before signing an informed consent form. The subject's average age (with±SE), height and body mass was 25.8±3.0 years, 174.8±3.0 cm, and 73.7±2.8 kg, respectively.

Exercise protocol: Prior to the exercise test for $\dot{V}O_2$ kinetics, all subjects performed an incremental exercise test to estimate the lactate threshold and maximal oxygen uptake. For $\dot{V}O_2$ kinetics determination an electrically-braked computer-controlled bicycle ergometer was used (Combi 232-C, Combi, Tokyo). The constant-load exercise test consisted of 5 min of rest followed by 5 min of one-legged pedaling at 50 W, using the subject s preferred leg and 5 min resting recovery. This exercise protocol was repeated at least 8 times for each subject to establish an ensemble average with a high signal-to-noise ratio. During the exercise test, ventilatory and gas exchange responses were measured with a computerized on-line breath-by-breath system and cardiac output ($\dot{Q}$) was determined continuously by computerized impedance technique (7, 13-16). The transient responses of both $\dot{V}O_2$ and $\dot{Q}$ were characterized by first-order kinetics. To determine response kinetics, a best-fit mono-exponential procedure with a time delay was employed (12).

Statistical Analysis: Data are expressed as means±SE. An analysis of variance (ANOVA) with repeated measures was used to analyze changes within each of the experiments. A post hoc t-test for a significant F value finding was applied to specify where significance occurred. A probability level of $P<0.05$ was accepted as significant.

RESULTS AND DISCUSSION

Both $\dot{V}O_2$ and $\dot{Q}$ during the transients were well characterized as a mono-exponential in all subjects. There was, however, a wide variation of time constant among the individual subjects, with τ ranging from 24 to 50 seconds for $\dot{V}O_2$ and from 19 to 43 for $\dot{Q}$ (Table 1).

The $\tau\dot{V}O_2$ at the on-transient (33.9±3.5 s, mean±SEM) and at the off-transient (37.2±2.9 s) were not significantly different. $\tau\dot{Q}$, however, was consistently longer at the off-transient than at the on (Table 1). That is, the ratio $\tau\dot{V}O_2/\tau\dot{Q}$, an important determinant of tissue oxygenation during the transient, was typically less than 1.0.

Table 1. Value for oxygen uptake ($\dot{V}O_2$) and cardiac output ($\dot{Q}$) for on-transient and off-transient of constant-load exercise

	$\dot{V}O_2$		$\dot{Q}$	
	Gain (ml·min^{-1})	Time constant (s)	Gain (%)	Time constant (s)
On-transient				
mean	472.2	33.88	62.5	29.43
± SEM	26.6	3.51	9.23	3.17
Off-transient				
mean	519.3	37.22	51.9	44.28
± SEM	25.9	2.91	4.91	3.58

In Fig 1 we explore this issue in greater detail. We plot the degree of hysteresis between $\dot{V}O_2$ and $\dot{Q}$ and its consequence upon the actual value of $\dot{V}O_2/\dot{Q}$ throughout the transient. For cardiac output, a t of 0.33 min and a steady-state gain derived from the equation $\dot{Q}$ (L/min) = $5\dot{V}O_2 + 5$ was used; for $\dot{V}O_2$ a t of 0.5 min and a steady-state gain derived from the equation $\dot{V}O_2$ (L/min) = $0.01W + 0.5$ was used, where W is the work rate in watts and 0.5 is the $\dot{V}O_2$ for unloaded cycling (i.e. "0" watts).

In panel A the lines radiating from the origin represent isopleths of $\dot{V}O_2/\dot{Q}$, with the line at 0.2 representing the 'critical' value, for subjects with a normal CaO_2 of 200ml/L, i.e. the value at which CvO_2 theoretically falls to zero. The line at 0.067 represents the value which obtains at unloaded cycling. The heavy black line (line of 0.14) gives the normal steady-state response between the variables. Note that, as a consequence of $\tau\dot{Q}$ being faster than $\tau\dot{V}O_2$ at the on transient but typically slower at the off transient, the ratio $\dot{V}O_2/\dot{Q}$ actually

Figure 1. The degree of hysteresis between $\dot{V}O_2$ and $\dot{Q}$ and its consequence upon the actual value of $\dot{V}O_2/\dot{Q}$ throughout the exercise transient, using modally representative values for the gains and time constants for $\dot{V}O_2$ and $\dot{Q}$. Panel A: the lines radiating from the origin represent isopleths of $\dot{V}O_2/\dot{Q}$, with the line at 0.2 representing the 'critical' value and the line at 0.067 represents the value which obtains at unloaded cycling. The line of 0.14 gives the steady-state response between the variables. Panel B: the ratio $\dot{V}O_2/\dot{Q}$ actually falls in both phases, representing better O_2 delivery.

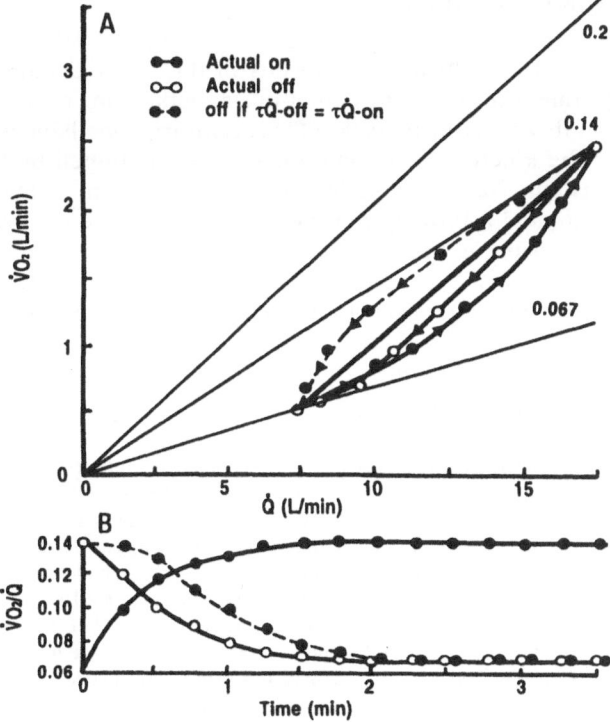

falls in both phases (Panel B), representing better O_2 delivery, presumably to the contracting muscle units.

In Fig 1 we also show the effects on $\dot{V}O_2/\dot{Q}$ had the on and off transient $\dot{Q}$ kinetics been effectively the same, i.e. as they are for $\dot{V}O_2$. Interestingly, $\dot{V}O_2/\dot{Q}$ even in this case does not rise (Panel B). The reason for this is evident in Panel A, that is, the slope of the actual steady-state $\dot{V}O_2$ versus $\dot{Q}$ plot is so steep relative to the steepest $\dot{V}O_2/\dot{Q}$ isopleth that the significant hysteresis resulting from the symmetrical $\dot{Q}$ dynamics (i.e. the solid dots joined by the dashed line, in Fig. 1) prevents $\dot{V}O_2/\dot{Q}$ from actually increasing. This effect tends to 'protect' the contractile units, on average, against inadequate O_2 delivery during the transients.

As stated above, the time-constant for the off-transient in $\dot{Q}$ was appreciably longer than that for its on-transient response. This serves to maintain oxygen flow to the muscle sufficiently high during recovery i.e. at a time when the metabolic oxygen exchange continuous at a high rate, assuming of course, that the relatively high cardiac output in this phase is a result of relatively high flow to the exercising muscles (2). This remains to be determined. It seems likely that this is the case, however, with metabolic factors maintaining the exercise-muscle component of the peripheral vascular conductance at higher levels. Potassium decreases very rapidly at the off-transient of exercise (13). In contrast, adenosine production is likely to remain high while the oxygen uptake is high. Similarly, osmolarity is likely to remain high during this phase, i.e. glycolytic and fatty acid fluxes continue at a high rate. Furthermore, the slow discharge of CO_2 from its capacitative storage within the muscle would maintain the demand for high muscle blood flow, and hence cardiac output, during this phase. This suggests that the hypothesis of Barstow et al. (1) that tissue oxygenation may be stressed more at the off-transient than at the on for constant load exercise appears to be falsified as a result of the marked increase in the off-transient $\tau\dot{Q}$.

CONCLUSION

In conclusion, the results of this study demonstrate that the time-constant for the on-transient of cardiac output during moderate constant load exercise is appreciably faster than that of oxygen uptake. This seems irreconcilable with the notion that the time-constant for the kinetics of oxygen uptake are determined by the limitations of blood flow in the transient. We have also demonstrated that the non-steady-state responses of cardiac output during and following constant-load exercise are not symmetrical. The time-constant at the off-transient is appreciably slower than that of the on-transient. Unlike O_2 uptake, at this work intensity, cardiac output manifests non-linear dynamics, the non-linearlity being advantageous to tissue gas exchange during recovery.

REFERENCES

1. Barstow, T.J., N, Lamarra, & B.J. Whipp. Modulation of muscle and pulmonary O_2 uptakes by circulatory dynamic during exercise. *J. Appl. Physiol.* 68: 979-989, 1990.
2. Clausen, J.P. Circulatory adjustments to dynamic exercise and effect of physical training in normal subjects and in patients with coronary artery disease. *Proc. Cardiovas. Dis.* 18: 459-495, 1976.
3. Davies. C.T.M., P.E. di Prampero, & P. Cerretelli. Kinetics of cardiac output and respiratory gas exchange during exercise and recovery. *J. Appl. Physiol.* 32: 615-625, 1972.
4. De Cort, S.C., J.A. Innes, T.J. Barstow, & A. Guz. Cardiac output, oxygen consumption and arteriovenous oxygen difference following a sudden rise in exercise level in humans. *J. Physiol.* 441: 501-512, 1991.
5. Griffiths, T.L., L.C. Henson, & B.J. Whipp. Influence of peripheral chemoreceptors on the dynamics of the exercise hyperpnea in man. *J. Physiol.* 380: 387-403, 1986.

6. Meuer, H.J., M. Ahrens, & C. Ranke. Distribution of local oxygen consumption in resting skeletal muscle. In: *Oxygen Transport to Tissue*, vol. VII. edited by S.M. Chain, Z. Turek & T.K. Goldstick, New York: Plenum, 1985, pp365-374.

7. Miyamoto, T., T. Hiura, T. Tamura, T. Nakamura, J. Higuchi, J. & T. Mikami. Dynamics of cardiac, respiratory, and metabolic function in men in response to step work load. *J. Appl. Physiol.* 52: 1198-1208, 1982.

8. Paterson, D.H. & B.J. Whipp. Asymmetries of oxygen uptake transients at the on- and off-set of heavy exercise in humans. *J. Physiol.* 443: 575-586, 1981.

9. Van Lieuw, H.D. Regional heterogeneity of PCO_2 and PO_2 in skeletal muscle. In: *Oxygen Transport to Tissue.* edited by H.I. Bicher & D.F. Bruley. New York: Plenum, 1973, pp457-462.

10. Whipp, B.J. The physiological and energetics basis of work efficiency. In: *Obesity in Perspective.* edited by G.A. Bray Section II,Chap 16, Washington: U.S. Government Printing Office, 1976, pp121-126.

11. Whipp, B.J. & S.A. Ward. Cardiopulmonary coupling during exercise. *Experiment. Physiol.* 100: 175-193, 1982.

12. Whipp, B.J., S.A. Ward, N. Lamarra, J.A, Davis, & K. Wasserman. Parameters of ventilatory and gas exchange dynamics during exercise. *J. Appl. Physiol.* 52: 1506-1513, 1982.

13. Yoshida, T., M. Chida, M. Ichioka, K. Makiguchi, J. Eguchi, & M. Udo. Relationship ventilation and arterial potassium concentration during incremental exercise and recovery. *Eur. J. Appl. Physiol.* 61:193-196, 1990.

14. Yoshida, T., K. Yamamoto, T. Naka, & M. Udo. Relationship between and cardiac output at the onset of exercise and recovery. *Proc. Int. Cong. Human-Environ. Sys.* 519-522, 1991.

15. Yoshida, T., M. Udo, T. Ohmori, Y. Matsumoto, T. Uramoto, & K. Yamamoto. Day-to-day changes in oxygen uptake kinetics at the onset of exercise during strenuous endurance training. *Eur. J. Appl. Physiol.* 64: 78-83, 1992.

16. Yoshida, T., K. Yamamoto, & M. Udo. Relationship between cardiac output and oxygen uptake at the onset of exercise. *Eur. J. Appl. Physiol.* 66: 155-160, 1993.

40

EXERCISE VENTILATION AND K⁺ IN PATIENTS WITH COPD

Positive and Negative Work

J. M. Rooyackers, P. N. R. Dekhuijzen, C. L. A. van Herwaarden, and H. Th. M. Folgering

Department of Pulmonary Diseases
University of Nijmegen
Medical Centre Dekkerswald
P.O. Box 9001, 6560 GB Groesbeek, The Netherlands

INTRODUCTION

Plasma potassium levels are closely related to ventilation during dynamic exercise, and potassium may be a potent stimulus of ventilation.(13) We studied the relationship between potassium and the exercise hyperpnoea during concentric (positive work, Wpos) and eccentric (negative work, Wneg) exercise in patients with chronic obstructive pulmonary disease (COPD).

As exercise ventilation is considerably less during Wneg in comparison with Wpos at similar work loads (5,6,8), we expected the plasma potassium level to be lower during Wneg. The aim of the present study was to test the hypothesis that the relationship between arterial plasma potassium concentration ($[K^+]a$) and ventilation is independent of the type of work, and that $[K^+]a$ equally contributes to the exercise hyperpnoea during both Wpos and Wneg.

METHODS

Twelve patients (10 male) with COPD according to ATS (1) criteria participated in the study. Exercise was performed at a pedalling rate of 60 RPM on a electrically braked cycle ergometer, which was adapted for positive and negative work. During Wneg the subjects had to brake the speed of the pedals, which were driven in backward direction by an electric motor.

All subjects cycled 6 minutes concentrically and 6 minutes eccentrically in a randomized order at a work load of 50% of the individuals' maximal work load, which had been measured previously. $[K^+]a$ was measured at one minute intervals during exercise and during 3 minutes of

Modeling and Control of Ventilation, Edited by S. J. G. Semple, L. Adams, and B. J. Whipp
Plenum Press, New York, 1995

Table 1. Exercise response (mean ± SD) to positive (Wpos) and negative work (Wneg) at 50% of maximal positive work capacity (n = 12)

	Wpos	Wneg
Resting values		
$\dot{V}E$ (L/min)	12 ± 3	11 ± 2
$[K^+]a$ (mM)	4.19 ± 0.3	4.14 ± 0.3
$\dot{V}CO_2$ (L/min)	0.26 ± 0.07	0.24 ± 0.06
Base-excess (mmol/L)	0.5 ± 1.9	0.2 ± 1.6
pH	7.437 ± 0.022	7.415 ± 0.020[†]
PaO_2 (kPa)	10.5 ± 1.5	10.3 ± 1.5
$PaCO_2$ (kPa)	4.6 ± 0.4	5.0 ± 0.5[*]
Exercise (6 minutes at a constant work load of 44 ± 14 Watts)		
$\dot{V}E$ (L/min)	29 ± 6	19 ± 7[†]
$[K^+]a$ (mM)	4.63 ± 0.3	4.50 ± 0.4
$\dot{V}CO_2$ (L/min)	0.83 ± 0.13	0.49 ± 0.14[†]
Base-excess (mmol/L)	0.3 ± 1.4	0.4 ± 1.8
pH	7.397 ± 0.030	7.409 ± 0.022
PaO_2 (kPa)	9.8 ± 2.1	10.2 ± 1.5
$PaCO_2$ (kPa)	5.3 ± 0.8	5.1 ± 0.6
Recovery (3 minutes)		
$\dot{V}E$ (L/min)	15 ± 5	12 ± 5[†]
$[K^+]a$ (mM)	4.36 ± 0.3	4.33 ± 0.3
$\dot{V}CO_2$ (L/min)	0.35 ± 0.10	0.28 ± 0.09
Base-excess (mmol/L)	0.1 ± 1.4	0.4 ± 1.8
pH	7.429 ± 0.029	7.419 ± 0.028
PaO_2 (kPa)	11.7 ± 1.6	10.3 ± 2.1[†]
$PaCO_2$ (kPa)	4.7 ± 0.6	5.1 ± 0.6[†]

Wpos versus Wneg: [*]$p<0.05$, [†]$p<0.01$.

recovery. Samples for blood gas analysis were taken at three minute intervals. $\dot{V}E$, $\dot{V}O_2$ and $\dot{V}CO_2$ were measured every 30 seconds by a mixing chamber ergospirometry unit.

RESULTS

The patients (mean (SD) age 56 (12) yrs.) had a FEV_1 of 1.5 (0.4) L (46 (16) % of predicted). $\dot{V}E$ increased by 17 L/min during Wpos versus 8 L/min during Wneg (p<0.01). $[K^+]a$ increased 0.44 mM and 0.36 mM, respectively (NS) (Table 1). Within subjects, the differences in ventilatory response between Wpos and Wneg did not correlate with differences in PaO_2 (r = -0.22), $PaCO_2$ (r = -0.14), base-excess (r = 0.10) or pH (r = 0.32).

The time course of $\dot{V}E$ was more flat during Wneg (Figure 1). $\dot{V}E$ had returned to resting values within 3 minutes of recovery after Wneg, but $[K^+]a$ was still increased after both types of exercise (Table 1). The time, in which $\dot{V}E$ and $[K^+]a$ had decreased by 50% of the difference between rest and the end of exercise, did not differ significantly after both Wpos and Wneg.

Significant correlations between mean $\dot{V}E$ and mean $[K^+]a$ were obtained during Wpos (r = 1.0, p = 0.02) and Wneg (r = 0.96, p = 0.02). However, for any change in $[K^+]a$ $\dot{V}E$ increased more during Wpos than during Wneg (Fig. 2a). In contrast, no differences were found in the relationship between mean $\dot{V}E$ and mean $\dot{V}CO_2$ (Fig. 2b).

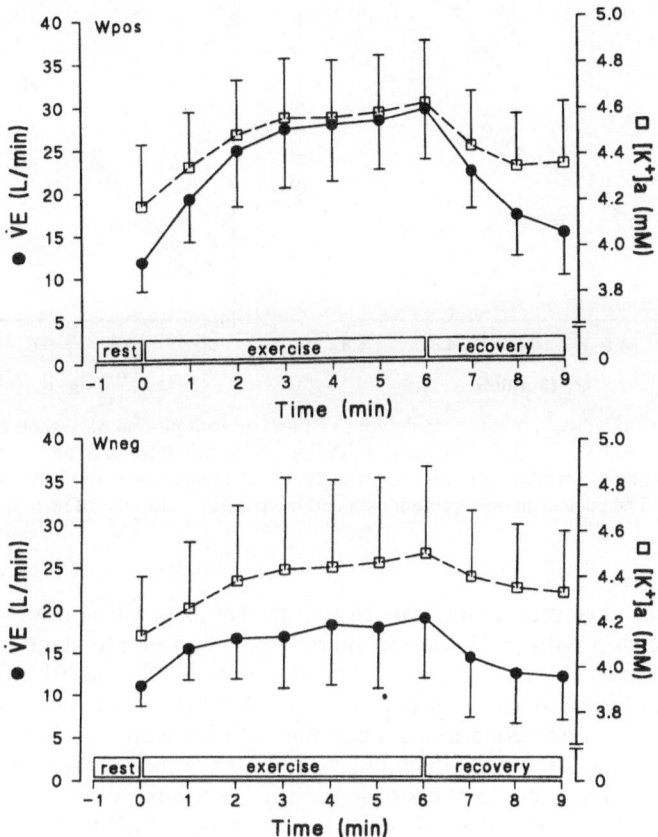

Figure 1. Time course of mean minute ventilation ($\dot{V}E$: closed circles) and mean arterial plasma potassium concentration ([K⁺]a: open squares) in 12 patients with COPD during 6 minutes of cycle exercise at a constant work load of 50% of the individual maximal (positive work) capacity and during 3 minutes of recovery. Figure 1a: concentric exercise (Wpos); Figure 1b: eccentric exercise (Wneg).

The means of the individual relationships between $\dot{V}E$ and [K⁺]a were described by the following equations:

$$\text{Wpos: } \dot{V}E = 42.7 \times [K⁺]a - 169$$

$$\text{Wneg: } \dot{V}E = 18.7 \times [K⁺]a - 67$$

The mean slope was more than two times steeper during Wpos than during Wneg (p = 0.01).

DISCUSSION

Ventilation during exercise is controlled by a complex of neural and humoral drives. Like potassium, lactate, $PaCO_2$ and catecholamines stimulate ventilation at the site of the

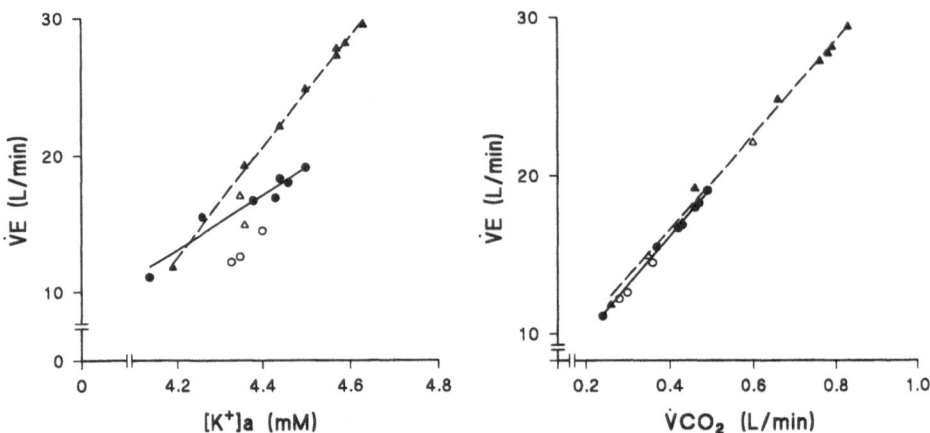

Figure 2. Relationship between minute ventilation ($\dot{V}E$) and arterial plasma potassium concentration ($[K^+]a$) (Figure 2a), and between $\dot{V}E$ and $\dot{V}CO_2$ (Figure 2b) during concentric (closed triangles) and eccentric (closed circles) exercise. The closed symbols indicate the mean values of 12 subjects at rest and at each minute during 6 minutes of exercise. The open symbols represent data obtained during 3 minutes of recovery.

arterial peripheral chemoreceptor.(3,10) In normal subjects $\dot{V}E$ and $\dot{V}O_2$ showed a close temporal relationship with $[K^+]a$ during incremental and single stage exercise.(2,11,12) Yoshida et al.(15) found similar results in patients with COPD, but for the same change in $[K^+]a$ $\dot{V}E$ increased less in patients than in normal subjects. However, exercise was not performed under the same conditions, since the patients were ventilatory limited during exercise and produced less lactate in comparison with healthy subjects.(15) The patients in our study performed sub-maximal exercise within the limits of their ventilatory capacity. Although small differences in PaO_2, $PaCO_2$, base-excess and pH between Wpos and Wneg were present, these did not correlate significantly with differences in the ventilatory response. Furthermore, $\dot{V}E$ was closely related to $\dot{V}CO_2$ in a similar way during Wpos and Wneg. So, the difference in the $\dot{V}E$ to $[K^+]a$ relationship between Wpos and Wneg suggests that potassium is not responsible for the difference in exercise ventilation between the two types of work.

A problem of sub-maximal exercise could be that the rise in $[K^+]a$ was too small to stimulate the arterial chemoreceptors.(10) In our study $[K^+]a$ increased by 0.44 mM during concentric exercise at a constant work load of 44 Watts. Yet, the increase in $\dot{V}E$ was the same as observed previously in normal subjects during maximal incremental exercise for the same change in $[K^+]a$.[12,15]

The release of potassium from working muscle is proportional to the exercise intensity.(14) We expected that at comparable work loads the rise in potassium during Wneg would be less than during Wpos, as fewer motor units are activated during Wneg.(9) Muscle damage, which is associated with Wneg and results in the release of muscle proteins (creatinine), may have contributed to extra loss of potassium from the active muscles.(4)

In normal subjects, $[K^+]a$ fell below resting values within three minutes after heavy exercise.(12) The delayed normalization of $[K^+]a$, as we found in patients with COPD, may be caused by detraining and the older age.(7)

In conclusion, the results of this study do not support the concept that potassium contribute to the ventilation during exercise at 50% of the maximal work load in patients with COPD.

REFERENCES

1. American Thoracic Society. Standards for the diagnosis and care of patients with chronic obstructive pulmonary disease (COPD) and asthma. Am. Rev. Respir. Dis. 136:225-244, 1987.
2. Busse, M.W., N. Maassen, and H. Konrad. Relation between plasma K$^+$ and ventilation during incremental exercise after glycogen depletion and repletion in man. J. Physiol. 443:469-76, 1991.
3. Cunningham, D.J.C. Studies on arterial chemoreceptors in man. J. Physiol. 384:1-26, 1987.
4. Ebbeling, C.B., and P.M. Clarkson. Muscle adaptation prior to recovery following eccentric exercise. Eur. J. Appl. Physiol. 60:26-31, 1990.
5. Hesser, C.M., D. Linnarsson, and H. Bjurstedt. Cardiorespiratory and metabolic responses to positive, negative and minimum-load dynamic leg exercise. Respir. Physiol. 30;51-67, 1977.
6. Hulsbosch, M.A.M., R.A. Binkhorst, and H.Th. Folgering. Interaction of CO$_2$ and positive and negative exercise stimuli on the ventilation in man. Pflügers Arch. 394:16-20, 1982.
7. Klitgaard, H., and T. Clausen. Increased total concentration of Na-K pumps in vastus lateralis of old trained human subjects. J. Appl. Physiol. 67:2491-2494, 1989.
8. Knuttgen, H.G., F.B. Petersen, and K. Klausen. Oxygen uptake and heart rate responses to exercise performed with concentric and eccentric muscle contractions. Med. Sci. Sports 3:1-5, 1971.
9. Komi,P.V., M. Kaneko, and O. Aura. EMG activity of the leg extensor muscles with special reference to mechanical efficiency in concentric and eccentric exercise. Int. J. Sports Med. 8:22-29S, 1987.
10. Linton, R.A.F., and D.M. Band. The effect of potassium on carotid chemoreceptor activity and ventilation in the cat. Respir. Physiol. 59:65-70, 1985.
11. McCoy, M., and M. Hargreaves. Potassium and ventilation during incremental exercise in trained and untrained men. J. Appl. Physiol. 73: 1287-1290, 1992.
12. Paterson, D.J., P.A. Robbins, and J. Conway. Changes in arterial plasma potassium and ventilation during exercise in man. Respir. Physiol. 78: 323-330, 1989.
13. Paterson, D.J. Potassium and ventilation in exercise. J. Appl. Physiol. 72:811-820, 1992.
14. Wilkerson, J.E., S.M. Horvath, B. Gutin, S. Molnar, and F.J. Diaz. Plasma electrolyte content and concentration during treadmill exercise in humans. J. Appl. Physiol. 53:1529-1539, 1982.
15. Yoshida, T., M. Chida, M. Ichioka, K. Makigushi, N. Tojo, and M. Udo. Ventilatory response and arterial potassium concentration during incremental exercise in patients with chronic airways obstruction. Clin. Physiol. 11:73-82, 1991.

41

PHASE-COUPLING OF ARTERIAL BLOOD GAS OSCILLATIONS AND VENTILATORY KINETICS DURING EXERCISE IN HUMANS

Phase Coupling and the Exercise Hyperpnoea

S. A. Ward, L. Swain, and S. Frye-Kryder

St. George's Hospital Medical School
London, SW17 0RE
United Kingdom and
UCLA School of Medicine
Los Angeles, California 90024

INTRODUCTION

The carotid bodies (CBs) have been implicated in the kinetic control of ventilation ($\dot{V}_E$) during moderate exercise (reviewed in Refs. 10 & 11). That is, augmenting CB responsiveness (eg. by hypoxia or dietary-induced metabolic acidaemia) shortens the time constant of the $\dot{V}_E$ response ($\tau\dot{V}_E$) to square-wave exercise forcings, while CB suppression (induced by hyperoxia, metabolic alkalaemia or intravenous dopamine infusion) or surgical CB resection is associated with a prolonged $\tau\dot{V}_E$. The present investigation considers the possibility that intra-breath oscillations of arterial PO_2 (PaO_2), PCO_2 ($PaCO_2$) and pH (pHa) may be involved in this CB modulation. Yamamoto (12) proposed that these oscillations, which have been measured in humans at rest and during exercise (1), could provide a component of respiratory control independent of their mean levels, mediated possibly via the rate of change of pHa in the falling phase of the oscillation (dpHa/dt↓) (3). The reflex efficacy of this humoral oscillation can be modulated by the "phase-coupling" (9) between the ensuing oscillation of CB discharge and the on-going respiratory cycle: stimuli presented to the CBs during inspiration evoke stimulation, while respiration is relatively unaffected (or may even be depressed) by stimuli arriving at the CBs during expiration (2-6). We therefore tested the following hypothesis: if, at exercise onset, the peaks of the oscillating CB discharge were to arrive at the brainstem respiratory integrating areas at a more optimal phase of the respiratory cycle than at rest (e.g., during inspiration, rather than expiration), then the kinetics of the exercise hyperpnoea would be faster compared to a situation where the phase coupling at exercise onset was unchanged or even became "pessimal" (4).

Modeling and Control of Ventilation, Edited by S. J. G. Semple, L. Adams, and B. J. Whipp
Plenum Press, New York, 1995

NORMAL BREATHING

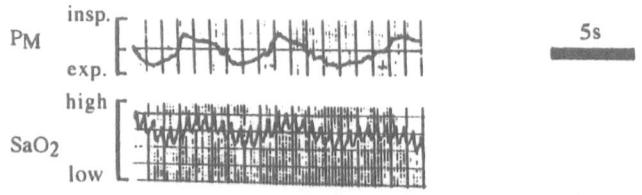

ALTERED BREATHING

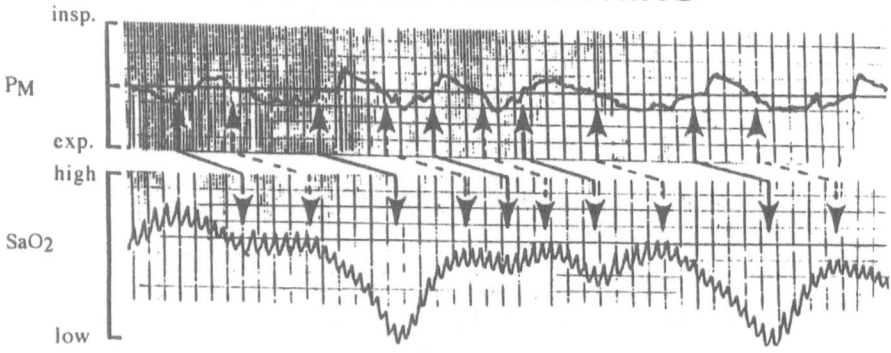

Figure 1. SaO$_2$ and Pm signals during spontaneous breathing (top panel) and during a period of deliberately irregular or "altered" breathing" (bottom panel).

METHODS

Five healthy subjects exercised on an electrically-braked cycle ergometer (Collins). Each performed an incremental test (10W.min^{-1}) for estimation of the lactate threshold (θ_L), using standard noninvasive criteria (9,11). On separate occasions, each then completed 2-4 sub-θ_L constant-power tests from "0" W to 80 W (6 min). They breathed through a low dead-space, low-resistance valve (Rudolph, 2700); expiratory flow was monitored by pneumotachography (Fleisch, 3; Validyne, MP45) and tidal volume derived by integration (Beckman); mouth pressure (Pm) was monitored at the mouthpiece (Validyne, MP45), and respired gas was sampled for measurement of PCO$_2$ and PO$_2$ (Beckman, LB2; Applied Electrochemistry, S3A); SaO$_2$ was recorded by high-gain, high-frequency ear oximetry (Waters, 351: $\tau < 0.2$ s). $\dot{V}_E$ and end-tidal PCO$_2$ (P$_{ET}$CO$_2$) and PO$_2$ (P$_{ET}$O$_2$) were determined breath-by-breath. The phase-coupling (PC) characteristic of the pHa oscillation at the CBs was estimated from the SaO$_2$ profile: successive minima of the SaO$_2$ signal were related to the phase of the on-going respiratory cycle, the inspiratory (I) and expiratory (E) phases of which were subdivided into early (I$_E$, E$_E$), mid (I$_M$, E$_M$) and late (I$_L$, E$_L$) phases of equal duration, via the Pm profile. Two assumptions were made: (a) the vascular transit delay from the lungs to the ear lobe (LET) closely approximated that to the CBs (7); (b) the minima of the SaO$_2$ and pHa oscillations were essentially coincident. LET was measured as the latency of the fall in SaO$_2$ caused by the abrupt, surreptitious introduction of 100% N$_2$ into the inspirate. For each test, $\tau\dot{V}_E$ for the "phase 2" ($\Phi2$) component of the response was estimated by standard least-squares regression (10,11):

$$\Delta\dot{V}_E(t) = \Delta\dot{V}_E(ss) \cdot (1 - e^{(t-\delta VE)/t})$$

where $\Delta\dot{V}_E(ss)$ is the steady-state $\dot{V}_E$ difference between "0" W and 80 W, $\Delta\dot{V}_E(t)$ is the $\dot{V}_E$ difference between the 80 W steady-state and time t, and δ is a delay term.

RESULTS

Fluctuations in SaO_2 with a respiratory periodicity were consistently observed: in Fig. 1 (above), for example, the SaO_2 minima typically coincided with the E_M phase. These fluctuations appeared to be determined by the intra-breath PaO_2 oscillations, rather than being artefacts of arterial blood flow or pressure (induced by respiratory fluctuations of intra-thoracic pressure). This contention is based upon the demonstration that the SaO_2 fluctuations measured at the ear are delayed by an interval consistent with the lung-CB body transit time, rather than being immediately transmitted to the ear (as would be expected for a simple hydraulic effect). That is, on several occasions during the steady state of exercise, subjects were requested to volitionally alter their normal respiratory rhythm for periods of

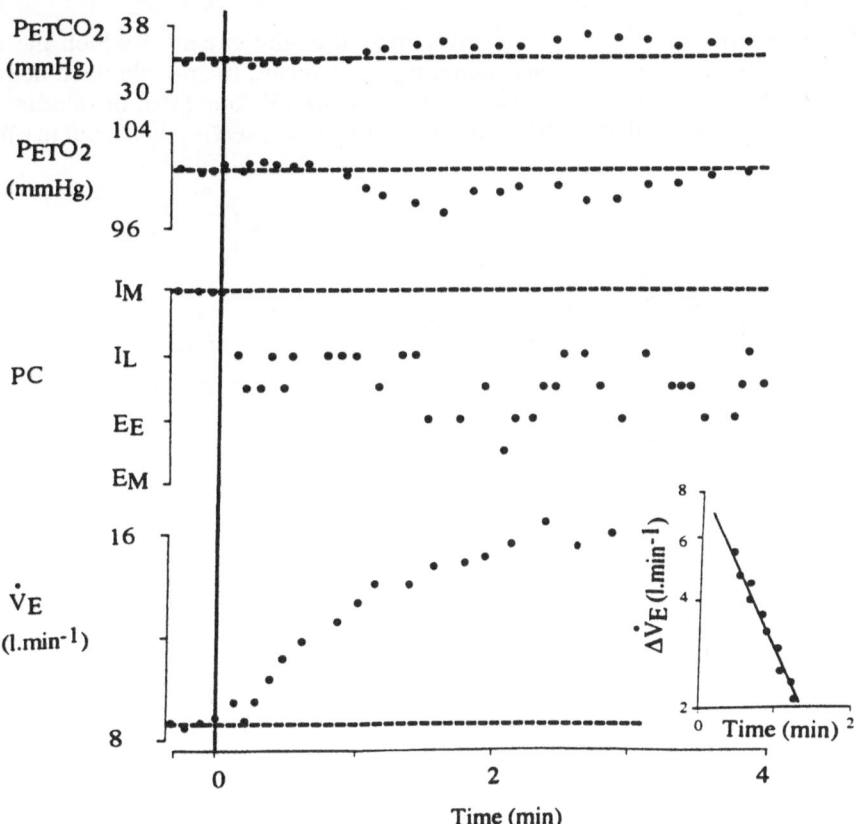

Figure 2. Representative response profiles.

1-2 min; this generated relatively unique and distinctive temporal sequences of consecutive T_I's and T_E's (Fig. 1, below). It is evident that the T_I-T_E sequence is preserved in the SaO_2 oscillation, with a delay corresponding to the LET.

The $\dot{V}_E$ and gas exchange responses to square-wave exercise (Fig. 2) cohere with previous descriptions (see Refs. 9 & 11). Following a small and sluggish increase at exercise onset (ie. Φ1), $\dot{V}_E$ rose slowly in Φ2 to attain a new steady state within 4-5 min. $P_{ET}CO_2$ and $P_{ET}O_2$ were reasonably stable during Φ1. In Φ2, $P_{ET}CO_2$ rose towards a new, higher steady-state level and $P_{ET}O_2$ fell (with a slight transient undershoot). The Φ2 $\dot{V}_E$ response was well fit by the exponential function, with a $\tau\dot{V}_E$ of 58 s in this particular example.

In this example, the PC profile changed systematically and abruptly at exercise onset from I_M towards the I-E transition, and subsequently spanned the I_L-E_E range (Fig. 3).

The PC characteristic varied widely across subjects, both in absolute terms and also relative to the corresponding $\dot{V}_E$ kinetics. One subject (Fig. 3) demonstrated a PC profile that was similar across tests: located in the I_M-I_L region at "0" W and moving into the E_E region during the exercise. However, $\tau\dot{V}_E$ was variable (Fig. 3): 56 s in one test, but 67 s for the other. This lack of consistency was evident also for the group as a whole, with no clear relationship between the PC characteristic and the corresponding $\tau\dot{V}_E$ (Fig. 4).

DISCUSSION

Previous Findings

There has been little systematic investigation of the extent to which the exercise hyperpnoea is influenced by this phase-coupling mechanism. In anaesthetized dogs undergoing electrically-induced hindlimb exercise, Cross *et al.* (3) found that the phase-coupling of the intra-breath pHa oscillation changed systematically at exercise onset, but in a direction

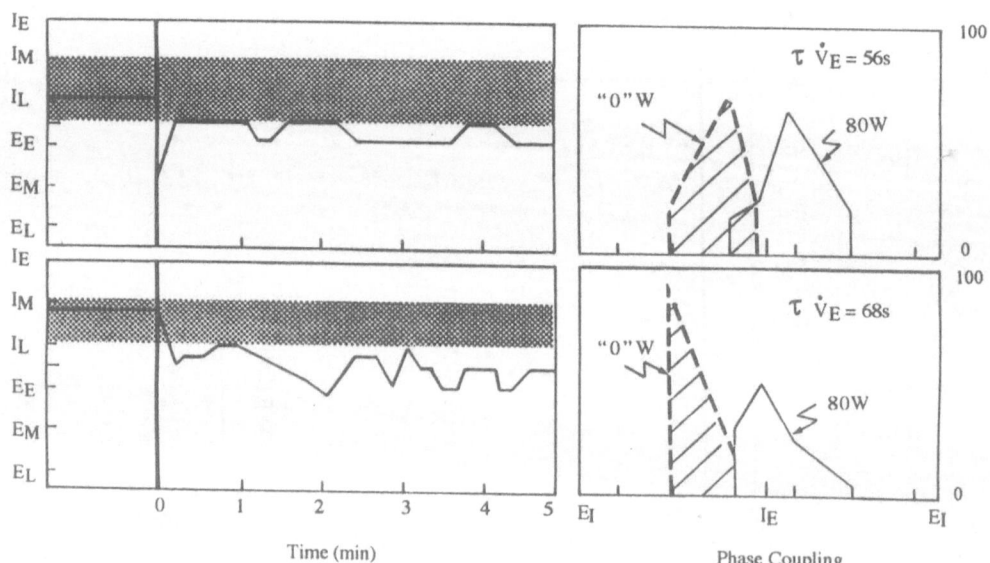

Figure 3. The PC responses for two exercise tests in the same subject: *left panels* show PC time course (shaded band indicates range of PC response at "0" W, and vertical line denotes exercise onset); *right panels* show the PC distribution histograms for "0" W (dashed line; hatched) and 80 W (solid lines; open).

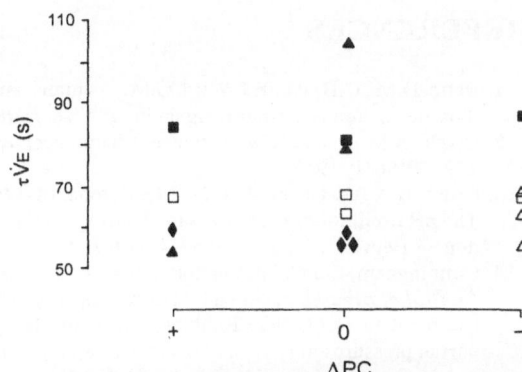

Figure 4. Response of $\tau\dot{V}_E$ expressed as a function of the change in PC occurring between "0" W and 80 W (ΔPC) for all five subjects (+, - and 0 represent "improved", "pessimal" and an unchanged PC characteristic, respectively).

that was opposite to that evoking respiratory stimulation. For humans in the steady state of moderate exercise, Petersen *et al.* (8) could discern no significant contribution to $\dot{V}_E$ from phase coupling: using slugs of N_2 to repeatedly estimate LET and therefore the phase-coupling, $\dot{V}_E$ was not systematically different for breaths in which the N_2 slug was assumed to arrive at the CBs during inspiration compared to expiration.

Present Findings

Despite the important role played by the CBs in establishing the dynamics of the exercise hyperpnoea (reviewed in Refs 10 & 11), we could find no support for phase coupling being the means of mediation. That is, in some instances $\tau\dot{V}_E$ was not different despite appreciable differences in the phase-coupling characteristic; in other instances, significant differences in $\tau\dot{V}_E$ were evident despite no difference in the PC characteristic (Fig. 4).

Timing Technique. Naturally, the validity of our conclusion depends on the validity of our timing estimates. The technique previously used by Petersen *et al.* (8) - i.e., the imposition of discrete inspiratory "slugs" of 100% N_2 - was not appropriate for our purposes as it would not allow the timing to be discerned on a breath-to-breath basis. Rather, we would be constrained to one or possibly two estimates that could reasonably be implemented during the entire nonsteady-state phase. Furthermore, the consequent transient hypoxaemia might alter the $\dot{V}_E$ time course under investigation. These problems were circumvented by our use of high-gain ear oximetry, which was sufficiently reliable to allow identification of the nadir of the intra-breath SaO_2 oscillation (assumed to coincide with that of the corresponding pHa oscillation).

Stimulus Characteristic. Our choice of the nadir of the SaO_2 oscillation as the index with which to define the phase coupling characteristic, rather than $d(SaO_2)/dt\downarrow$ (ie. equivalent to $dpH/dt\downarrow$: Ref 3), was made on the basis of practical expediency: the slope of the falling phase of the SaO_2 oscillation could not be defined consistently with adequate resolution.

Conclusion

It seems, therefore, that any CO_2-linked drive to ventilation during the nonsteady-state of exercise that has been suggested, based upon the close temporal co-relationship between ventilation and $\dot{V}CO_2$ (9,11), is unlikely to involve respiratory phase-coupling of intra-breath humoral oscillations in its mediation.

REFERENCES

1. Band, D.M., C.B. Wolff, J. Ward, G.M. Cochrane, and J.G. Prior. Respiratory oscillations in arterial carbon dioxide tension as a control signal in exercise. *Nature* 283: 84-85, 1980.
2. Black, A.M.S., and R.W. Torrance. Chemoreceptor effects in the respiratory cycle. *J. Physiol. (Lond.)* 189: 59P-61P, 1967.
3. Cross, B.A., A. Davey, A. Guz, P.G. Katona, M. Maclean, K. Murphy, S.J.G. Semple, and R.P. Stidwell. The pH oscillations in arterial blood during exercise; a potential signal for the ventilatory response in the dog. *J. Physiol. (Lond.)* 329: 57-73, 1982.
4. Cunningham, D.J.C. Introductory remarks. Some problems in respiratory physiology. In: *Modelling and Control of Breathing*, edited by B.J. Whipp and D.M. Wiberg. New York: Elsevier, 1983, p. 3-19.
5. Cunningham D.J.C., P.A. Robbins, C.B. Wolff. Integration of respiratory responses to changes in alveolar partial pressures of CO_2 and O_2 and in arterial pH. In: *Handbook of Physiology: The Respiratory System, vol. II, Control of Breathing, pt. 2*, edited by N.S. Cherniack and J.G. Widdicombe. Bethesda: Am. Physiol. Soc., 1986, p. 475-528.
6. Eldridge, F.L. and T.G. Waldrop. Neural control of breathing. In: *Pulmonary Physiology and Pathophysiology of Exercise*, edited by B.J. Whipp and K. Wasserman. New York: Dekker, 1991, p. 309-370.
7. Jain, S.K., S. Subramanian, D.B. Julka, and A. Guz. Search for evidence of lung chemoreflexes in man: study of respiratory and circulatory effects of phenyldiguanide and lobeline. *Clin. Sci.* 42: 163-177, 1972.
8. Petersen, E.S., B.J. Whipp, D.B. Drysdale, and D.J.C. Cunningham. Testing a model. In: *Control of Respiration During Sleep and Anesthesia*, edited by R. Fitzgerald, S. Lahiri, and H. Gautier. New York: Plenum, 1978, p. 335-342.
9. Whipp, B.J. The control of exercise hyperpnea. In: *The Regulation of Breathing*, edited by T. Hornbein. New York: Dekker, 1981, p. 1069-1139.
10. Whipp, B.J. Peripheral chemoreceptor control of the exercise hyperpnea in humans. *Med. Sci. Sports Ex.* 26: 337-347, 1994.
11. Whipp, B.J., and S.A. Ward. The coupling of ventilation to pulmonary gas exchange during exercise. In: *Pulmonary Physiology and Pathophysiology of Exercise*, edited by B.J. Whipp and K. Wasserman. New York: Dekker, New York, p. 271-307, 1991.
12. Yamamoto, W.S. Mathematical analysis of the time course of alveolar CO_2. *J. Appl. Physiol.* 15: 215-219, 1960.

42

OPTIMIZATION OF RESPIRATORY PATTERN DURING EXERCISE

G. Benchetrit,[1] T. Pham Dinh,[2] and J. Viret[3]

[1] Département de Physiologie (PRETA)
[2] Département de Mathématique
   Université de Grenoble
[3] Centre de Recherche des Services de Santé des Armées
   Grenoble, France

INTRODUCTION

In a previous study (3), we analyzed airflow profiles from 11 healthy subjects, at rest and during steady state exercise (at 50% of their maximal O2 consumption) both at sea level and simulated altitude (4,500 m in a hypobaric chamber). The mean values of ventilation were: 7.97 l/min (rest normoxia), 9.24 l/min (rest hypooxia), 31.57 l/min (exercise normoxia) and 43.91 l/min (exercise hypoxia). Breath-by-breath analysis of airflow profiles were performed in all 4 conditions and statistical tests were used to compare the within-individual to between-individual variations. The findings of this study were (i) hypoxia does not significantly change the individual flow pattern at rest, (ii) there existed a diversity in flow patterns among individuals during exercise and (iii) hypoxia does not significantly change the individual flow pattern during exercise.

Several studies (2, 7, 8, 4, 5), considering various optimization criteria have shown that a rectangular pattern of airflow during exercise was an optimal pattern. Our results show also a more rectangular airflow shape during exercise, with however persistence of diversity of the pattern among individuals. This suggested that, one possibility would be that the optimization of breathing pattern should take into account the <u>characteristics of the modifications</u> of the shape from rest to exercise, rather than the characteristics of the shape itself. Furthermore, these characteristics of modifications would be common to all subjects.

Model 1

The envisaged model could be:

Exercise profile = *Basic profile* + α (Resting profile - *Basic profile*)

with *Basic profile* and α being common to all individuals, exercise and resting profile varying between individuals.

Modeling and Control of Ventilation, Edited by S. J. G. Semple, L. Adams, and B. J. Whipp
Plenum Press, New York, 1995

Flow profile were quantified by harmonic analysis (1), the fundamental and the first 3 harmonics are enough to quantify the shape of the airflow profile ; they contain more than 95% of the power of the original signal. Thus, for each breath this analysis provides four amplitude and four phase angles or four vectors (8 cartesian coordinates) which describe the shape of the airflow of each breath. In addition, this quantification was normalized for differences in breath duration and amplitude, so that the analysis concerned solely the shape.

The model hypotheses are:

- Basic profile: a rectangular profile was reported by all previous studies

  In our study, the sum of the differences between exercise airflow profiles is much less than the sum of the differences between resting profiles. Although there persists a diversity among individuals, there is a tendency for greater similarity between exercise profiles than between resting profiles;
- linear transformation
- Harmonic component-by-component transformation

Lets us denote by H_k^e, $k = 1,...,4$, the harmonic components of the signal "exercise" (represented as *complex* numbers)

H_k^r, $k = 1,...,4$, the harmonic component of the signal "at rest"

Our hypotheses relate H_k^e to H_k^r (for each k) according to the formula

$$H_k^e, k = H_k^b + \alpha_k(H_k^r - H_k^b)$$

Here H_k^b representing the "basic" shape common to all "exercise" signals, α_k is a coefficient also common to all signals: This is a *complex* number and the multiplication of $(H_k^r - H_k^b)$ with it is the complex multiplication (equiv. to scaling and a rotation).

We estimated the H_k^b and α_k by least squares fitting based on the signals recorded at sea level at rest and at sea level during exercise :

$$\text{Minimize} \sum_{i=1}^{N} \sum_{k=1}^{4} | H_{k,i}^e - H_{k,i}^b - \alpha_i(H_{k,i}^r - H_{k,i}^b) |^2.$$

The index i refers to these signals associated with each individual in the study (there are N = 11 individual).

We found:

$H_1^b = -0.952, -0.145,$	$\alpha_1 = 0.0492$ (scaling) $- 7.1°$ (rotation)
$H_2^b = -0.140, -0.030,$	$\alpha_2 = 0.2079$ (scaling) $- 47.9°$ (rotation)
$H_3^b = -0.209, 0.058,$	$\alpha_3 = 0.4660$ (scaling) $- 25.3°$ (rotation)
$H_4^b = -0.094, -0.209,$	$\alpha_4 = 0.4331$ (scaling) $- 18.1°$ (rotation)

We also computed the quantities

$$\sqrt{\sum_{k=1}^{4} | H_{k,i}^e - H_{k,i}^b - \alpha_i(H_{k,i}^r - H_{k,i}^b) |^2} .$$

These represent the relative errors in the reconstruction formula: the signals have been normalized so that

$$\sum_{k=1}^{4} |H_k^e|^2 = 1.$$

We calculated the basic profile and α from the set of data recorded at rest and during exercise at sea level. We then applied the model to the data recorded at rest during hypoxia to calculate the exercise airflow profiles. We then compared these theoretical profiles to those obtained experimentally i.e. exercise in hypoxia. The mean differences between theoretical and experimental flow profiles were 15% (ranging from 6 to 25%).

This model implies that there exists one resting pattern and one exercise pattern for an individual. The more the resting pattern is close to the basic shape, the less there will be changes from rest to exercise.

MODEL 2

A topological formalization issued from catastrophy theory (6, 9) may illustrate qualitatively the above hypothesis. Such formalization enables multivariate description of various phenomenological changes in one single representation. We chose the simplest cusp catastrophy which, in a first attempt, is convenient in this study. In figure 1 the axes represent (i) the individual diversity of the flow patterns, (ii) an exercise intensity variable - these two

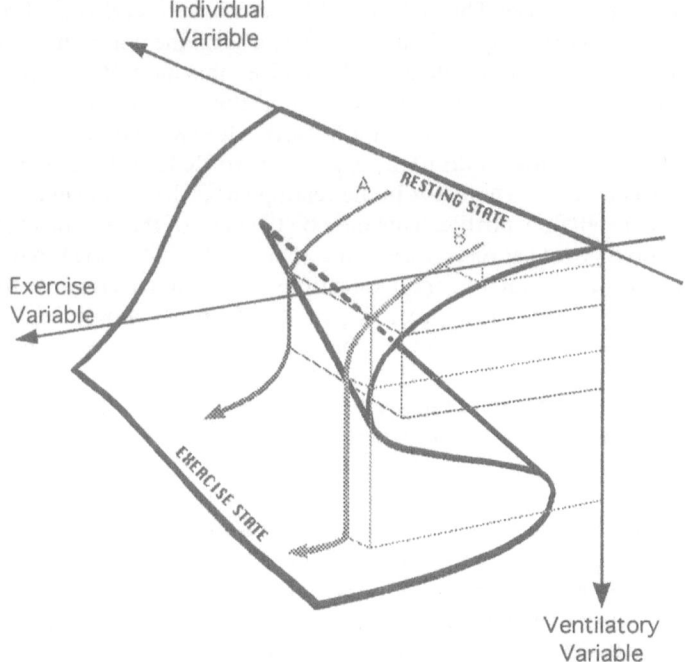

Figure 1. Flow profiles at rest and during exercise in a topological (cusp) formalization.

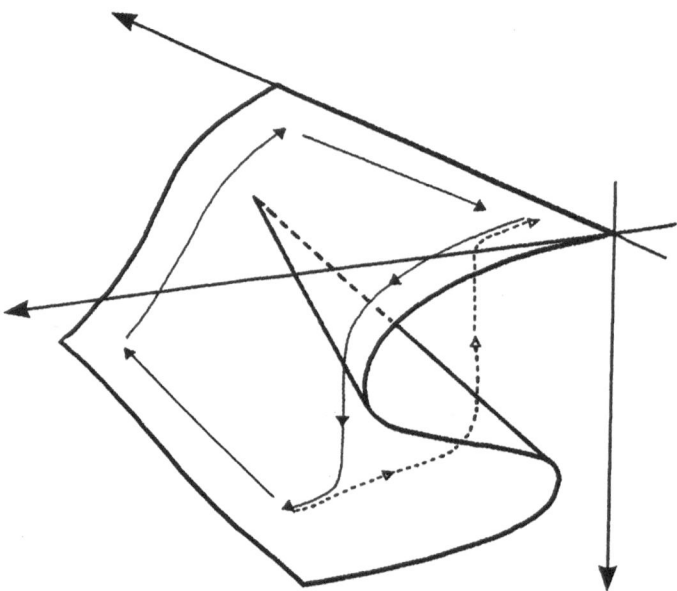

Figure 2. Recovery trajectories of the airflow profile from exercise to rest.

variables define the control plan - and (iii) a ventilatory variable which is the state variable. Thus, the upper sheet of the cusp is the resting state and the lower sheet the exercise state.

According to the model, individuals airflow profile will follow a trajectory when the level of exercise and ventilation increase. The state variable (ventilatory variable) should have the following characteristics : slight changes while the trajectory is on the upper sheet, then a sudden change (i.e. a drop) at some value of the intensity of exercise - this intensity being different from one individual to another- and again slight changes when the trajectory is on the lower sheet.

A and B on figure 1 are two examples of trajectories for two individuals. It can be seen that there are less changes (i.e. smaller drop) in the airflow profile for subject A than for subject B, and for a subject placed behind subject A in the resting sheet there may even be no change from rest to exercise in the airflow profile. This may be the case of those subjects who already have a rectangular airflow profile at rest, close to the basic profile defined above.

Following this formalization two main questions arise. First, as regards the individual variable, is there optimal resting flow profiles per se? In other words, do those subject already having an airflow profile close to the basic profile have an optimal flow profile because there will be little change in their flow profile during exercise? Or on the contrary, could the fact that there will be no changes from rest to exercise be detrimental. This individual variable axis may then be considered as a "fitness" axis with the increasing direction to be determined.

The second question pertains to the recovery. Figure 2 shows two possible recovery trajectories. This raises the question of ventilatory recovery from exercise. Does the flow pattern have a trajectory with an hysteresis as is represented by the dashed line, or does it follows the solid line with progressive changes in flow profile? Which trajectory is optimal? A more developed model with a time dimension would then be necessary to answer this question. However, additional exprimental data would be needed before envisaging more complicate models.

REFERENCES

1. Bachy, J. P., A. Eberhard , P. Baconnier, and G. Benchetrit . A programm for cycle-by-cycle shape analysis of biological rhythms. Application to respiratory rythm. *Comput Meth. Prog. in Biomed.* 23: 297 - 307. 1986.
2. Bretschger H. J. Die Geschwindigkeitskurve der menschlichen Atemluft (Pneumotachogramm). *Pflügers Arch.. f. d. Ges. Physiol.* 210: 134 - 148. 1925.
3. Eisele J H, B. Wuyam, G Savourey, J. Eterradossi, JH Bittel JH and Benchetrit G. The individuality of breathing patterns during hypoxia and exercise. *J. Appl. Physiol.* 72 : 2446-2453, 1992.
4. Hämäläinen, R. P. Optimisation of respiratory airflow. In: *Modeling and control of breathing,* edited by B.J. Wipp and D.M. Wiberg. New York; Elveiser, 1983, p. 181-188.
5. Lafortuna, c. l., A. E. Minetti, and P. Mognoni. Inspiratory flow patterns in humans.*J. Appl. Physiol.* 57: 1111 - 1119, 1984.
6. Thom, R. Stabilité structurelle et morphogénèse. Benjamin, New York, 1972.
7. Yamashiro, S. M., and F. S. Grodins. Optimal regulation of respiratory airflow. *J. Appl Physiol.* 30: 597 - 602, 1971.
8. Yamashiro, S. M., and F. S. Grodins. Respiratory cycle optimisation in exercise.*J. Appl. Physiol.* 35: 522 - 525, 1973.
9. Zeeman, EC. Catastrophe Theory. *Proc. Roy. INSTN,* 49 : 77-92, 1976

43

BREATHING IN EXERCISING QUADRUPEDS

There Ain't No Such Thing as a Free Breath!

D. M. Ainsworth,[1] C. A. Smith,[2] K. S. Henderson,[2] and J. A. Dempsey[2]

[1] Department Clinical Sciences
Cornell University
Ithaca, New York 14850
[2] Department Preventive Medicine
University of Wisconsin
Madison, Wisconsin 53705

INTRODUCTION

Recently, Bramble and Jenkins (1993) advanced a novel mechanism for pulmonary ventilation in exercising dogs. They proposed that locomotory-driven oscillations of the diaphragm, rather than active diaphragmatic contractions *per se*, were responsible for airflow generation during high frequency breathing in exercising dogs. They based this tenet on (1) the observation of strict locomotory:respiratory coupling ratios in dogs exercising on the treadmill (2:1) as well as upon (2) cineradiographic analysis of diaphragmatic silhouettes in exercising dogs. Using kinematic profiles - in which the rate of diaphragmatic displacements were calculated as a function of locomotion - they concluded that the function of the primary muscle of inspiration, the diaphragm, was to modulate locomotory-induced movements of the viscera, rather than to actively produce the intrathoracic pressure changes required for ventilation. Unfortunately, conclusions of diaphragmatic physiological activity - contracting or relaxing - were based upon cineradiography and lacked confirmatory electromyographic (EMG) or sonomicrometry measurements.

Previous work in our laboratory demonstrated that inspiratory and expiratory muscles contribute actively to the hyperpnea of exercise in dogs[1]. However, the tightly-fitting breathing mask worn by these dogs may have constrained their breathing frequencies to less than one Hz and may thereby have altered the respiratory muscle recruitment patterns. Thus, one objective of this study was to determine if airflow generated by unencumbered stationary or exercising dogs breathing at high frequencies was dependent upon phasic inspiratory and expiratory muscle activity. A second objective was to determine the contribution of locomotion, i.e. footplant activity, upon airflow generation in exercising dogs.

Modeling and Control of Ventilation, Edited by S. J. G. Semple, L. Adams, and B. J. Whipp
Plenum Press, New York, 1995

231

MATERIALS AND METHODS

Five chronically instrumented dogs with bipolar electromyographic (EMG) electrodes implanted in the costal (CS) and crural (CR) diaphragm and in the transverse abdominal (TA) muscles were examined. EMG signals were processed through a fourth-order Butterworth filter, with a band pass at 50-1000 Hz. The dogs were studied at rest and during moderate treadmill exercise speeds ranging from 3.0 km/hr (walking) to 6.4 km/hr 10% grade (trotting). During these trials, breathing patterns were not constrained by the use of a face mask. Breathing frequency and transpulmonary pressure changes (PTP) were measured by an esophageal balloon catheter, filled with 1 ml of air and attached to a pressure transducer (Validyne model MP45, range ± 50 cm H_2O). Hindlimb impact and thus stride frequency, were assessed by an accelerometer affixed to the left hind leg. In two additional dogs, surgically-implanted sonomicrometry transducers enabled measurements of both phasic and tonic shortening of the crural and costal diaphragm at rest and during exercise.

RESULTS

During our analysis, two distinct types of breathing patterns emerged which we classified as Type A and Type B. Type A, characterized by pure high frequency (2-6 Hz) oscillations in the transpulmonary pressure, was found in most of the dogs at rest and in all of the dogs during some portion of the exercise trials. The maximum change in transpulmonary pressure (peak positive to peak negative) ranged from 3-10 cm H_2O but was variable both within dogs as well as between dogs. In this breathing pattern, decrements in transpulmonary pressure were always associated with electrical activation of the costal and crural diaphragm, whether the dog was standing quietly on the treadmill or was trotting briskly at 6.4 km/hr. Similarly, increases in the transpulmonary pressure in the standing dog (expiration) were associated with phasic activation of the transversus abdominis muscle. However, during exercise the activation pattern of the TA was more variable: in two of the dogs, transversus abdominis activity coincided with expiration (i.e. increases in PTP), but in the

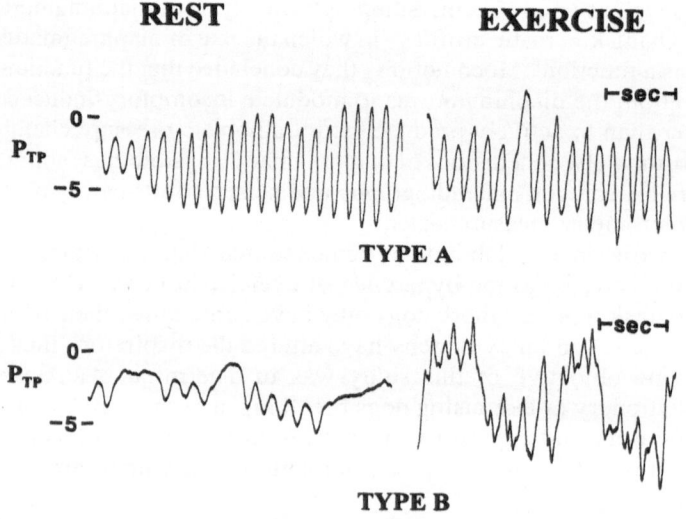

Figure 1. Transpulmonary pressure changes (PTP) which characterize type A and type B breathing patterns in resting or exercising dogs. (See text for description of activation patterns for the respiratory muscles.)

remaining three dogs, TA activity was correlated with hind foot impact. When dogs utilized the type A breathing pattern, strict coupling between breathing frequency and stride frequency was never observed. For example, when dogs trotted up a 10% grade, respiratory:locomotory ratios ranged from 1.2 to 3.2. Furthermore, an obvious mechanical effect of footplant on transpulmonary pressure (and hence airflow) generation was not detected during this type of breathing.

Type B breathing pattern was a mixed pattern characterized by high frequency oscillations in the transpulmonary pressure ranging from 4-6 Hz, superimposed upon a slower inherent breathing rate, ranging from 0.5-1 Hz. This type of pattern was found in all dogs during quiet breathing as well as in all dogs during some phase of the exercise trials. The maximum change in transpulmonary pressure ranged from 4-16 cm H_2O, and although these pressure swings were variable both within and between dogs, the transpulmonary pressure excursions tended to increase with increasing exercise intensity. During the inspiratory phase of the slower breathing pattern, decrements in the oscillating transpulmonary pressures were always associated with phasic activation of the costal and crural diaphragm. Similar such oscillations in the PTP were absent during the expiration in the resting dog. Furthermore, the gradual increase in the transpulmonary pressure during this time was always accompanied by phasic activation of the transversus abdominis muscle. When the dog exercised and utilized a type B pattern, high frequency oscillations were noted to occur during both the slower inspiratory phase as well as during the slower expiratory period. When they occurred during inspiration, decrements in the oscillating PTP were again associated with phasic activation of the costal and crural diaphragm. When they occurred during expiration (1) the frequencies of these oscillations in transpulmonary pressure ranged from 2.6 Hz in walking dogs to 3.6 Hz during the trotting gaits and (2) were not related to phasic activation of the diaphragm but rather correlated best with footfall. Thus, in type B pattern of breathing, transversus abdominis activity exhibited both a locomotory-related activation pattern, coinciding with footfall, as well as an expiratory-related activation pattern, with EMG activity further increasing during the slower expiratory period of the breath. The footplant associated alterations in transpulmonary pressure averaged 2.3 for walking trials and 3.2 cm H_2O for the trotting trials. Based upon estimates of dynamic pulmonary compliance[3] for dogs of 0.097 L/cm H_2O, between 0.2 and 0.3 liters of the total expired tidal volume may have been generated from locomotion. Undoubtedly, positive footplant related alterations in the transpulmonary pressure during the slower inspiratory phase were apposed by diaphragm-induced decrements in P_{TP}, making it difficult to accurately assess the impact of locomotion on airflow generation during inspiration in type B breathing patterns.

Sonomicrometry measurements of the costal or of the crural diaphragm in two dogs resting or exercising allowed assessment of electrical-mechanical events. In these two dogs which exhibited a type A breathing pattern (breathing frequencies ranged from 1.5 to 3 Hz), phasic shortening of either the costal or crural diaphragm occurred synchronous with EMG activation and approximated 3% of the resting diaphragmatic length. During high frequency breathing, tonic (end-expiratory diaphragmatic) length also decreased (shortened) suggesting that end-expiratory lung volume had increased. More importantly, phasic changes in the costal or crural muscle did not occur in the absence of phasic EMG activity, suggesting that locomotory-induced oscillations of the diaphragm are insignificant.

DISCUSSION

Unencumbered stationary or exercising dogs use at least two distinct types of breathing patterns to actively generate transpulmonary pressures and effect airflow into and out of the lungs. The type A pattern is characterized by a pure high frequency oscillation in the transpulmonary pressure whereas type B pattern is a mixed pattern consisting of a slower inherent breathing rate with superimposed hi frequency oscillations in the transpulmonary pressure.

These present studies, as well as earlier ones from our laboratory, confirm the hypothesis that breathing in quadrupeds is a neurally-dependent event. These conclusions are based upon the finding (1) that decrements in transpulmonary pressure are correlated one for one with phasic activation of the costal and crural diaphragm; (2) that electrical activation of the costal and crural diaphragm is clearly linked with phasic shortening of that specific muscle segment and (3) that breathing frequency is not strictly linked with locomotion during exercise. While footplant-related changes in the transpulmonary pressure during type B breathing may contribute to expired tidal volume excursions ranging from 0.2 -0.3 liters, a similar mechanical effect of footplant on ventilation could not be demonstrated when dogs utilized type A breathing patterns. Furthermore, the actual effect of footplant on inspired tidal volume generation during type B breathing was difficult to assess as locomotory-asso-ciated alterations in intrathoracic pressures were undoubtedly offset by diaphragmatic contractions. In summary, whether the dog is at rest or during exercise or whether it elects a slow or a fast breathing frequency, all breaths require active contributions by respiratory muscles: There are no free lunches when it comes to breathing!

ACKNOWLEDGMENTS

This work was supported by a grant from the National Heart Lung Blood Institute.

REFERENCES

1. Ainsworth, D.M., C.A. Smith, S.W. Eicker, K.S. Henderson, and J.A. Dempsey. The effects of locomotion on respiratory muscle activity in the awake dog. *Resp Physiol* 78: 145-162, 1989.
2. Bramble, D.M., and F.A. Jenkins Jr. Mammalian locomotor-respiratory integration: implications for diaphragmatic and pulmonary design. *Science* 262: 235-240, 1993.
3. Gillespie, D.J., and R.E. Hyatt. Respiratory mechanics in the unanesthetized dog. *J Appl Physiol* 36: 98-102, 1974.

44

DYNAMIC CHEMORECEPTIVENESS STUDIED IN MAN DURING MODERATE EXERCISE BREATH BY BREATH

A. K. Datta and Annabel Nickol

Neuroscience Critical Care Unit, Box 167
Addenbrooke's Hospital
Hills Road, Cambridge, CB2 2QQ, United Kingdom

INTRODUCTION

During exercise ventilation matches the metabolic production of carbon dioxide such that the mean arterial level of carbon dioxide remains constant. One hypothesis for the mechanism of exercise hyperpnoea is that there is a dynamic sensitivity to carbon dioxide, i.e. that ventilation responds to the rate of change to carbon dioxide. The aim of this study was to test this hypothesis directly in man during steady moderate exercise by administering the same pulse of inhaled carbon dioxide either early or late in inspiration.

METHODS

Nine fit male subjects (age 18-30) unaware of the experiment's aims were studied. They cycled on an electromagnetically braked bicycle ergometer (Instrumentum Lode) at a steady sub-anaerobic workload. Mellifluous opera (Mozart's 'Cosi fan Tutti') played through headphones masked external noise. Subjects breathed into a low resistance, non-re-breathing valve (Hans Rudolph) where tidal PCO_2 was measured with infra red (Beckman LB2). Expired air was collected in a series of Douglas bags every 3 minutes to determine expiratory minute volume ($\dot{V}_E$). Unknown to subjects, the inspirate could be changed for a 6 minute epoch from air to 5%CO_2/95% air at 0-300 msec ("test gas" = CO_2 Early) or 300-600 msec ("test gas" = CO_2 Late) after the onset of inspiration, using apparatus constructed for the purpose (1). To provide appropriate control epochs (AirEarly, AirLate), air was delivered throughout with the apparatus left operational. Each epoch was repeated two to four times and the order randomized. An opaque screen shielded subjects' view of apparatus and experimenters. In eight subjects, the PCO_2 in arterialized venous blood, sampled from a vein in the dorsum of the hand, was measured($P_{AV}CO_2$) every three minutes; for this purpose the subjects' left forearm was kept in a heated waterbath at 44°C. The ventilatory response to the early or late CO_2 pulse, was determined using both $P_{ET}CO_2$ and $P_{AV}CO_2$. $\propto$ = Arc tan

Modeling and Control of Ventilation, Edited by S. J. G. Semple, L. Adams, and B. J. Whipp
Plenum Press, New York, 1995

Table 1. Median(standard deviation) of group results.(p) = probability of null hypothesis in paired
t-test. There is a significant respiratory phase dependency to the ventilatory response to inhaled
CO_2 during steady moderate non-hypoxic exercise in normal human subjects

	Air early	CO_2 early	(p)	Air late	CO_2 late	(p)
V_E (l/min)	31.8 (3.0)	39.1 (5.1)	.0001	31.6 (3.2)	32.5 (3.2)	.594
$P_{ET}CO_2$ (mmHg)	43.1 (4.4)	45.1 (5.1)	.001	44.9 (4.7)	45.7 (4.9)	.002
$P_{AV}CO_2$ (mmHg)	41.3 (1.4)	42.9 (1.7)	.001	40.7 (1.7)	42.1 (1.9)	.003
$\propto_{AV}(°)$	74.4 (7.7)			-16.5 (55.2)		.010
$\propto_{ET}(°)$	70.1 (13.6)			-16.7 (50.1)		.007

$((\dot{V}_E[CO_2]-\dot{V}_E[Air])/(PCO_2[CO_2]-PCO_2[Air]))$. A negative value of $\propto$ represents ventilatory
depression by CO_2. In four subjects a CUSUM of $P_{ET}CO_2$, V_T, T_{TOT} and T_E was constructed
for a three minute period when "test gas" was air and then changed for a three minute period
when "test gas" was CO_2Early or CO_2Late. Mean baseline level for each CUSUM is taken
from a twenty breath period when subjects breathed air throughout (i.e. "test gas" = AirEarly
or AirLate); for any given breath number the CUSUM represents the cumulative sum of the
difference between the value of any variable minus mean baseline value. CUSUM time zero
is the breath at which changeover of "test" gas occurred.

RESULTS

The group results for ventilation, $P_{ET}CO_2$ and $P_{AV}CO_2$.are given in the table. A
comparable increase in median $P_{ET}CO_2$ and $P_{AV}CO_2$ was achieved when inhaled carbon
dioxide was given either early or late in inspiration. Ventilation increased by some 7.3 l/min
when carbon dioxide was administered early in inspiration. In contrast when carbon dioxide

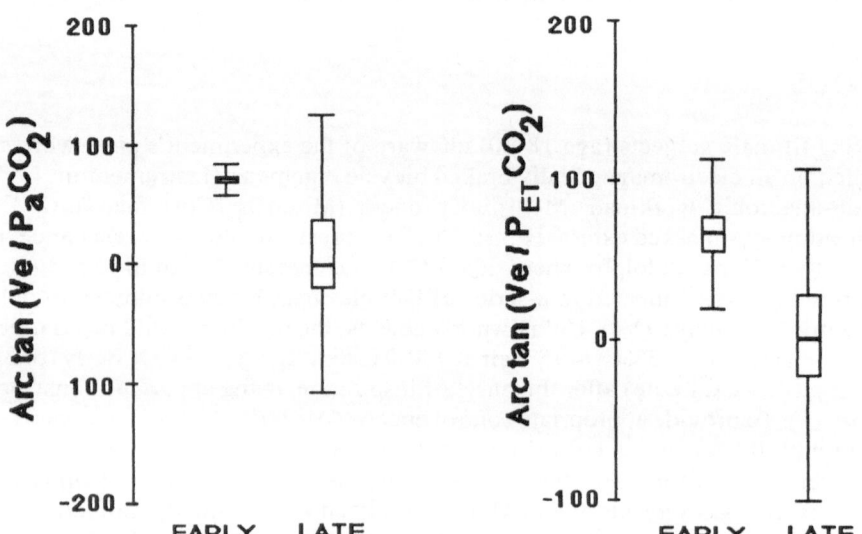

Figure 1. Box and Whisker plots for ventilatory response to CO_2, alpha, for early and late pulses of inhaled CO_2.
The central line represents the median, the box ±1SD and the bars represent the range of values.

was administered late in inspiration, median ventilation rose by only 0.9 l/min; this rise was not statistically significant.

The calculated ventilatory response to carbon dioxide was much greater when CO_2 was given early in inspiration than when it was given late in inspiration. Negative alpha values for late inspiratory CO_2 pulses indicate ventilatory depression by the late pulse of carbon dioxide. The results are displayed graphically in Figure 1.

CUSUMs showed that the lag for the V_T response to inhaled CO_2 pulses was about two to three breaths, consistent with the response being mediated by peripheral chemoreceptors. $P_{ET}CO_2$ CUSUMs showed a similar profile for CO_2Early and CO_2Late. In general CUSUMs were linear i.e. the response was reproducible for each breath during the whole of the epoch studied. A greater V_T response was seen for CO_2Early than for CO_2Late in all subjects confirming a significant dynamic chemoreceptiveness. A typical example is shown in Figure2.

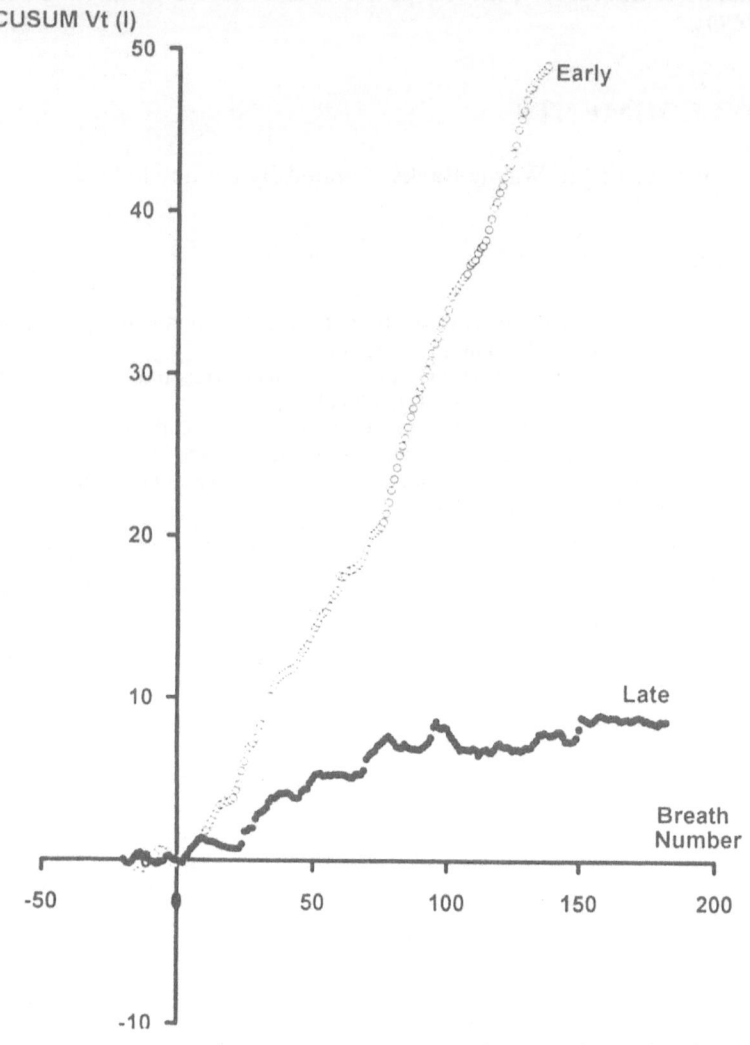

Figure 2. CUSUM of tidal volume during moderate exercise in a single subject during the administration of CO_2 Early and Late in inspiration.

In two subjects a reduction in V_T was seen after CO_2 Late - an extreme example of dynamic chemoreceptiveness, confirming the group results that inhaled pulses of CO_2 depress ventilation.

DISCUSSION

The results of this study indicate that there is a respiratory phase dependent ventilatory response to inhaled CO_2 during moderate aerobic exercise in man; a greater ventilatory response is seen when the same pulse of CO_2 is given early in inspiration than when it is given late.The results are analogous to those seen in the anaesthetised cat (2). These results have been ascribed in the cat to a dynamic sensitivity of carotid body chemoreceptors (3,4). The results of the current study suggest that a similar mechanism exists in conscious man.during exercise. The finding of respiratory depression by late pulses of CO_2 suggest that this dynamic chemoreceptiveness is powerful and may be greater than the steady state response to CO_2.

ACKNOWLEDGMENTS

The assistance of Mr. Wayne Banks is gratefully acknowledged

REFERENCES

1. Datta A.K., Nickol Annabel and Band D.M. (1994) New technique for studying dynamic chemoreceptiveness in man breath by breath. J.Physiol., 475, 10P.
2. Band D,M., Cameron I.R. and Semple S.J.G. (1970) The effect on respiration of abrupt changes in carotid artery pH and PCO_2 in the cat. J.Physiol., 211, 479-494.
3. Band D.M., McClelland M., Phillips D.L., Saunders K.B. and Wolff C.B. (1978) Sensitivity of the carotid body to within-breath changes in arterial PCO_2. J.Appl. Physiol., 45, 768-777
4. Black A.M.S. and Torrance R.W. (1971) Respiratory oscillations in chemoreceptor discharge in the control of breathing. Respir. Physiol., 13, 221-237.

45

CO$_2$ RETENTION DURING EXERCISE

A Role for the Carotid Chemoreceptors?

J. H. Howell and B. A. Cross

University College London
Gower St., London WC1E 6BT
United Kingdom

INTRODUCTION

It is generally accepted that during moderate intensity exercise arterial partial pressure of CO_2 (PaCO$_2$) is maintained at close to normal resting levels in man. Dempsey et al. (1) reported that although the mean change in PaCO$_2$ on transition from rest to exercise within a group of subjects was small, there was considerable inter-individual variation: some subjects became hypercapnic on exercise while others became hypocapnic.

We have recently been studying the cardiorespiratory responses to moderate intensity exercise among a group of sportsmen, including a number of oarsmen and have noticed that a number of the sportsmen (eg. 9 of the 15 oarsmen tested) allowed their end-tidal partial pressure of CO_2 (PETCO$_2$) to rise on exercise to a level in excess of that which can be accounted for simply in terms of the increase in the slope of the plateau phase of the expired CO_2 partial pressure profile (unpublished observations, see Figure 1).

There is evidence to suggest that the ventilatory response to moderate intensity exercise is smaller in patients who have undergone bilateral carotid body resection than in normal subjects matched for pulmonary function (2, 3). One consequence of this blunted ventilatory response to exercise is that PaCO$_2$ rises on exercise.

We hypothesised that the CO_2 retention seen in some of the sportsmen may be a manifestation of a reduced carotid body sensitivity and therefore would be associated with a reduced ventilatory response to changes in inspired O$_2$ tension (PIO$_2$).

METHODS

9 male subjects performed the experiment. Their anthropometric data are presented in Table 1. 3 of the subjects were elite oarsmen, 3 were college oarsmen, 2 played rugby football for their college and 1 was generally fit. All the subjects had refrained from eating and drinking for 2 hours prior to the experiment.

Modeling and Control of Ventilation, Edited by S. J. G. Semple, L. Adams, and B. J. Whipp
Plenum Press, New York, 1995

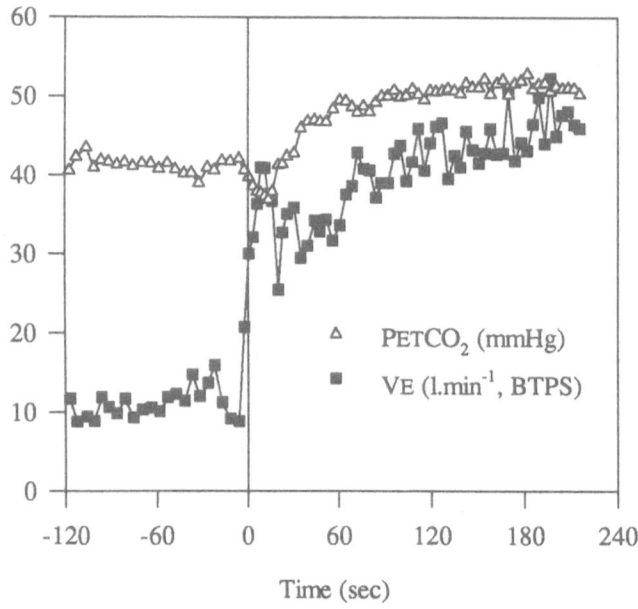

Figure 1. Typical changes in $\dot{V}E$ (BTPS, l.min^{-1}) and PETCO$_2$ (mmHg) for an oarsmen following the onset (Time 0) of 150 W cycle ergometer exercise. Data points are breath-by-breath.

Table 1. Anthropometric data for the subjects used in this study

	Mean ± S.E.
Age (Years)	23 ± 1
Height (m)	1.84 ± 0.02
Weight (Kg)	80.5 ± 1.8

The experimental protocol consisted of 2 tests in a randomised crossover design. The test design is presented in fig. 2.

Each test consisted of 8 min rest followed by 9½ min exercise (125 W @ 75 rpm). The rest and exercise periods both included 2 x 30 sec periods during which the subject was exposed to either a hypoxic (FIO$_2$ = 0.12) or a hyperoxic (FIO$_2$ = 0.50) gas mixture. Test order was randomised between the subjects.

Subjects exercised on a Löde electrically braked cycle ergometer. Throughout the test subjects breathed through a mouthpiece (Hans-Rudolph 2700 series) and wore a noseclip. Expiratory flow was measured using a Fleisch head (No. 3) connected to a Gould pneumotachograph. Respired PO$_2$ and PCO$_2$ were measured using an Airspec 2600 mass

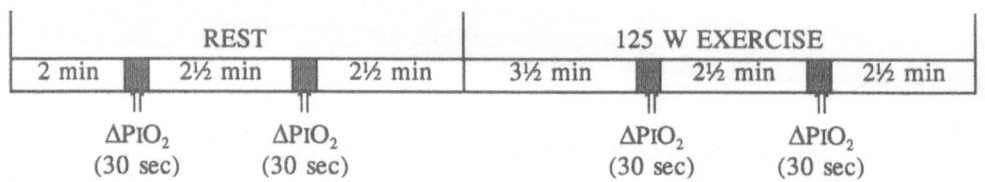

Figure 2. Schematic representation of the protocol design. FIO$_2$ = 0.21 except for shaded portions, when FIO$_2$ = either 0.12 or 0.50.

spectrometer. Minute ventilation ($\dot{V}E$, BTPS), PETCO$_2$ and PETO$_2$ were calculated and saved to disc by a custom-written data acquisition program.

RESULTS

Administration of the hyperoxic gas mixture was associated with a fall in $\dot{V}E$ both at rest and during exercise. The converse was true of administration of the hypoxic gas mixture (see Figure 3).

The % change in $\dot{V}E$ from normoxic levels as a result of administration of the gas mixtures both at rest and during exercise was calculated for each subject. These results are

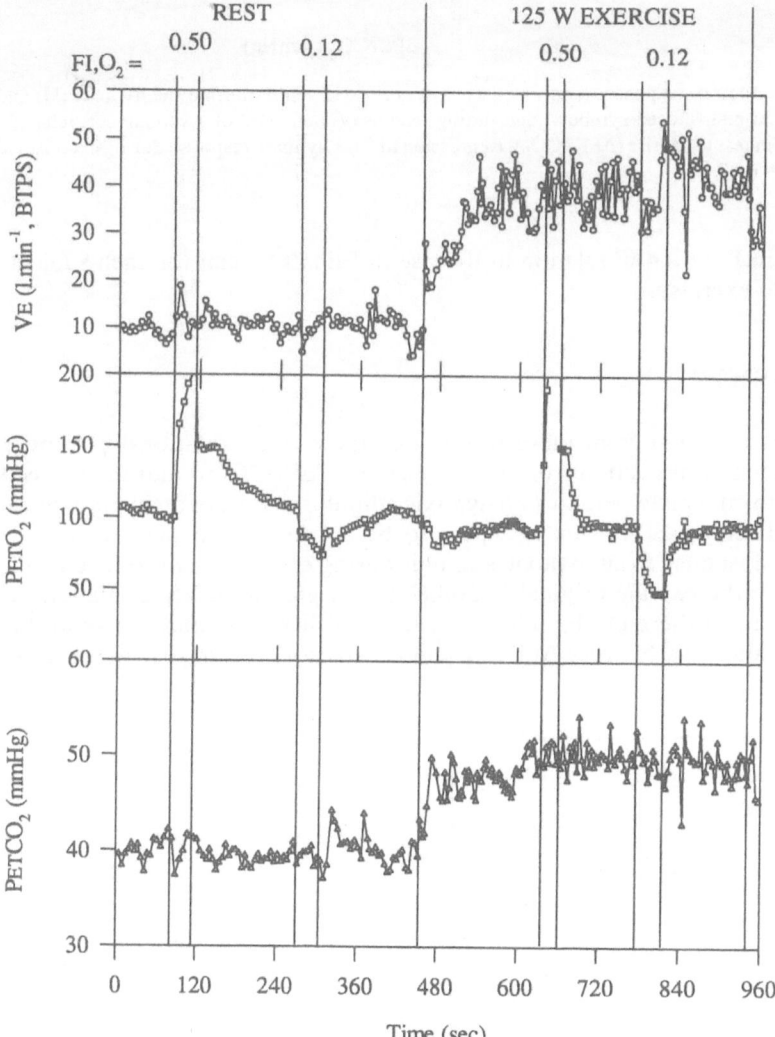

Figure 3. Representative example of breath-by-breath changes in $\dot{V}E$ (top graph, l.min^{-1}), PETO$_2$ (middle graph, mmHg) and PETCO$_2$ (bottom graph, mmHg) throughout a test. Lines denote changes in FIO$_2$ or changes in exercise state.

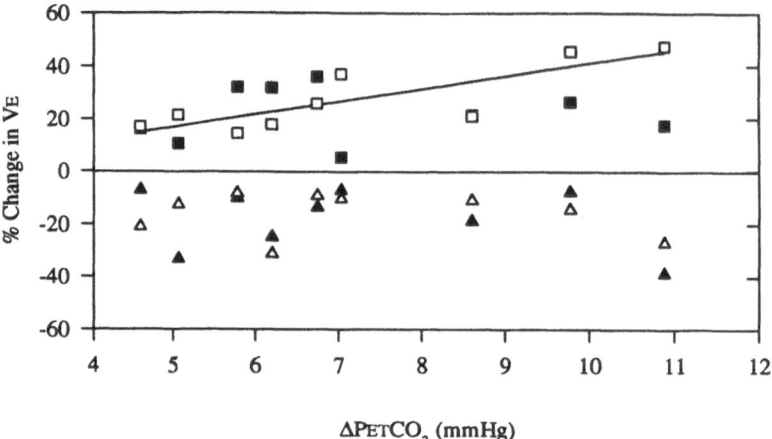

Figure 4. Ventilatory responses (expressed as % of normoxic ventilation) to the hypoxic (■) and hyperoxic (▲) stimuli both at rest (closed symbols) and during exercise (open symbols) vs. subjects' change in $PETCO_2$ on transition from rest to exerise ($\Delta PETCO_2$). Regression line for hypoxic response during exercise (□) vs. $\Delta PETCO_2$ is shown. R = 0.83.

presented in Figure 4 in relation to the rise in $PETCO_2$ seen for each subject on transition from rest to exercise.

DISCUSSION

It is apparent from these results that there is no relationship between the rise in $PETCO_2$ seen on transition from rest to exercise ($\Delta PETCO_2$) and the subjects' ventilatory responses to either a hyperoxic or a hypoxic stimulus at rest, or an hyperoxic stimulus during exercise. There does, however, appear to be a relationship between $\Delta PETCO_2$ and the ventilatory response to an hypoxic stimulus during exercise. Unfortunately, the relationship is positive, whereas the original hypothesis predicted a negative relationship. The initial hypothesis must therefore be rejected, but it still leaves a question as to the cause of the positive relationship between $\Delta PETCO_2$ and ventilatory response to hypoxia during exercise.

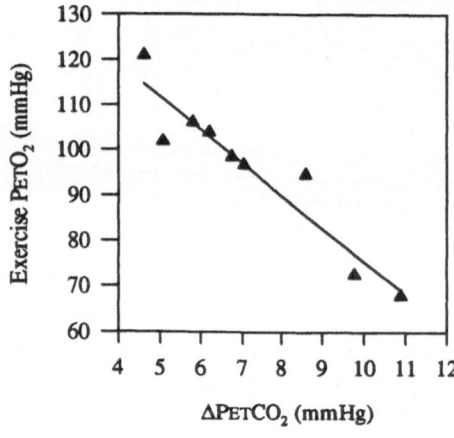

Figure 5. Relationship between the subject's change in $PETCO_2$ seen on transition from rest to exercise and the subject's exercise $PETO_2$. Regression line is shown (R = 0.95).

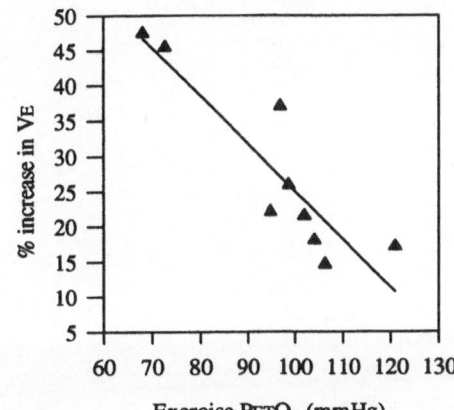

Figure 6. Relationship between the subjects' exercise PETO$_2$ and the % increase in $\dot{V}E$ associated with breathing the hypoxic (FIO$_2$ = 12 %) gas mixture). Regression line is shown (R = 0.90).

A plausible explanation is presented by analysis of the end-tidal O$_2$ concentration during exercise(PETO$_2$) in these subjects (Figure 5).

There is a negative correlation between ΔPETCO$_2$ and PETO$_2$ during exercise, i.e. the baseline from which the hypoxic stimulus is imposed is not consistent between the subjects. The profile of the O$_2$ dissociation curve means that this does not affect the ventilatory response to the hyperoxic stimulus, but will influence the response to the hypoxic stimulus. Plotting the hypoxic ventilatory response during exercise against exercise PETO$_2$ supports this view (Figure 6): there is a stronger correlation between the hypoxic ventilatory response and exercise PETO$_2$ than for the hypoxic ventilatory response during exercise and ΔPETCO$_2$.

CONCLUSIONS

There appears to be a positive correlation between ΔPETCO$_2$ and the ventilatory response to an hypoxic stimulus during exercise. This could be accounted for by the reduced PETO$_2$ associated with CO$_2$ retention on exercise. These results do not support the hypothesis that CO$_2$ retention on exercise is due to a reduced carotid body sensitivity.

JHH was supported by an MRC studentship.

REFERENCES

1. Dempsey, J. D., G. S. Mitchell and C. A. Smith. Exercise and chemoreception. *Am. Rev. Respir. Dis.* 129: Suppl S31-S34, 1984.
2. Honda, Y., S. Myojo, T. Hasegawa and J. Severinghaus. Decreased exercise hyperpnea in patients with carotid body chemoreceptor resection. *J. Appl. Physiol.* 46(5): 908-912, 1979.
3. Honda, Y. Role of carotid chemoreceptors in control of breathing at rest and in exercise: studies on human subjects with bilateral carotid body resection. *Jap. J. Physiol.* 35: 535-544, 1985.

46

HYPOXIC EXERCISE DOES NOT ELICIT LONGTERM MODULATION OF THE NORMOXIC EXERCISE VENTILATORY RESPONSE IN GOATS

Duncan L. Turner,[1] Patricia A. Martin,[2] and Gordon S. Mitchell[2]

[1] Department of Physiology
University of Leeds
Leeds LS2 9NQ, United Kingdom
[2] Department of Comparative Biosciences
University of Wisconsin
Madison, Wisconsin 53706

INTRODUCTION

Ventilation typically increases during exercise to an extent whereby $PaCO_2$ is maintained at or below resting levels (4).Traditionally, attempts to describe the neural mechanisms responsible for the exercise hyperpnoea have invoked a combination of feed-forward and feedback signals (5,8).

Recently, we have proposed that additional mechanisms must be considered that impart a degree of modulation and plasticity to the exercise ventilatory response (6,9). For example, we hypothesize that unusual stimuli such as increased respiratory dead space: 1) reversibly alter the exercise ventilatory response within an exercise trial *via* spinal serotonergic mechanisms (*short term modulation*; 2); and 2) alter future exercise ventilatory responses following repeated exposures (*long term modulation*, LTM; 6). LTM, which changes the degree to which $PaCO_2$ is regulated during exercise, can be elicited by repeated hypercapnic exercise (6) and/or repeated bouts of hypercapnia at rest (7). We speculate that LTM may play an important role in shaping the exercise ventilatory response in the face of changing conditions of acid-base balance, pulmonary mechanics or hormonal status.

The aims of the present studies were: 1) to extend our knowledge of the stimuli that elicit LTM; specifically, we determined if hypoxia (*versus* hypercapnia) elicits LTM and thereby 2) to provide insight concerning the identity of the relevant chemoreceptor groups (ie. peripheral *versus* central chemoreceptors).

Modeling and Control of Ventilation, Edited by S. J. G. Semple, L. Adams, and B. J. Whipp
Plenum Press, New York, 1995

METHODS

Adult female goats were prepared with a translocated carotid artery, thus allowing repeated arterial blood sampling. They were familiarised with running on a treadmill whilst wearing a tight-fitting respiratory mask (6,8). Each goat undertook each of three training protocols: A) 6 bouts of normocapnic, hypoxic exercise in one day (2.5 mph, 5% grade; 5-7 min; PaO_2 = 40 Torr); B) 10 bouts of isocapnic hypoxia at rest (3 min; PaO_2 = 47 Torr) and C) 15 bouts of poikilocapnic, hypoxic exercise over two days (2.5 mph, 5% grade; 5-7 min, PaO_2 = 40 Torr). Isocapnic or normocapnic conditions were maintained in protocols A and B by adding CO_2 to the inspired gas as necessary to maintain $PaCO_2$ at normoxic resting levels (*i.e.* isocapnic), or at the same level of $PaCO_2$ as observed during normoxic exercise (*i.e.* normocapnic). In poikilocapnic exercise trials, $PaCO_2$ was controlled by the goat. After each training protocol, one hour was allowed for the goat to recover while standing quietly on the treadmill. Each series was complemented by an equivalent number of exercise (or resting) trials performed in poikilocapnic, normoxic (sham) conditions. Ventilatory responses to subsequent poikilocapnic, normoxic exercise and the steady-state ventilatory response to increased $PaCO_2$ at rest (*S*) were measured 1-6 hours after the training protocols. Significant differences between experimentally paired, sham and hypoxic protocols were tested with Student's T-tests (p<0.05).

RESULTS

The $PaCO_2$ decrease from rest to exercise was not significantly different 1-6 hrs following hypoxic *versus* sham training in any of the three training protocols (fig. 1); *i.e.*, there were no significant differences in $\Delta PaCO_2$ between sham and hypoxia. Furthermore, there were no significant changes in exercise ventilatory responses or breathing patterns, or in *S* at rest (not shown). Model estimates for each goat (8) indicated that the feedforward exercise ventilatory stimulus (or stimuli; GEX) was also unchanged by hypoxic training. Thus, there was no evidence for LTM following any of the hypoxic training trials since

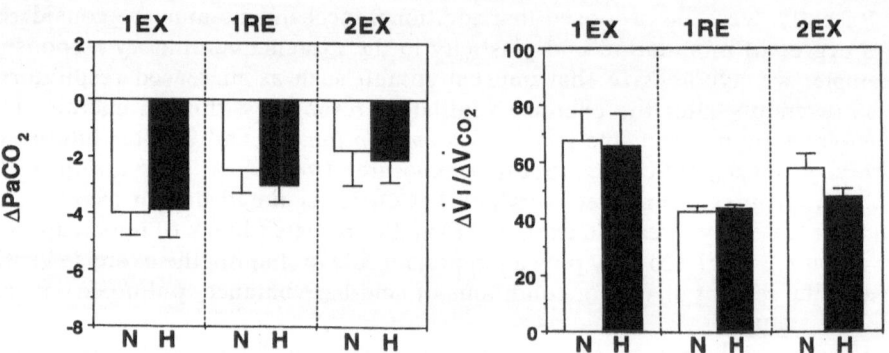

Figure 1. Changes in $PaCO_2$ from rest to exercise ($\Delta PaCO_2$; Torr) or the exercise ventilatory response, characterized by the slope of the relationship between ventilation and CO_2-production ($\Delta\dot{V}I/\Delta\dot{V}CO_2$; BTPS/STPD), during normoxic, poikilocapnic exercise 1-6 hrs following exercise training protocols of one (1EX; protocol A) and two days duration (2EX; protocol C), or 1-6 hrs following hypoxic training at rest (1RE; protocol B). Post-hypoxic training is indicated by an H (solid bars) whereas post-normoxic (sham) training is indicated by an N (open bars).

responses were not significantly different from those encountered when subjecting the goats to a sham training protocol.

DISCUSSION

These studies indicate that hypoxic episodes, either alone or in association with exercise, do not elicit LTM expressed as an augmentation of future ventilatory responses to normoxic exercise. This result contrasts with similar protocols utilizing hypercapnia, which alone (7) or with exercise (6) elicit a degree of LTM, expressed as a greater $PaCO_2$ decrease from rest to exercise after hypercapnic training. These findings suggest that peripheral chemoreceptor stimulation alone is not sufficient to elicit LTM. The stimulus for LTM may be linked to CO_2 chemoreception per se, or to some variable only indirectly related to increased $PaCO_2$ but not decreased PaO_2, such as a greater increase in tidal volume or the activation state of brainstem respiratory neurons.

The increase in ventilation and the decrease in $PaCO_2$ during hypoxic exercise trials are greater than in normoxia, an effect that is dependent on intact peripheral chemoreceptors (3). We have hypothesized that this response results from a combination of short term modulation, and an additional peripheral chemoreceptor stimulus that is unique to hypoxic exercise (9). The "extra" stimulus to ventilation during hypoxic exercise is unknown, but is unlikely to require increases in circulating potassium or norepinephrine, at least in goats (9). Whatever the identity of the unique stimulus during hypoxic exercise, it does not appear to be present 1-6 hrs following repeated hypoxic exercise.

Although there was no evidence for LTM following any of the training protocols utilized in this study, it remains possible that protocols utilizing additional training trials of (or more severe) hypoxic exercise may elicit LTM. If this is so, we are still left with the conclusion that hypoxia is less potent as a stimulus to LTM than hypercapnia. Alternately, repeated hypoxic episodes may elicit other, inhibitory mechanisms that override the expression of LTM.

During our two-day protocol utilizing poikilocapnic hypoxia as a training stimulus (protocol C), it may be that recurrent hypocapnia attendant to hypoxic exercise offset LTM elicited by hypoxic peripheral chemoreceptor stimulation. For this reason, experiments utilizing normocapnic hypoxia (protocol A) were conducted with a similar result. Furthermore, in these experiments using 6 trials of normocapnic hypoxic exercise, the exercise ventilatory response during training trials was significantly greater than during poikilocapnic exercise (not shown), and approached the level of ventilation observed during hypercapnic exercise. Thus, a greater ventilatory response alone is not sufficient to reveal LTM, at least with six training trials. At this time, the mechanism of LTM elicited by repeated hypercapnic exercise is not known. However, the ability to augment the exercise ventilatory response based on experience (eg. repeated hypercapnic exercise) indicates a previously unsuspected degree of plasticity in the ventilatory control system. These types of experiments may help in explaining how the control of breathing during exercise is adjusted to accommodate the constantly changing physiological conditions that characterize development and ageing, or during unnatural circumstances such as progressive chronic lung or neurological disease. The occurrence of LTM in humans has been demonstrated (1) and is presently being studied in greater detail by us (supported by: NIH HL36780).

REFERENCES

1. Adams, L., Moosavi, S. & Guz, A. (1992) Ventilatory response to exercise in man increases by prior conditioning of breathing with added dead space. American Review of Respiratory Disease, 145, A882.

2. Bach, K.B., Lutcavage, M.E. & Mitchell, G.S. (1993) Serotonin is necessary for short-term modulation of the exercise ventilatory response. Respiration Physiology, 91, 57-70.
3. Bisgard, G.E., Forster, H.V., Mesina, J. & Sarazin, R.G. (1982) Role of the carotid body in hyperpnoea of moderate exercise in goats. Journal of Applied Physiology, 52, 1216-1222.
4. Dempsey, J.A., Vidruk, E.H. & Mitchell, G.S. (1985) Pulmonary control systems in exercise: Update. Federation Proceedings, 44, 2260-2270.
5. Grodins, F.S. (1950) Analysis of factors concerned in regulation of breathing in exercise. Physiological Reviews, 30, 220-239.
6. Martin, P.A. & Mitchell, G.S. (1993) Long-term modulation of the exercise ventilatory response in goats. Journal of Physiology (London), 470, 601-617.
7. Martin, P.A., Bloomer, S.E.M. & Mitchell, G.S. (1993) Repeated hypercapnia at rest augments future exercise ventilatory responses in goats. FASEB Journal, 7, A858.
8. Mitchell, G.S. (1990) Ventilatory control during exercise with increased respiratory dead space in goats. Journal of Applied Physiology, 69, 718-727.
9. Mitchell, G.S., Bach, K.B., Martin, P.A. & Foley, K.T. (1993) Modulation and plasticity of the exercise ventilatory response. Funktionsanalyse biologischer Systeme, 23, 269-277.

Chemical Control of Breathing

47

RESPIRATORY RESPONSES TO HYPOXIA PERIPHERAL AND CENTRAL EFFECTS

Chairman's Introductory Communication

A. Berkenbosch, C. N. Olievier, and J. DeGoede

Department of Physiology
University of Leiden
PO Box 9604, 2300 RC Leiden
The Netherlands

INTRODUCTION

The ventilatory response following a stepwise change into hypoxia shows an overshoot i.e. a fast increase in ventilation followed by a slow decline. This overshoot in the ventilatory response is observed in newborns and adults, humans as well as animals, awake as well as anaesthetized (see Berkenbosch et al. (4)) although it is not a universal finding (1,16).

The ventilatory response to an isocapnic step out of hypoxia is less well investigated. In awake man ventilation often falls to or slightly below the original euoxic value (9, 15). Recent studies in awake cats also suggest that there is no manifest undershoot in the ventilatory response upon a step out of hypoxia (11,12). In contrast in lightly anaesthetized humans an undershoot in the ventilatory response upon sudden relief of hypoxia is observed. This is illustrated in Fig. 1 which summarizes data from a study by Dahan and coworkers (8) who measured the time course of the ventilatory response upon stepwise changes in end-tidal PO_2 in awake healthy man and man lightly anaesthetized with halothane.

MECHANISMS OF HYPOXIC DEPRESSION

Various mechanisms have been proposed to play a role in the time course of the ventilatory response to hypoxia. Some of these mechanisms will be discussed in the light of experimental findings in anaesthetized cats.

Modeling and Control of Ventilation, Edited by S. J. G. Semple, L. Adams, and B. J. Whipp
Plenum Press, New York, 1995

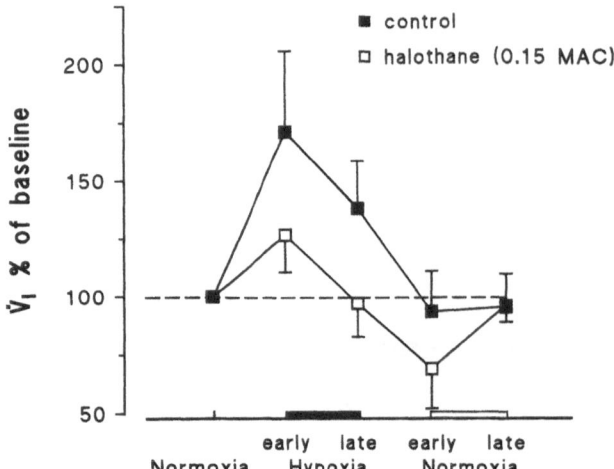

Figure 1. Ventilatory responses to sustained hypoxia in 9 subjects awake and lightly anaesthetized. Means with SD.

RESULTS

Adaptation of the Peripheral Chemoreflex Loop

In the anaesthetized cat the early rise in ventilation is due to stimulation of the peripheral chemoreceptors, while the secondary fall is of central origin. This is illustrated in Fig. 2. The over- and undershoot in the ventilatory response to stepwise changes in end-tidal PO_2 in the intact cat are absent when the brain stem is kept hyperoxaemic using the technique of artificial brain stem perfusion (ABP)(6). This indicates that moderate levels of hypoxia do not have a time dependent effect on the neuromechanical link between brain stem respiratory centres and respiratory muscles. Furthermore our experiments show that the ventilatory response due to stimulation of the peripheral chemoreceptors by hypoxia does not adapt (5) in agreement with measurements of the carotid sinus nerve activity (1,18). However, in awake man (and also in awake animals) there is some evidence that there is adaptation of the peripheral chemoreflex loop, but it is unknown whether this occurs at the level of the peripheral chemoreceptors or at the level of the central processing of the peripheral chemoreceptor signals (2,3).

Neuromodulators

A much discussed hypothesis is that the central depressant effect of hypoxia is due to changes in the synthesis or release of neuromodulators or neurotransmitters. In this regard GABA is frequently mentioned since Melton and coworkers (14) reported that a significant part of the ventilatory depression during hypoxia can be attributed to the inhibitory effect of gamma-aminobutyric acid (GABA) on respiratory neurons. They progressively decreased the arterial O_2 content of cats to about 50 per cent by means of CO inhalation and observed that infusion of the $GABA_A$ antagonist bicuculline reverses hypoxic respiratory depression. Inspired by these observations we performed experiments to investigate the effects of this blocking agent using the ABP technique (6).

In 4 cats the depressant effect of brain stem hypoxaemia on ventilation was evaluated by changing the PO_2 of the blood perfusing the brain stem from a hyperoxic level (about 50 kPa) to a hypoxic level of about 6 kPa (range 5.7-6.8 kPa between cats) while the PCO_2 was kept constant. This procedure was performed twice. Thereafter bicuculline was added to the

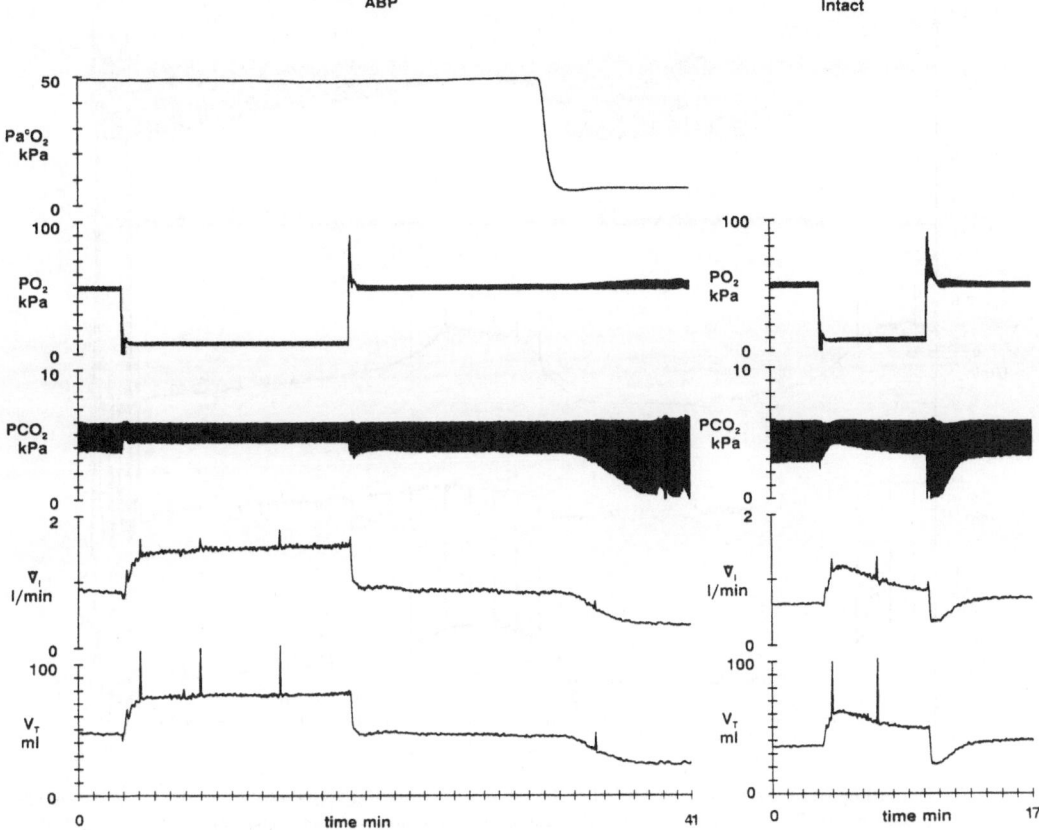

Figure 2. Ventilatory responses to isocapnic step-wise changes in end-tidal PO_2 in a cat. Right panel: 'intact' cat. Left panel: during artificial perfusion of the brain stem. Note the absence of an over- and undershoot; lowering the PaO_2 of the blood perfusing the brain stem (Pa^cO_2) shows the central depressant effect of hypoxia.

hypoxic blood perfusing the brain stem at a rate of 4 µg/min per kg bodyweight. After 10 minutes the infusion of bicuculline was stopped and the PO_2 brought back to its hyperoxic value and the increase in ventilation measured. The brain stem was then made hypoxaemic again and the infusion of bicuculline resumed for another 10 min after which the increase in ventilation due to the relief of brain stem hypoxaemia was again measured. During this procedure the blood gas tensions in the systemic circulation were kept constant by manipulating the O_2 and CO_2 concentration of the inspired gas. Thereafter during brain stem hypoxaemia the infusion of bicuculline was resumed till convulsions occurred and reliable measurements of the ventilatory parameters no longer could be made.

In all 4 cats we observed that perfusing the brain stem with hypoxic blood, keeping the other blood gas tensions constant, caused a decrease in ventilation. A recording of the second part of an experiment is shown in Fig. 3. Lowering the central PO_2 caused an appreciable ventilatory depression, although bicuculline was already given. A second infusion of bicuculline into the hypoxic blood perfusing the brain stem did not lead to an increase in ventilation. The subsequent increase in PO_2 to hyperoxic level indicates that the hypoxaemic depression was not manifestly changed. This was also observed in the other cats. The mean (with SEM) ventilatory depression due to brain stem hypoxaemia before and after infusion of bicuculline was 0.63±0.25 and 0.71±0.27 l/min respectively.

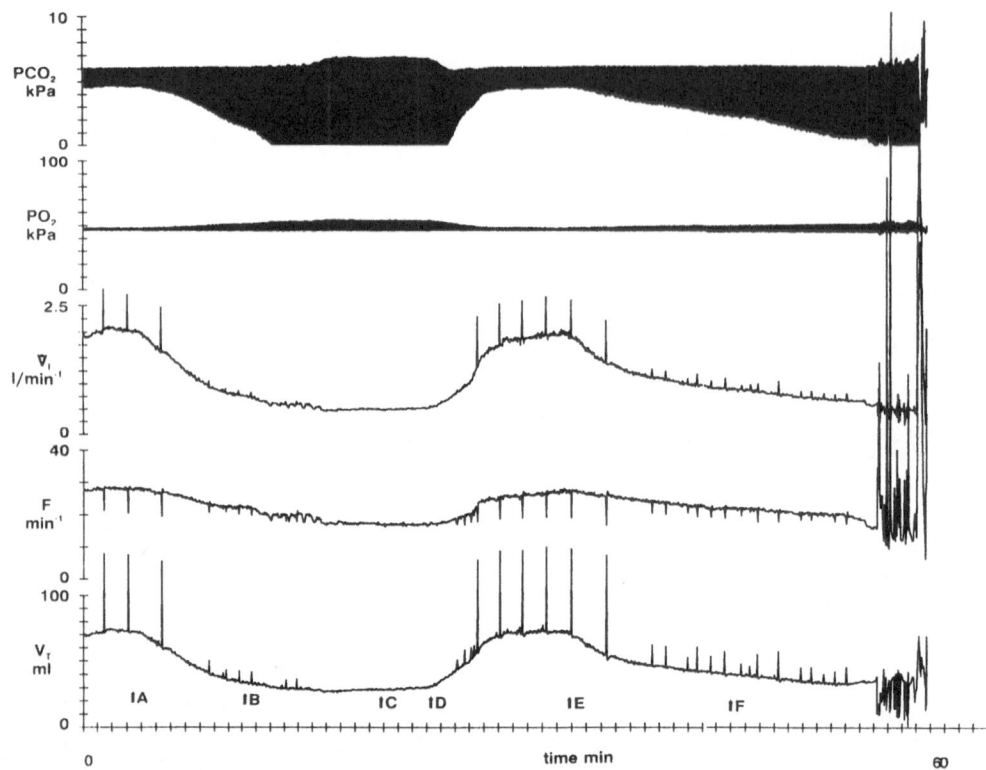

Figure 3. Recording of an ABP experiment. At A the PaO$_2$ of the blood perfusing the brain stem was changed from 48.5 to 6.4 kPa. At B bicuculline was added to the blood perfusing the brain stem at a rate of 4 μg/min per kg bodyweight and at C the infusion was ended. At D the blood was made hyperoxic again and at E hypoxic (6.8 kPa). At F bicuculline infusion was resumed till convulsions occurred accompanied with very irregular breathing.

Melton and coworkers (14) lowered the O$_2$ content of the arterial blood by CO inhalation. It is known that CO in addition to reducing the amount of active haemoglobin affects the position and shape of the oxygen dissociation curve. Zwart et al. (20) reported that reducing the arterial O$_2$ content with CO to 50 % of the normal value diminishes the oxygen half-saturation pressure with 1.8 kPa. Although we lowered the O$_2$ content of the arterial blood to about the same value as Melton did, brain tissue PO$_2$ levels in their study must be much lower than in ours. This severe brain hypoxia in all probability would have led to GABA formation and does not occur at the moderate levels of hypoxaemia we applied. Weyne et al. (19) reported that, brain GABA concentrations only increase at severe levels of hypoxaemia (PaO$_2$ 3.6 kPa) while there is no change at a PaO$_2$ of 7.2 kPa. Our ABP experiments in which the hypoxia was restricted to the pons and medulla oblongata show that the central depressant effect on ventilation by moderate levels of brain stem hypoxaemia is not influenced by blocking the GABA receptor at its α-site with bicuculline.

Brain Blood Flow

Another mechanism to explain the central depressant effect of hypoxia is that hypoxic cerebral vasodilatation increases cerebral blood flow followed by the washout

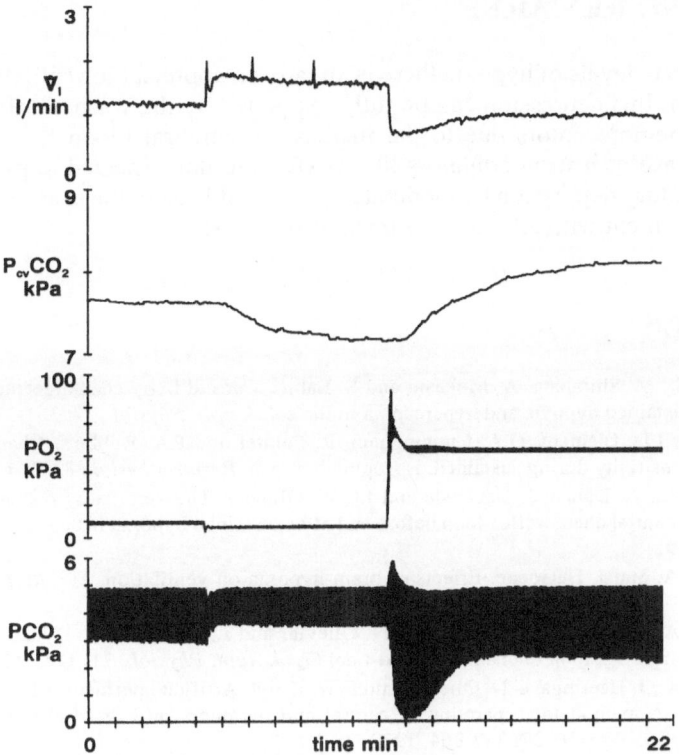

Figure 4. Recording of the breath-by-breath ventilatory response and cerebral venous PCO_2 upon isocapnic step-wise changes in end-tidal PO_2. Note that the step out of hypoxia shows a manifest undershoot in the ventilatory response together with an increase in $PcvCO_2$ of about 1 kPa.

of CO_2 at the site of the central chemoreceptors and resulting in a reduced ventilatory drive.

By administering the vasodilator papaverine to the blood perfusing the brain stem we have shown that cerebral vasodilatation indeed has a large effect on ventilation (7). This indicates that there is an appreciable gradient between brain tissue PCO_2 at the site of the central chemoreceptors and arterial PCO_2. Changes in brain tissue PCO_2 are reflected in the cerebral venous PCO_2 ($PcvCO_2$). In 7 cats we therefore measured at different levels of $PaCO_2$ and PaO_2 continuously the PCO_2 in cerebral venous blood drawn from the superior sagittal sinus. As illustrated in Fig. 4 isocapnic changes in end-tidal PO_2 induces appreciable changes in cerebral venous PCO_2.

The mean decrease in $PcvCO_2$ at a $PaCO_2$ of 5.1 kPa was 0.46 kPa going from hyperoxia (PaO_2 50 kPa) to a moderate hypoxic level of 6.6 kPa and 0.90 kPa at a PaO_2 of 4 kPa. This decrease in $PcvCO_2$ was independent of the level of $PaCO_2$ in the $PaCO_2$ range of 4.5 to 7.5 kPa. This strongly suggests that the change in brain tissue PCO_2 due to a hypoxic challenge is the same at different levels of $PaCO_2$ and is of sufficient magnitude to explain the central hypoxaemic depression in ventilation observed in the anaesthetized cat. Furthermore these results show that it is to be expected that the ventilatory CO_2 sensitivity of the central chemoreceptors is independent of the arterial PO_2 as indeed is found (10,12,13,17).

CONCLUDING REMARKS

At moderate levels of hypoxia there is already an appreciable ventilatory depression. The magnitude of this depression can be fully explained by the washout of CO_2 at the site of the central chemoreceptors due to the increase in cerebral blood flow. Therefore the synthesis or release of neuromodulators like GABA are not expected to play a prominent role in the ventilatory depression by moderate hypoxia, at least in the anaesthetized cat. This picture is in agreement with all our experimental findings.

REFERENCES

1. Andronikou, S., M. Shirahata, A. Mokashi and S. Lahiri. Carotid body chemoreceptor and ventilatory responses to sustained hypoxia and hypercapnia in the cat. *Respir. Physiol.*, 72: 361-374, 1988.
2. Bascom, D.A., I.D. Clement, D.A. Cunningham, R. Painter and P.A. Robbins. Changes in peripheral chemoreflex sensitivity during sustained, isocapnic hypoxia. *Respir. Physiol.*, 82: 161-176, 1990.
3. Berkenbosch, A., A. Dahan, J. DeGoede and I.C.W. Olievier. The ventilatory response to CO_2 of the peripheral and central chemoreflex loop before and after sustained hypoxia in man. *J. Physiol. (London)*, 456: 71-83, 1992.
4. Berkenbosch, A. and J. DeGoede. Effects of brain hypoxia on ventilation. *Eur. Respir. J.*, 1: 184-190, 1988.
5. Berkenbosch, A., J. DeGoede, D.S. Ward, C.N. Olievier and J. VanHartevelt. Dynamic response of the peripheral chemoreflex loop to changes in end-tidal O_2. *J. Appl. Physiol.*, 71: 1123-1128, 1991.
6. Berkenbosch, A., J. Heeringa, C.N. Olievier and E.W. Kruyt. Artificial perfusion of the ponto-medullary region of cats. A method for separation of central and peripheral effects of chemical stimulation of ventilation. *Respir. Physiol.*, 37: 347-364, 1979.
7. Berkenbosch, A., C.N. Olievier, J. DeGoede and E.W. Kruyt. Effect on ventilation of papaverine administered to the brain stem of the anaesthetized cat. *J. Physiol. (London)*, 443: 457-468, 1992.
8. Dahan, A., M.J.L.J. VandenElsen, A. Berkenbosch, J. DeGoede, I.C.W. Olievier, J.W. VanKleef and J.G. Bovill. Effects of Subanesthetic Halothane on the Ventilatory Responses to Hypercapnia and Acute Hypoxia in Healthy Volunteers. *Anesthesiol.*, 80: 727-738, 1994.
9. Easton, P.A., L.J. Slykerman and N.R. Anthonisen. Ventilatory response to sustained hypoxia in normal adults. *J. Appl. Physiol.*, 61: 906-911, 1986.
10. Georgopoulos, D., S. Walker and N.R. Anthonisen. Effect of sustained hypoxia on ventilatory response to CO2 in normal adults. *J. Appl. Physiol.*, 68: 891-896, 1990.
11. Long, W.Q., G.G. Giesbrecht and N.R. Anthonisen. Ventilatory response to moderate hypoxia in awake chemodenervated cats. *J. Appl. Physiol.*, 74: 805-810, 1993.
12. Long, W.Q., D. Lobchuk and N.R. Anthonisen. Ventilatory responses to CO2 and hypoxia after sustained hypoxia in awake cats. *J. Appl. Physiol.*, 76: 2262-2266, 1994.
13. Melton, J.E., J.A. Neubauer and N.H. Edelman. CO2 sensitivity of cat phrenic neurogram during hypoxic respiratory depression. *J. Appl. Physiol.*, 65: 736-743, 1988.
14. Melton, J.E., J.A. Neubauer and N.H. Edelman. GABA antagonism reverses hypoxic respiratory depression in the cat. *J. Appl. Physiol.*, 69: 1296-1301, 1990.
15. Painter, R., S. Khamnei and P. Robbins. A mathematical model of the human ventilatory response to isocapnic hypoxia. *J. Appl. Physiol.*, 74: 2007-2015, 1993.
16. Suzuki, A., M. Nishimura, H. Yamamoto, K. Miyamoto, F. Kishi and Y. Kawakami. No effect of brain blood flow on ventilatory depression during sustained hypoxia. *J. Appl. Physiol.*, 66: 1674-1678, 1989.
17. VanBeek, J.H.G.M., A. Berkenbosch, J. DeGoede and C.N. Olievier. Effects of brain stem hypoxaemia on the regulation of breathing. *Respir. Physiol.*, 57: 171-188, 1984.
18. Vizek, M., C.K. Pickett and J.V. Weil. Biphasic ventilatory response of adult cats to sustained hypoxia has central origin. *J. Appl. Physiol.*, 63: 1658-1664, 1987.
19. Weyne, J., F. VanLeuven and I. Leusen. Brain amino acids in conscious rats in chronic normocapnic and hypocapnic hypoxemia. *Respir. Physiol.*, 31: 231-239, 1977.
20. Zwart, A., G. Kwant, B. Oeseburg and W.G. Zijlstra. Human whole-blood oxygen affinity: effect of carbon monoxide. *J. Appl. Physiol.*, 57: 14-20, 1984.

48

HYPOXIC VENTILATORY DEPRESSION MAY BE DUE TO CENTRAL CHEMORECEPTOR CELL HYPERPOLARIZATION

John W. Severinghaus

Department of Anesthesia and
Cardiovascular Research Institute, Box 0542
University of California
San Francisco California

INTRODUCTION

This paper concerns the possible relationship of the acid secretion found to be localized primarily over the medullary ventral surface CO_2 chemosensitive regions to the phenomenon of hypoxic ventilatory depression, HVD, or "roll-off". HVD occurs between the 5th and 25th minutes of acute steady isocapnic hypoxia in normal man at sea level, eliminating about half of the hypoxic ventilatory stimulus without reducing the slope of the CO_2 response curve or central response to a constant stimulus from peripheral chemoreceptors [1-4], recovery requiring similar times [5]. We have demonstrated hypoxic ventilatory depression (HVD) of subjects breathing ambient air, with SaO_2 - 80-90% [6,7]. After 25 min at SpO_2 = 75%, when ventilation was apparently depressed by HVD, subjects had normal or above-normal HVR (computed as the differential slope) when SpO_2 was rapidly lowered from 75% to 65%.

Methods

Six normal human volunteers were studied at 3810m altitude (Pb = 488 Torr). CO_2 response (HCVR) was determined with step hyperoxic hypercapnia. Hypoxic ventilatory responses (HVR) were determined at the end of a 5 min step of isocapnic hypoxia at SaO_2 values from 65-85%, and computed as $\Delta \dot{V}_I / \Delta SaO_2$, determined in a new way, beginning with a period of either 20 min of moderate hypoxia, or after several days of altitude exposure. For this purpose, $P_{ET}CO_2$ was elevated during purported HVD to the level expected at that saturation from non-hypoxic tests on that subject. HVD was computed in two ways: 1) The reduction in HVR slope was tested beginning with ambient saturation. Prior to the acute added hypoxia, $P_{ET}CO_2$ was increased for 6-8 min to a predefined control PCO_2 (yielding a standard minute ventilation during hyperoxia of 140 ml/kg/min) while keeping SpO_2

Modeling and Control of Ventilation, Edited by S. J. G. Semple, L. Adams, and B. J. Whipp
Plenum Press, New York, 1995

257

constant at its ambient value, and then SaO_2 was suddenly reduced to about 70-75%, and the slope, $\Delta\dot{V}_1/\Delta SaO_2$ was computed. 2) The difference in ventilation was determined between that found chronically while breathing ambient air, and that found after 25 min of oxygenation, when SaO_2 was abruptly lowered to the ambient level, after Pco_2 had been set at the level observed while breathing ambient air.

RESULTS

HVR rose from -1.13 +/-0.23 at sea level to -2.17 +/-0.13 by day 12 at altitude. Hypoxic depression failed to significantly reduce HVR slope when measured without preoxygenation on any altitude day. If all altitude data were combined, HVR slope averaged 18% lower without preoxygenation (p<0.05). On altitude day 2, without preoxygenation, ventilation was 13.3 +/-2.4 L/min lower than after preoxygenation at equal levels of SaO_2 and $Paco_2$. This difference of ventilation with and without preoxygenation at the same SaO_2 and Pco_2 diminished as SaO_2 improved with acclimatization. Assuming this ambiently decreased ventilation to be due to altered transmembrane pH gradient, we estimated the hypoxic shift of Pco_2 setpoint compared with that found after preoxygenation to be 9.2 +/-2.1 torr on the 2nd day at altitude when mean SaO_2 was 86.2%, 6.1 +/-2.4 torr on day 6 with $SaO_2 = 89.4+/-0.9\%$, and only 4.5+/-1.3 torr on day 12 with $SaO_2 = 91.0+/-0.6\%$. This shift of set point Pco_2 was calculated at ambient SpO_2 as the ratio of the ambient-induced ventilatory decrease divided by the slope of the CO_2 response, HCVR.

DISCUSSION

HVD is more significant when assessed as a fall of ventilation at constant SaO_2 and Pco_2 than when measured as a change of HVR slope, tested with and without preoxygenation. This supports the proposal of Van Beek et al [1] that HVD is an alteration of central CO_2 chemosensitivity set point, and not a decrease in either peripheral or central chemosensitivity. HVD is not blocked by naloxone, ruling out an endorphin cause [8]. Adenosine or GABA may contribute [9-10]. HVD can be prevented by either systemic or ventral medullary topical use of dichloroacetate which blocks the degradation of pyruvate dehydrogenase and thus increases pyruvate entry into the citric acid cycle in mitochondria, thereby reducing hypoxic lactic acidosis [11].

Sato et al [12] have been able to stain cells in this ventral surface chemoreceptor by use of a c-fos immunocytochemical method after stimulating the cells with 1 hour of 13-15% CO_2 inhalation in rats. Increased numbers of c-fos stained cells were found in the areas identified as topically chemosensitive by Loeschcke, Mitchell and their co-workers long ago, together with increased staining of the NTS, but surprisingly, no general labelling in other medullary areas thought to be involved in respiratory integration.

These neurons appear to express lactic acid during hypoxia detectable by surface pH electrodes [13,14]. The magnitude of this acid is greater over the known chemosensitive areas of the medulla than over pons, spinal cord or cortex in goats with implanted electrodes and is closely related to the known areas of chemosensitivity [14]. The degree of acid shift over the rostral area averaged -0.12+/-0.09 pH units at a mean SaO_2 of 48+/-10%, whereas the surface of adjacent brain stem areas, spinal cord and cortex pH fell only 0.02+/-0.01. The slope, $\Delta pH/\Delta SaO_2$, was 3 times greater on the rostral area than on cortex.

The data indicating easily detectable acid production from a very few scattered cells suggest either a very high local metabolic rate of the specific chemosensor cells or an unusual sensitivity to hypoxia of other non-neuronal cells in the region. This may indicate that the

CO_2 and acid detection process involves significant [metabolic] work. The ECF acid found with hypoxia together with HVD implies that normal stimulation of medullary respiratory chemoreceptors by CO_2 or acid is therefore not solely ECF acidification.

The topographic identity of these CO_2-stimulated cells, this hypoxic acid generation and the classic chemosensitive region suggests a connection between hypoxic ventilatory depression, especially when measured as an upward shift of the CO_2 setpoint, and intracellular acidosis of chemoreceptor cells. If chemoreceptor neuronal intracellular acid secretion causes HVD by membrane hyperpolarization, it suggests a new model in which chemoreceptor neurones respond to changes in transmembrane H^+ gradient, like a glass pH electrode. ICF acid depresses neuronal activity. Electrogenic inward flux of H^+ would depolarize (stimulate) the chemoreceptor, whereas outward H^+ flux would hyperpolarize and quiet (stabilize). Rising Pco_2 acidifies ECF more than ICF due to ICF protein buffering. ICF hypoxic acid generation shifts the Pco_2 setpoint upward like an increase in ECF HCO_3^- concentration until the Pco_2 rises sufficiently to normalize the transcellular H^+ gradient. A hypothetic change of transmembrane pH due to hypoxia can be computed from the upward shift of Pco_2 setpoint as $\Delta pH = 0.6 \Delta log Pco_2$ assuming an intracellular CO_2 buffer factor of 0.6. The 9.2 Torr shift we computed using HCVR on day 2 at altitude (with $SaO_2 = 86\%$) is equivalent to a 0.054 fall of ICF pH compared to ECF pH.

SUMMARY

By re-examining the results of various studies of HVD, of the localization of medullary CO_2 chemosensory cells, and of their acid secretion, an hypothesis has been developed suggesting that the neurones which detect increased CO_2 or CSF acid respond to decreased transmembrane H^+ gradient, i.e. a greater fall in ECF than in ICF pH. Hypoxic lactic acid generated within these cells depresses activity, which can be restored by an appropriate rise of $Paco_2$, disclosing both normal peripheral chemoreceptor hypoxic sensitivity and normal medullary integrative response.

REFERENCES

1. Van Beek, J.H.G.M., A. Berkenbosch, J. DeGoede and C.N. Olievier. Effects of brain stem hypoxaemia on the regulation of breathing. Respiration Physiol 57:171-188, 1984.
2. Melton, J.E., J. A. Neubauer and N.H. Edelman. CO_2 sensitivity of cat phrenic neurogram during hypoxic respiratory depression. J Appl Physiol 65: 736-743, 1988.
3. Kiwull-Schöene, H and P. Kiwull. Hypoxic modulation of central chemosensitivity. In: Central Neurone Environment , edited by M.E.Schläfke, H.P.Köpfchen and W.R.See. Berlin: Springer Verlag, p.88-95, 1983.
4. Georgopoulos, D., S. Walker, and N. R. Anthonisen. Effect of sustained hypoxia on ventilatory response to CO_2 in normal adults. J Appl Physiol 68:891-896, 1990.
5. Easton, P.A., L.J. Slykerman, and N.R. Anthonisen. Recovery of the ventilatory response to hypoxia in normal adults. J Appl Physiol 64:521-528, 1988.
6. Sato, M., J.W. Severinghaus, F.L. Powell, F.D. Xu, and M.J. Spellman, Jr. Augmented hypoxic ventilatory response in men at altitude. J Appl Physiol 73: 101-107, 1992.
7. Sato, M, J.W.Severinghaus and P. Bickler. Time course of augmentation and depression of hypoxic ventilatory responses at altitude. J Appl Physiol 76:313-316, 1994.
8. Kagawa, S, M.J.Stafford, T.B.Waggener and J.W.Severinghaus. No effect of naloxone on hypoxia-induced ventilatory depression in adults. J Appl Physiol 52: 1030-1034, 1982.
9. Georgopoulos D, S.G. Holtby, D.Berezanski and N.R.Anthonisen. Aminophylline effects on ventilatory response to hypoxia and hyperoxia in normal adults. J Appl Physiol 67: 1150-1156, 1989.
10. Melton, J.E., J.A. Neubauer and N.H. Edelman. GABA antagonism reverses hypoxic respiratory depression in the cat. J Appl Physiol 69: 1296-1301, 1990.

11. Neubauer J.A., A. Simone and N.H.Edelman. Role of brain lactic acid in hypoxic depression of respiration. J Appl Physiol 65:1324-1331, 1988.

12. Sato M, J.W.Severinghaus, A.I.Basbaum. Medullary CO_2 chemoreceptor neuron identification by c-fos immunocytochemistry. J Appl Physiol 73: 96-100, 1992.

13. Xu, F., M.J. Spellman,Jr, M. Sato, J.E. Baumgartner, S. F. Ciricillo and J.W.Severinghaus. Anomalous hypoxic acidification of medullary ventral surface. J. Appl Physiol 71: 221-2217, 1991.

14. Xu, F., M. Sato, M.J. Spellman, Jr, R. A. Mitchell, and J.W. Severinghaus. Topography of cat medullary ventral surface hypoxic acidification. J Appl Physiol 73: 2631-2637, 1992.

49

CENTRAL HYPOXIC CHEMORECEPTORS IN THE VENTROLATERAL MEDULLA AND CAUDAL HYPOTHALAMUS

P. C. Nolan[1], G. H. Dillon,[2] and T. G. Waldrop[1]

[1] Department of Physiology and Biophysics, Neuroscience Program and
    College of Medicine
University of Illinois at Urbana/Champaign
524 Burrill Hall, 407 S. Goodwin Ave
Urbana, Illinois 61821
[2] Department of Pharmacology
University of North Texas Health Science Center at Fort Worth
3500 Camp Bowie Blvd., Fort Worth, Texas 76107

INTRODUCTION

Even though hypoxia is known to depress the activity of many central nervous system neurons (11,12,18), CNS hypoxia increases ventilation and cardiovascular drive under some experimental and clinical conditions (1,5,8). Central hypoxia elicits the cerebral ischemic response which involves an increase in sympathetic nerve activity resulting in elevated arterial pressure, heart rate and ventricular contractility (1,8). Moreover, perfusion of hypoxic blood to the brain of the awake goat while maintaining the isolated carotid body circulation normoxic results in a tachypnea (2). In addition, hyperventilation is evoked by systemic hypoxia in the awake rat (15), cat (13), dog (3), pony (10) and goat (2) after peripheral chemoreceptor denervation. These studies suggest that central receptors exist which increase cardiorespiratory drive when stimulated by hypoxia. The recent focus of this laboratory has been to determine if hypoxia exerts a direct effect upon neurons in brain sites involved in control of the cardiovascular and respiratory systems. Our approach was to examine the *in vivo* and *in vitro* electrophysiological responses of single neurons to hypoxia (6,7,14). Our findings support the hypothesis that the inherent responses of neurons in the ventrolateral medulla (VLM) and caudal hypothalamus (CH) are involved in the coordinated response to hypoxia observed in the intact animal.

METHODS

In vivo studies were used to investigate extracellular responses of ventrolateral medullary and caudal hypothalamic neurons in rats and cats which were anaesthetized with

Modeling and Control of Ventilation, Edited by S. J. G. Semple, L. Adams, and B. J. Whipp
Plenum Press, New York, 1995

a combination of chloralose and urethane and allowed to breathe spontaneously (7,14). Single unit activity, arterial pressure, heart rate and respiratory activity (diaphragm electromyographic activity-rat, phrenic nerve activity-cat) were recorded. Computer signal averaging techniques were utilized to determine if neurons in the VLM or CH display spontaneous discharges related to the cardiovascular (arterial pressure pulse) and/or the respiratory cycles (phrenic or diaphragmatic activity). These correlations were determined by comparing 3-5 minute bins of neuronal discharge and the respective cycle and testing for a significant relationship. Each neuron was then tested for a response to systemic hypoxia (10% O_2, 60-90 seconds) and to hypercapnia (5% CO_2, 3-5 minutes). Neuronal activity was also examined in some animals after bilateral transection of the carotid sinus, vagus and aortic depressor nerves to eliminate input from peripheral chemoreceptors and baroreceptors in the mediation of the responses to hypoxia.

In order to determine if hypoxia exerts a direct excitatory effect on VLM and CH neurons, an *in vitro* brain slice preparation was utilized (6,14). These studies used 400 μm slices of acutely dissected brain tissue (rat medulla or caudal hypothalamus) maintained in a chamber perfused with nutrient medium and bubbled with 95% O_2/5% CO_2. Both

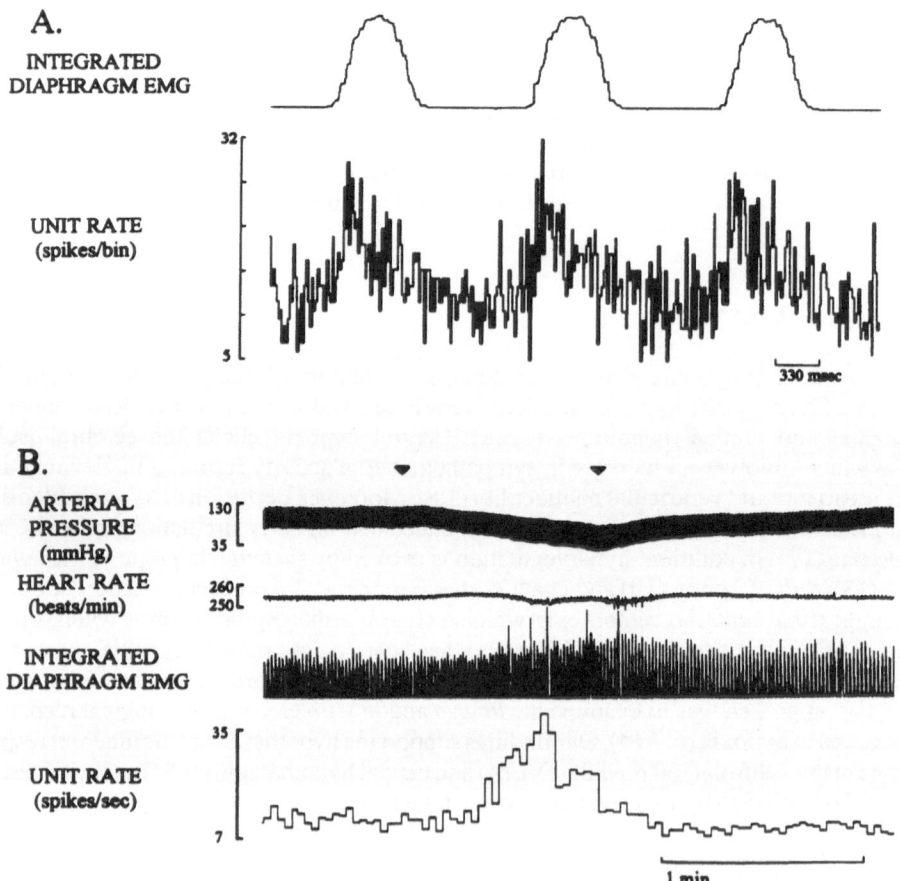

Figure 1. An example of a ventrolateral medullary neuron from an intact rat which responded to hypoxia (B). The averaged discharge frequency of this neuron relative to the diaphragmatic discharge (257 inspirations) is shown in A. Cardiorespiratory responses to hypoxia in this animal are also shown in B. Averages were compiled in 10 ms bins.

intracellular (whole cell patch) (6) and extracellular recordings (6,14) were made from individual neurons in various regions of the brain slice during presentation of a hypoxic gas (10% O_2/5% CO_2/ 85% N_2 or 5% CO_2/95% N_2, 90 sec). Some neurons were studied before and during blockade of classical synaptic transmission (high Mg^{2+}/low Ca^{2+}). This medium has been shown previously to reversibly block synaptic neurotransmission (4,6).

RESULTS

Extracellular recordings were made from thirty-nine neurons located in the ventro-lateral medulla of anaesthetized, spontaneously breathing rats (14). Inhalation of a hypoxic gas (10% O_2) elicited an increase in the discharge frequency of the majority (64%) of the neurons studied (14). Seventy-six percent of these neurons had a basal discharge related to the cardiac and/or respiratory rhythms. Figure 1B shows an example of one VLM neuron which displayed an elevated discharge rate during systemic hypoxia. In addition, this neuron had a basal rhythm linked to the respiratory cycle (Fig. 1A). Retrograde labelling with rhodamine microspheres demonstrated that the neurons stimulated by hypoxia were located in a region of the ventrolateral medulla that has extensive projections to the intermediolateral columns of the thoracic spinal cord.

Other studies in our laboratory have shown that hypoxia exerts similar effects on neurons in the caudal hypothalamus (7). Hypoxia significantly increased the discharge rate of twenty-one percent of the hypothalamic neurons recorded in the anaesthetized cat. Ninety percent of these had a cardiovascular and/or respiratory-related rhythm. It is interesting that hypercapnia stimulated thirty-two percent of the neurons tested; however, only thirteen percent of the neurons were stimulated by both hypoxia and hypercapnia.

Since both the caudal hypothalamus and ventrolateral medulla are known to receive input from peripheral chemoreceptors, the observed excitation could have resulted from peripheral input. Therefore, the effects of hypoxia upon VLM and hypothalamic neurons were also tested after denervation of peripheral chemoreceptors and baroreceptors (7,14). The excitation elicited by hypoxia was still present in neurons located in both the ventrolateral medulla and the caudal hypothalamus. In addition, the hypoxia-stimulated neurons in the denervated animals possessed basal discharges entrained to the cardiovascular and respiratory rhythms as observed in the intact cats and rats. An example of a ventrolateral medullary neuron stimulated by hypoxia in a denervated animal is shown in Figure 2. Note that this neuron also had a basal discharge related to the cardiac cycle.

One consideration with the above studies is that the excitation produced by hypoxia could have been due to input from some unidentified peripheral sites or from other brain regions. Therefore, our next experiments utilized a brain slice preparation to determine if hypoxia exerts a direct effect upon VLM and CH neurons (6,14). Perfusing a hypoxic gas over the surface of the brain slice evoked an increase in the discharge frequency of 73% and 88% of the ventrolateral medullary and caudal hypothalamic neurons, respectively. The magnitude of the neuronal excitation produced by hypoxia was related to the severity of the hypoxic stimulus. The response to hypoxia observed in the brain slice preparation persisted during perfusion of the tissue with a low calcium medium to block synaptic transmission. Neurons in adjacent regions of the slices were either unaffected or depressed by hypoxia (14). These neurons were located in brain nuclei (inferior olive, cochlear nucleus, medial vestibular nucleus) that are not known to be involved in cardiorespiratory regulation.

Whole cell patch recordings revealed that the hypoxia-induced excitation involves a membrane depolarization and an increased conductance which persist throughout the hypoxic period (6). Figure 3 shows an example of a whole cell patch recording of a caudal hypothalamic neurons which was excited by hypoxia.

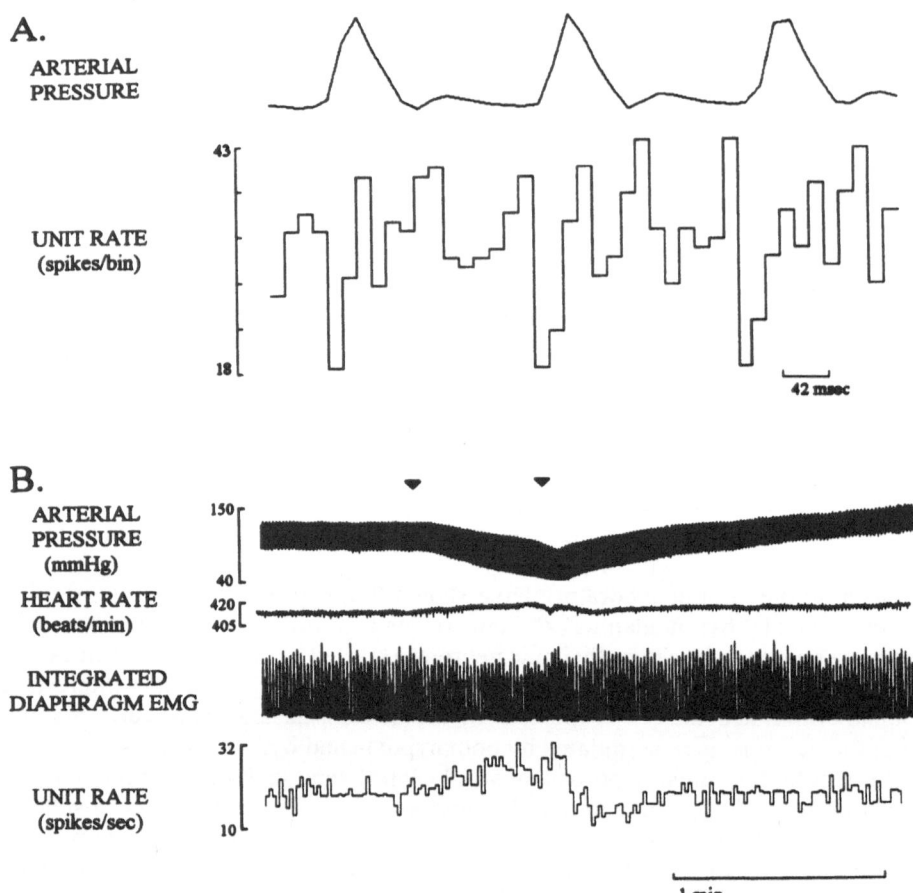

Figure 2. An example of a ventrolateral medullary neuron from a peripherally chemodenervated rat that was excited by hypoxia (B). The averaged discharge frequency of this neuron relative to the cardiac cycle (323 arterial pressure pulses) is shown in A. The cardiorespiratory responses to hypoxia in this denervated rat are shown in B. Averages were compiled with a bin duration of 10 ms.

DISCUSSION

Our observations demonstrate that the majority of neurons in the ventrolateral medulla and caudal hypothalamus, two regions involved in cardiorespiratory regulation, are inherently sensitive to hypoxia. The response to this stimulus involves a modest, prompt depolarization increasing basal discharge levels (6). This stimulation persists following removal of peripheral chemoreceptor afferent input and during a similar hypoxic stimulus *in vitro* in absence of synaptic interactions (6,7,14). These studies have also shown that hypoxia either has no effect or depresses neurons in other brain sites not known to influence cardiorespiratory function. Based on the present evidence, this suggests that the excitatory response to hypoxia is specific to areas involved in cardiorespiratory integration.

The excitatory response observed in VLM and CH neurons may underlie some of the cardiorespiratory responses elicited by brain hypoxia. A number of studies have provided evidence that central hypoxia can stimulate the cardiovascular and respiratory systems (1,5,8). Central hypoxic chemoreceptors may mediate this increased cardiorespiratory

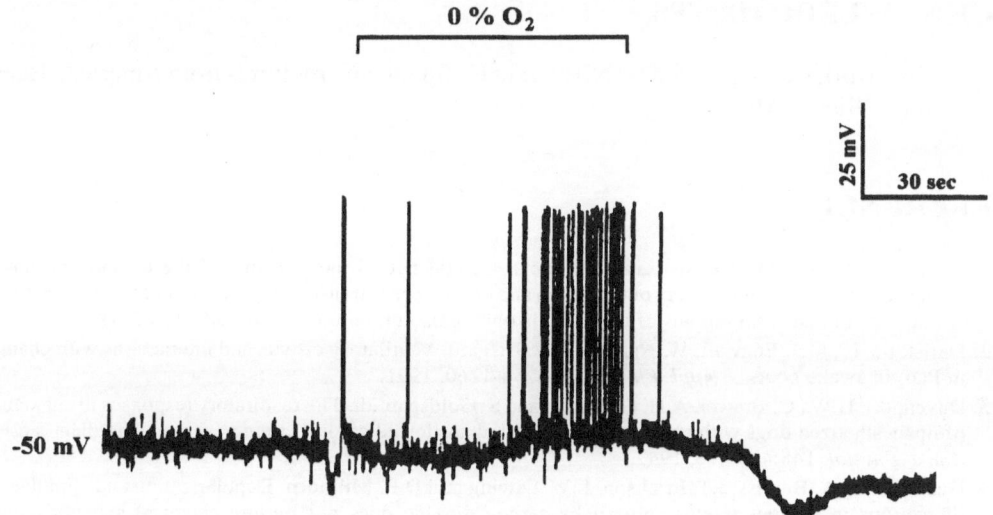

Figure 3. Whole-cell patch recording of a caudal hypothalamic neuron during a hypoxic stimulus.

activity. Moreover, several laboratories have shown electrical or chemical activation of both the VLM and CH increase ventilation and sympathetic nerve activity (16,20). Thus, activation of these cardiorespiratory neurons during hypoxia may be responsible for this excitatory effect upon the cardiovascular and respiratory systems.

It is especially interesting that the excitatory effects of hypoxia were only seen in areas known to be involved in cardiorespiratory regulation. Our studies demonstrate that most VLM and CH neurons stimulated by hypoxia *in vivo* possess basal rhythms correlated temporally with the cardiovascular and/or respiratory cycles suggesting that these cells may play a role in cardiorespiratory integration. In other areas of the brain, such as the cortex, electrical activity is depressed during hypoxia (11,18). These observations suggest there may be a hierarchical organization of hypoxia responsiveness with autonomic regions specifically excited by hypoxia and higher centres being depressed. Decreased cortical activity may result in a disinhibition of diencephalic and brainstem neurons resulting in a stimulation of ventilation. This hypothesis is supported by Tenney and Ou's finding that decortication potentiates the respiratory response to hypoxia in awake, peripherally chemodenervated cats (19).

It is also likely that the excitation evoked by hypoxia is potentiated at several levels of the diencephalon, midbrain and medulla. The direct effects of hypoxia upon hypothalamic neurons as well as additional input from the carotid body result in a strong descending drive. We have shown recently that some of the neurons in the caudal hypothalamus which are stimulated by hypoxia project to the periaqueductal grey (PAG) region of the midbrain which, in turn, projects to the ventrolateral medulla (17). These projections and possibly direct hypothalamic input to ventrolateral medullary neurons would have an additive effect on VLM neurons inherently sensitive to hypoxia and which receive excitatory input from peripheral chemoreceptors. This descending amplification has the potential to evoke a powerful stimulatory response during conditions of systemic hypoxia.

Thus, ventrolateral medullary and caudal hypothalamic neurons which are excited by the direct action of hypoxia may act as central hypoxic chemoreceptors. Mechanisms responsible for this unique response and the changes that occur during neonatal development are important areas to investigate as possible causes of autonomic dysfunction (9). Additional studies are needed to further define the role these neurons play in the coordinated response to hypoxia.

ACKNOWLEDGMENTS

This work was supported by NIH Grant HL38726 and by funds from American Heart Association-Illinois Affiliate.

REFERENCES

1. Dampney, R.A.L., M. Kumada, and D.J. Reis. Central neural mechanisms of the cerebral ischemic response: characterization, effect of brainstem and cranial nerve transections, and simulation by electrical stimulation of restricted regions of the medulla oblongata in rabbit. *Circ. Res.* 45: 48-62, 1979.

2. Daristotle, L., M.J. Engwall, W. Niu, and G.E. Bisgard. Ventilatory effects and interactions with change in Pao$_2$ in awake goats. *J. Appl. Physiol.* 71:1254-1260, 1991.

3. Davenport, H.W., G. Brewer, A.H. Chambers and S. Goldschmidt. The respiratory responses to anoxemia of unanesthetized dogs with chronically denervated aortic and carotid chemoreceptors and their causes. *Am. J. Physiol.* 148:406-416, 1947.

4. Dean, J.B, D.A. Bayliss, J.T. Erickson, L.W. Lawing and D.E. Millhorn. Depolarization and stimulation of neurons in nucleus tractus solitarii by carbon dioxide does not require chemical synaptic input. *Neuorsci.* 36:207-216, 1990.

5. Dempsey, J.A. and R.B. Schoene. Pulmonary system adaptations to high altitude, In: Pumonary and Critical Care Medicine, Vol I (C), edited by D. Dantzker, R.B. George, R.A.Matthay and H.Y. Reynolds, New York: Mosby Year Book, 1993.

6. Dillon, G.H. and T.G. Waldrop. *In vitro* responses of caudal hypothalamic neurons to hypoxia and hypercapnia. *Neurosci.* 51:941-950, 1992.

7. Dillon, G.H. and T.G. Waldrop. Responses of feline caudal hypothalamic cardiorespiratory neurons to hypoxia and hypercapnia. *Exp. Brain Res.* 96:260-272, 1993.

8. Downing, S.E., J.H. Mitchell and A.G. Wallace. Cardiovascular responses to ischemia, hypoxia, and hypercapnia of the central nervous system. *Am. J. Physiol.* 204:H881-H887, 1963.

9. Filiano, J.J. and H.C. Kinney. A perspective on neuropathologic findings in victims of the sudden infant death syndrome: the triple risk model. *Biol. Neonate.* 65:194-197, 1994.

10. Forster, H.V., G.E. Bisgard, B. Rasmussen, J.A. Orr, D.D. Buss and M. Manohar. Ventilatory control in peripheral chemoreceptor denervated ponies during chronic hypoxemia. *J. Appl. Physiol.* 41:878-885, 1976.

11. LeBlond, J. and K. Krnjevic. Hypoxic changes in hippocampal neurons. *J. Neurophysiol.* 62:15-30, 1989.

12. Luhmann, H.J. and U. Heinemann. Hypoxia-induced functional alterations in adult rat neocortex. *J. Neurophysiol.* 67:798-811, 1992.

13. Miller, M.J. and S.M. Tenney. Hypoxia-induced tachypnea in carotid-deafferented cats. *Respir. Physiol.* 23:31-39, 1975.

14. Nolan, P.C. and T.G. Waldrop. *In vivo* and *in vitro* responses of neurons in the ventrolateral medulla to hypoxia. *Brain Res.* 630:101-114, 1993.

15. Olson, E.B. Jr. and J.A. Dempsey. Rat as a model for humanlike ventilatory adaptation to chronic hypoxia. *J. Appl. Physiol.* 44:763-769, 1978.

16. Ross, C.A., D.A. Ruggiero, D.H. Park, T.H. Joh, A.F. Sved, J. Fernandez-Pardal, J.M. Saavedra and D.J. Reis. Tonic vasomotor control by the rostral ventrolateral medulla: effect of electrical or chemical stimulation of the area containing C1 adrenergic neurons on arterial pressure, heart rate and plasma catecholamines and vasopressin. *J. Neurosci.* 4:474-494, 1984.

17. Ryan, J.W. and T.G. Waldrop. Hypoxic sensitive neurons in the caudal hypothalamus project to the periaqueductal gray. *Resp. Physiol.* 100: 185-194, 1995.

18. Suzuki, R., T. Yamaguchi, Y. Inaba and H. Wagner. Microphysiology of selectively vulnerable neurons. *Prog. Brain Res.* 63:59-68, 1985.

19. Tenney, S.M. and L.C. Ou. Ventilatory response of decorticate and decerebrate cats to hypoxia and CO$_2$. *Resp. Physiol.* 29:81-92, 1977.

20 Waldrop, T.G. and J.P. Porter. Hypothalamic involvement in respiratory and cardiovascular regulation, In: *Regulation of Breathing*, 2nd ed., edited by J.A. Dempsey and A.I. Pack, New York: Marcel Dekker, Inc., 1995.

50

VENTILATORY RESPONSES TO ISOCAPNIC HYPOXIA IN THE EIGHTH DECADE

W. D. F. Smith, D. A. Cunningham, M. J. Poulin, and D. H. Paterson

Centre for Activity and Ageing
University of Western Ontario
London, Ontario, N6A 3K7 Canada

INTRODUCTION

The ventilatory response to hypoxia has been intensively studied in the young healthy adult, however, comparatively little attention has been directed towards investigating the responses to acute and sustained hypoxia in the older adult. With older age, changes in physiology and pathophysiology may alter the ventilatory response to a given stimulus and understanding these changes may have significant clinical implications.

Kronenberg & Drage (6) demonstrated a significantly reduced ventilatory response to both hypoxia and hypercapnia, and a reduced heart rate response to hypoxia, in a group of eight untrained subjects aged 70 years. More recently Ahmed et al. (1) demonstrated no change in the ventilatory response to acute or sustained hypoxia in a group of comparatively young subjects aged 62.

Poulin et al. (7), using a computer-controlled end-tidal forcing system, demonstrated an impaired ventilatory response to hypercapnia in a background of hypoxia in the eighth decade, but no difference in backgrounds of euoxia or hyperoxia.

The purpose of this study was to examine the ventilatory response to sustained isocapnic hypoxia in physically active men in their eighth decade. The precision of a computerised end-tidal forcing system has been used to study changes in the response as changes in neuromuscular drive, central timing and kinetics.

METHODS

Subjects: Two groups of five male subjects, whose day-to-day physical activities were comparable, were recruited. Group 1 were undergraduate students or members of the university staff; none was a highly trained athlete. Group 2 were elderly male subjects who exercised for one hour three mornings weekly as part of an exercise programme. All the older subjects were screened medically prior to study to exclude disease, particularly cardio-respiratory disease. All young and older subjects were non-smokers and life-long low altitude

Modeling and Control of Ventilation, Edited by S. J. G. Semple, L. Adams, and B. J. Whipp
Plenum Press, New York, 1995

dwellers. All subjects gave written informed consent for the study which had been approved by the University Ethics Committee in Human Research.

Precise breath-by-breath control of respiratory gases was achieved using a computer-controlled end-tidal forcing system (3, 4, 8). Prior to each experiment, resting end-tidal CO2 was estimated and the experimental CO_2 clamp set 1-2 torr higher. With the subject relaxed, reading and listening to soft music, each experiment began in isocapnic euoxia. Hypoxia was introduced abruptly after ten minutes and maintained for 20 minutes before returning abruptly to euoxia. Isocapnia was maintained throughout.

The protocol was repeated to reduce the signal-to-noise ratio, reduce the effect of day-to-day variation and to allow studies of recovery from hypoxic ventilatory depression not reported here. All subjects performed 12 active protocols in hypoxia and three control protocols in euoxia. All experiments were performed in random order as directed by blind ballot drawn by the subject. Data from each experiment were aligned on time signals and overlaid for plotting and curve fitting. Using DB-stats software (Ashton-Tate/SPSS), individual test data were analyzed using Student's t-test to assess within-group differences and mean group data (i.e. for each group N=5) used to assess between-group differences. The adequacy of the analysis was determined by constructing a normal probability plot from the model residuals; in some cases a log, square root or reciprocal transformation of the data were required. Statistical significance was taken as p equal or less than 0.05.

RESULTS

The mean ages (SD) for subjects in groups 1 and 2 were 29.8 (8.2) and 73.8 (2.8) years respectively; their descriptive data are shown in Table 1. From euoxia ($PETO_2$ 99.43-99.54 torr) all subjects reached their target end-tidal PO_2 ($PETO_2$ 49.07-49.39 torr) promptly at the onset of hypoxia and this was maintained throughout the 20 minutes of hypoxia. The end-tidal CO_2 clamp was maintained throughout the protocol, with slight imperfections at the onset of hypoxia which were minimised by use of the gain controls. Ventilation during the control protocols did not change significantly during the course of the protocol.

Both groups mounted a brisk response at the onset of hypoxia which reached a peak response within 2 minutes (within groups p=0.001, between groups NS). Mean group data

Table 1. Subject descriptive and ventilatory data

Variable	Group 1, Men N = 5	Group 2, Men N = 5
Age, years	29.8 (8.2)	72.8 (2.8)
Height, m	1.78 (0.07)	1.78 (0.05)
Mass, kg	78.4 (7.2)	80.2 (7.4)
FeV$_1$, 1 BTPS	4.53 (0.99)	3.06 (0.41)
Ventilation, baseline	10.79 (1.99)	11.88 (0.91)
Ventilation, peak	22.58 (2.60)	24.56 (2.54)
Ventilatory change	11.79 (2.49)	12.68 (1.84)
peak, % of baseline	215.6 (30.9)	208.7 (12.1)
Ventilation, end hypoxia	14.29 (1.92)	16.85 (2.34)
Ventilatory depression	8.28 (2.00)	7.71 (1.89)
Ventilation, nadir euoxia	7.23 (1.32)	8.41 (0.96)
Ventilation, 2' euoxia	9.42 (1.18)	10.62 (1.00)

Descriptive data values are means (standard deviation) and ventilations l.min^{-1} (standard deviation).

are shown in Table 1. In both groups, the increment in ventilation was produced by substantial changes in tidal volume (VT) and mean inspiratory flow (VT:TI) reflecting increased neuromuscular drive. Changes in central timing, demonstrated by changes in TI, TI:TTOT and f, were less marked in both groups, contributing approximately 32% and 30%, respectively, to the increase in ventilation. Examination of the kinetics of the ventilatory response gave time constants for 63% of the response (tau) of 16.0 and 18.5 seconds with delays of 4.8 and 4.6 seconds, respectively (NS). Ventilation declined during sustained isocapnic hypoxia though it remained above resting euoxic values in all subjects. The decline, achieved by reduction in neuromuscular drive, was significant within both groups (p = 0.001) though between-group differences were not significant. In both groups, the acute ventilatory response was significantly correlated with hypoxic ventilatory depression; group 1, r = 0.67 (p = 0.001), group 2, r = 0.72 (p = 0.001).

At the off-transient, ventilation fell in both groups (p = 0.001) undershooting the baseline but by the second minute of euoxia it had settled at 9.42 l.min^{-1} and 10.62 l.min^{-1}, respectively. In both groups, this latter change in ventilation was significant (p = 0.001) (between groups NS).

DISCUSSION

Both groups were well matched for body size and the older study group had well preserved lung function within the age-predicted range for non-smokers. The acute ventilatory response was similar in both groups in absolute or standardised terms, as were the neuromuscular drive, central timing components and kinetics.

Kronenberg & Drage (6), however, found a significant difference in the acute ventilatory response between their young and older subjects. Their subjects were of similar stature to the subjects in this study though their older group appears to have been physically inactive. Slower kinetics associated with the sedentary detrained state and the progressive onset of hypoxia may have delayed the peak response which also may have been masked by the onset of hypoxic ventilatory depression. However, the most marked difference between this study and Kronenberg's lies not with the older subjects but with his young subjects who demonstrated a response much greater than that observed in this or other studies.

Hypoxic ventilatory depression observed by Easton et al. (2) in young adults and later by Ahmed et al. (1) in young active retirees is confirmed in these physically active septuagenarians. The ventilatory changes at the on- and off-transients were asymmetric, similar in both groups and consistent with Khamnei & Robbins' (5) modelling which suggested a resetting of the peripheral chemo-receptor during hypoxia.

Poulin et al. (7) found an impaired ventilatory response to combined hypercapnia and hypoxia in his active older group, five of whom also participated in this study. In this study, using the same technique and adequate repetitions, no between-group difference nor trend was demonstrated in isocapnic hypoxia. This suggests that in most clinical circumstances a normal ventilatory response to hypoxia can be expected in healthy active men in their eighth decade.

REFERENCES

1. Ahmed, M. Giesbrecht, G.G. Serrette, C. Georgopoulos, D. Anthonisen, N.R. Ventilatory response to hypoxia in elderly humans. Respiration Physiology 1991;83:343-352.
2. Easton, P.A. Slykerman, L.J. Anthonisen, N.R. Ventilatory response to sustained hypoxia in normal adults. Journal of Applied Physiology 1986;61(3):906-911.

3. Howson, M.G. Khamnei, S. O'Connor, D.F. Robbins, P.A. The properties of a turbine device for measuring respiratory volumes in man. Journal of Physiology 1986;382:12P.
4. Howson, M.G. Khamnei, S. McIntyre, M.E. O'Connor, D.F. Robbins, P.A. A rapid computer-controlled binary gas-mixing system for studies in respiratory control. Journal of Physiology 1987;394:7P.
5. Khamnei, S. Robbins, P.A. Hypoxic depression of ventilation in humans: alternative models for the chemoreflexes. Respiration Physiology 1990;81:117-134.
6. Kronenberg, R.S. Drage, C.W. Attenuation of the ventilatory and heart rate responses to hypoxia and hypercapnia with aging in normal men. Journal of Clinical Investigation 1973;52:1812-1819.
7. Poulin, M.J. Cunningham, D.A. Paterson, D.H. Kowulchuk, J.M. Smith, W.D.F. Ventilatory sensitivity to CO2 in hyperoxia and hypoxia in older aged humans. Journal of Applied Physiology 1993;75(5):2209-2216.
8. Robbins, P.A. Swanson, G.D. Howson, M.G. A prediction correction scheme for forcing alveolar gases along certain time courses. Journal of Applied Physiology 1982;52:1357-1362.

51

HYPOXIC VENTILATORY RESPONSE NEAR NORMOCAPNIA

Lindsey C. Henson, John A. Temp, Andrea A. Berger,
and Denham S. Ward

Department of Anesthesiology
University of Rochester School of Medicine
601 Elmwood Avenue, Box 604, Rochester, New York 14642

INTRODUCTION

A classic study demonstrated a CO_2 threshold or "dog leg" in the CO_2 response curve when ventilation was measured after exposure to prolonged (>30 min) hypoxia and the existence of a component of ventilation during hypoxia which does not vary with CO_2 (1). However, ventilation after prolonged hypoxia is determined by the combination of acute hyperventilation mediated by the carotid bodies and a slower decline in ventilation originating in the central nervous system. More recently, Rebuck and Slutsky (2) described a simple linear relationship without a "dog leg" between acute hypoxic sensitivity (the carotid body response) and P_{CO2}.

In previous studies, we noted an abrupt increase in the acute hypoxic sensitivity near resting CO_2 (unpublished). The purpose of these experiments was to investigate the presence of a CO_2 threshold or "dog leg" in acute hypoxic sensitivity. The experiments were designed to measure only the carotid body component of the response and to systematically examine the hypoxic-hypercapnic interaction at a range of $P_{ET}CO_2$ near each subject's individual resting value.

METHODS

10 healthy subjects (8 females, 2 males, 26 ± 2 yr, mean ± SEM) fasted and refrained from respiratory stimulants for 8 hr before testing. All testing was completed on one day within <5 hr. Subjects breathed through a face mask and listened to classical music with headphones to minimize distraction from laboratory noise.

Each subject breathed room air for 20 min to determine resting $P_{ET}CO_2$ (34.4 + 0.6 Torr) and resting ventilation (7.7 + 0.3 l/min), followed by a 10 min rest period. Each subject then performed 8 separate hypoxic step experiments, using dynamic end-tidal forcing to control end-tidal gas concentrations (3). Each experiment consisted of 7 min of normoxia

Modeling and Control of Ventilation, Edited by S. J. G. Semple, L. Adams, and B. J. Whipp
Plenum Press, New York, 1995

($P_{ET}O_2 = 100$ Torr), 6 min of acute hypoxia ($P_{ET}O_2 = 42$ Torr), 1 min of hyperoxia ($F_1O_2 = 0.3$), and a 10 min rest period. The CO_2 controller was set to target values of -1, -3, -5, -7, +1, +3, +5, +7 Torr relative to the resting value. The order of $P_{ET}CO_2$ was randomized among subjects.

$P_{ET}CO_2$ was referenced by subtracting the resting value from actual normoxic and hypoxic values for each subject. Breath-by-breath data (3) for minute ventilation, $P_{ET}CO_2$, and $P_{ET}O_2$ were averaged over the last min of normoxia and min 3-4 of hypoxia. Hypoxic sensitivity was calculated as the change in ventilation divided by the change in S_aO_2 between these two periods (2). S_aO_2 was calculated from end-tidal gas concentrations (4).

Curves relating hypoxic sensitivity to referenced hypoxic $P_{ET}CO_2$ were divided into three segments; the slopes of each segment were calculated using linear regression and were compared using one-way ANOVA with post-hoc Newman-Keuls testing. A P value of 0.05 was considered significant.

RESULTS

For target $P_{ET}CO_2$ below resting, actual $P_{ET}CO_2$ did not reach the target value when ventilation was insufficient (during normoxia in all subjects and during hypoxia in a few subjects for the lowest target $P_{ET}CO_2$) (Table 1). This resulted in greater variability in control of $P_{ET}CO_2$ across subjects at lower values. In addition, the hypoxic sensitivity during hypocapnia may have been underestimated, since the CO_2 during normoxia was higher than that during hypoxia.

A CO_2 threshold (the point below which ventilation appears to be independent of P_{ETCO2}) was apparent during normoxia in every subject. During hypoxia, a CO_2 threshold was apparent in every subject but one, in whom acute hypoxic ventilation was quite variable. The CO_2 threshold occurred at a lower CO_2 during hypoxia than during normoxia (Table 1, Figure 1).

The relation between hypoxic sensitivity and $P_{ET}CO_2$ was not linear over the full range of CO_2 studied in any subject. A CO_2 threshold in hypoxic sensitivity was apparent in every subject; moreover, above this CO_2 threshold the rate of increase in hypoxic sensitivity lessened at higher CO_2 levels (Figure 2). The mean slope of the mid-portion of the curve relating hypoxic sensitivity to $P_{ET}CO_2$ was significantly steeper than the mean slope of the

Table 1. Mean end-tidal CO_2 (referenced to resting end-tidal CO_2) and minute ventilation during normoxia and hypoxia and mean hypoxic sensitivity

Target $P_{ET}CO_2$ (Torr)*	Normoxic $P_{ET}CO_2$ (Torr)*	Hypoxic $P_{ET}CO_2$ (Torr)*	Normoxic ventilation (L/min)	Hypoxic ventilation (L/min)	Hypoxic sensitivity (L/min/%sat)
-7	-1.4 (0.5)	-4.8 (0.6)	8.1 (0.3)	11.0 (0.5)	0.26 (0.05)
-5	-1.2 (0.3)	-4.1 (0.4)	8.1 (0.3)	11.1 (0.6)	0.26 (0.06)
-3	-1.1 (0.5)	-3.0 (0.1)	8.2 (0.3)	13.0 (1.8)	0.37 (0.12)
-1	-1.1 (0.2)	-0.9 (0.0)	8.9 (0.3)	16.6 (1.5)	0.52 (0.08)
+1	1.0 (0.1)	1.0 (0.0)	9.7 (0.6)	21.7 (2.1)	0.74 (0.12)
+3	2.8 (0.1)	3.0 (0.0)	13.0 (0.9)	30.3 (3.0)	1.00 (0.14)
+5	4.9 (0.1)	5.0 (0.0)	13.9 (0.7)	33.6 (2.9)	1.05 (0.12)
+7	6.8 (0.1)	7.0 (0.0)	20.5 (1.3)	41.8 (3.2)	1.05 (0.12)

*Change from resting $P_{ET}CO_2$. All values are mean (SEM), n = 10.
Values expressed as 0.0 are less than 0.05.

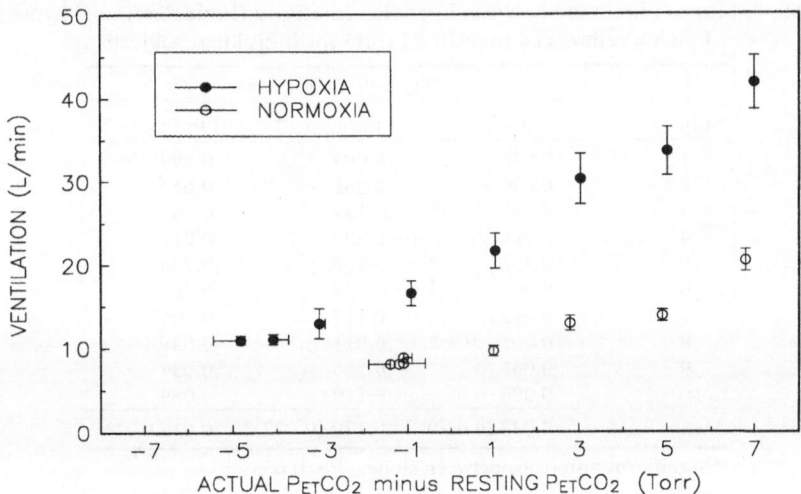

Figure 1. Mean ventilation during normoxia (open symbols) and hypoxia (closed symbols), plotted as a function of mean $P_{ET}CO_2$. Error bars represent SEM.

upper portion of the curve (Table 2). This suggests a "leveling off," or an altered, slightly decreased hypoxic-hypercapnic interaction, at the higher levels of hypercapnia.

DISCUSSION

Although the hypoxic-hypercapnic interaction has previously been studied over a wide range of P_{CO_2}, we have found no prior study which systematically examined the interaction at a number of $P_{ET}CO_2$ values near resting. Our data demonstrate a "dog leg" in

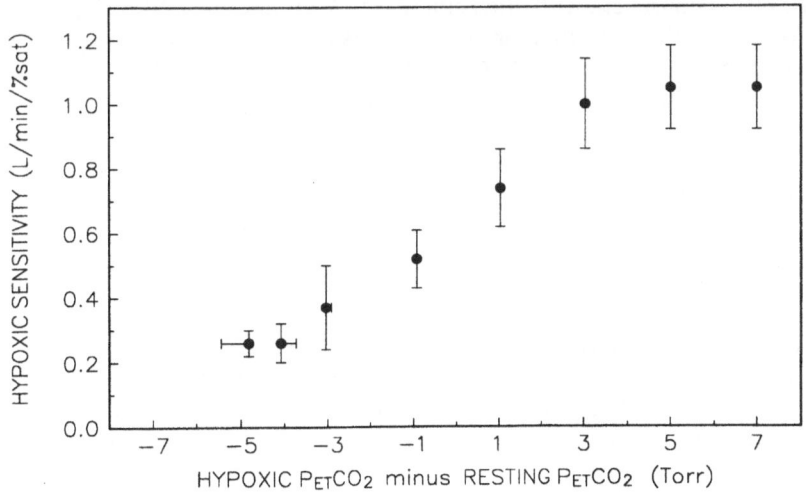

Figure 2. Mean hypoxic sensitivity, plotted as a function of mean hypoxic $P_{ET}CO_2$. Error bars represent SEM.

Table 2. Slopes of relation between hypoxic sensitivity (l/min/%sat) and hypoxic $P_{ET}CO_2$ referenced to resting (Torr) for individual subjects

Subject	Target $P_{ET}CO_2$ (Torr)		
	-7 through -1	-1 through +3	+3 through +7
1	0.063	0.097	0.099
2	0.039	0.088	0.053
3	0.095	0.139	0.041
4	0.084	0.228	-0.017
5	0.238	0.160	-0.128
6	0.061	0.034	0.006
7	0.012	0.117	0.005
8	0.065	0.011	-0.049
9	0.055	0.220	0.037
10	0.003	0.110	0.089
Mean (SEM)	0.072 (0.020)	0.120 (0.022)*	0.014 (0.021)*

*Significant difference between slopes, $P < 0.05$.

acute hypoxic sensitivity near resting CO_2 and suggest that hypoxic sensitivity is not increased in a strictly linear fashion by increases in P_{CO2} up to 7 Torr above resting.

We have demonstrated a threshold in both acute hypoxic and normoxic ventilatory responses to CO_2 which occurs at a lower CO_2 during hypoxia than during normoxia (Figure 1). These data are similar to those of Nielsen and Smith (1), who demonstrated in two subjects that pulmonary ventilation after prolonged (~30 minutes) hypoxia was not changed by the addition of CO_2 to inspired air until a threshold $P_A CO_2$ was reached; above this threshold ventilation increased linearly over a range of $P_A CO_2$ up to ~10 Torr above the threshold value. In addition, we found a threshold in acute hypoxic sensitivity (the carotid body-mediated response), which cannot be deduced from the data of Nielsen and Smith because they employed prolonged hypoxic exposure. Interestingly, it can be seen in their published data that one of their two subjects exhibited marked hypoxic ventilatory decline, while the other did not.

We observed a measurable hypoxic sensitivity at hypocapnic CO_2 in all but one of our subjects; this is consistent with other studies (5-7) comparing poikilocapnic to isocapnic hypoxia. However, other investigators have not examined the intermediate levels of CO_2 between resting and poikilocapnia. $P_{ET}CO_2$ was lower during hypoxia than during normoxia for our hypocapnic experiments. This might theoretically lead to underestimation of the difference between normoxic and hypoxic ventilation (*i.e.*, hypoxic sensitivity), particularly if the CO_2 threshold during normoxia were not achieved. However, we observed clear thresholds in normoxic ventilation in all ten subjects. Moreover, the direction of the theoretical error would tend to obscure a "dog leg", rather than to demonstrate one that did not exist.

It is possible that, due to the brief duration of hypoxia (and the correspondingly brief duration of hyperventilation and increased cerebral blood flow), P_{CO2} tension in the medulla lagged behind that in the arterial blood and the lung, where we measured $P_{ET}CO_2$. However, data relating to changes in jugular venous P_{CO2} during hypoxia (8) suggest to us that the time course of these experiments was sufficient to achieve P_{CO2} change at the medulla.

We studied the interaction between P_{CO2} and hypoxic sensitivity in spontaneously breathing subjects, in whom wakefulness drive maintains ventilation during hypocapnia (9). Whether there is some degree of hypocapnia which abolishes the ventilatory response to acute hypoxia cannot be determined using our experimental design. An experimental model using passive hyperventilation to produce hypocapnia before hypoxic exposure (10) may be more appropriate for addressing this question.

At the upper end of our range of $P_{ET}CO_2$, we observed a "leveling off" of hypoxic sensitivity in all but two of our subjects. Rebuck and Slutsky (2) described hypoxic sensitivity as a linear function of P_{CO_2} in four subjects, but they studied only two levels of isocapnic end-tidal P_{CO_2}, each subject's resting and mixed venous values. In earlier work (11), Rebuck and Woodley studied seven subjects at a minimum of three levels of isocapnic CO_2; analysis of their published data reveals that, in five of these subjects, hypoxic sensitivity did not increase linearly with increasing $P_{ET}CO_2$. A possible explanation for this finding is the inverse relationship between cerebral blood flow and central P_{CO_2}, since there should be a greater increase in cerebral blood flow during hypercapnic than normocapnic hypoxia, leading to attenuation of ventilation and an apparently decreased hypoxic-hypercapnic interaction.

Further experiments should include analyzing higher P_{CO_2} tensions in order to confirm or negate our result of the nonlinear "leveling off" effect at high levels of CO_2, as well as experiments to determine the reproducibility of the hypoxic-hypercapnic interaction from day to day. If confirmed, our data will have implications for the design of experiments testing effects of various treatments (*e.g.*, drugs, level of consciousness, visual stimulation) on the hypoxic response. Our data suggest that choosing a P_{CO_2} in relation to each subject's resting value may be more relevant than choosing the same P_{CO_2} for all subjects. Moreover, if the shape of the curve relating hypoxic sensitivity to P_{CO_2} were to change from day to day, or if the entire curve were to shift from day to day, the interpretation of data obtained on different days may be difficult.

REFERENCES

1. Nielsen, M., and H. Smith. Studies on the regulation of respiration in acute hypoxia. *Acta Physiol. Scand.* 24: 292-313, 1952.
2. Rebuck, A.S., and A.S. Slutsky. Measurement of ventilatory responses to hypercapnia and hypoxia. In: *Regulation f Breathing,* edited by T.F. Hornbein. New York: Dekker, 1981, p. 745-772.
3. Temp, J.A., Henson, L.C., and D.S. Ward. Effect of a subanesthetic minimum alveolar concentration of isoflurane on two tests of the hypoxic ventilatory response. *Anesthesiology* 80: 739-750, 1994.
4. Severinghaus, J.W., and K.H. Naifeh. Accuracy of response of six pulse oximeters to profound hypoxia. *Anesthesiology* 67: 551-558, 1987.
5. Moore, L.G., Huang, S.Y., McCullough, R.E., Sampson, J.B., Maher, J.T., Weil, J.V., Grover, R.F., Alexander, J.K., and J.T. Reeves. Variable inhibition by falling CO_2 of hypoxic ventilatory response in humans. *J. Appl. Physiol.* 56: 207-210, 1984.
6. Reynolds, W.J., and H.T. Milhorn. Transient ventilatory response to hypoxia with and without controlled alveolar PCO_2. *J. Appl. Physiol.* 35: 187-196, 1973.
7. Easton. P.A., and N.R. Anthonisen. Carbon dioxide effects on the ventilatory response to sustained hypoxia. *J. Appl. Physiol.* 64: 1451-1456, 1988.
8. Suzuki, A., Nishimura, M., Yamamoto, H., Miyamoto, K., Kishi, F., and Y. Kawakami. No effect of brain blood flow on ventilatory depression during sustained hypoxia. *J. Appl. Physiol.* 66: 1674-1678, 1989.
9. Fink, B.R. Influence of cerebral activity associated with wakefulness on the regulation of breathing. *J. Appl. Physiol.* 16: 15-20, 1961.
10. Datta, A.K., Shea, S.A., Horner, R.L., and A. Guz. The influence of induced hypocapnia and sleep on the endogenous respiratory rhythm in humans. *J. Physiol.* 440: 17-33, 1991.
11. Rebuck, A.S., and W. Woodley. Ventilatory effects of hypoxia and their dependence on PCO_2. *J. Appl. Physiol.* 38: 16-19, 1975.

52

A COMPARISON BETWEEN THE EFFECTS OF 8 HOURS OF ISOCAPNIC HYPOXIA AND 8 HOURS OF POIKILOCAPNIC HYPOXIA ON RESPIRATORY CONTROL IN HUMANS

L. S. G. E. Howard and P. A. Robbins

University Laboratory of Physiology
Parks Road, Oxford OX1 3PT
United Kingdom

INTRODUCTION

The ventilatory response to a 20 min exposure to isocapnic hypoxia is biphasic. There is an initial increase in ventilation associated with stimulation of the peripheral chemoreceptors, followed by a subsequent fall known as hypoxic ventilatory depression or decline (HVD) (8,13). In contrast to the decline in ventilation observed with 20 min of isocapnic hypoxia, there is a progressive increase in ventilation when humans are exposed to many hours/days of poikilocapnic hypoxia, such as occurs on going to altitude (1,7). The mechanisms underlying both the fall in ventilation associated with HVD and the rise in ventilation associated with acclimatization to altitude (or hypoxia, VAH) are incompletely understood. There are, however, two major differences between the hypoxic exposures. First, HVD develops over a period of minutes, whereas VAH takes hours/days. Secondly, HVD occurs under isocapnic conditions, whereas VAH occurs during hypocapnia when there is a respiratory alkalosis. The purpose of the current study was to undertake a more effective comparison of isocapnic and poikilocapnic conditions. In particular, we wished to extend the time scale of the observations with isocapnic hypoxia in order to determine whether ventilation continues to decrease (as with HVD), whether ventilation remains stable (at the level observed 20-30 min after the induction of hypoxia) or whether ventilation begins to rise (as is seen with VAH).

There is, however, a major methodological problem with maintaining isocapnia in humans over a period of hours as compared with a period of minutes. For experiments which do not last too long, it is possible to ask the subject to breathe through a mouthpiece with the nose occluded. This enables the composition of the inspired gas to be varied easily. However, for the longer periods, the mouthpiece and nose-clip arrangement is too restrictive for the subject, and becomes uncomfortable. In order to overcome these restrictions, a chamber was developed in which the atmospheric composition could be altered in order to

Modeling and Control of Ventilation, Edited by S. J. G. Semple, L. Adams, and B. J. Whipp
Plenum Press, New York, 1995

hold the subject's own end-tidal gas compositions (measured via a fine nasal catheter) at the desired values.

METHODS

Subjects and Protocols

Ten subjects were studied, and each undertook three protocols. The first protocol consisted of an 8 hour exposure to isocapnic hypoxia (Protocol I). End-tidal P_{CO2} (PET_{CO2}) was held at the subject's air breathing control value measured prior to the start of the exposure. End-tidal P_{O2} (PET_{O2}) was held at 55 Torr. The second protocol consisted of an 8 hour exposure to poikilocapnic hypoxia (Protocol P). PET_{CO2} was allowed to vary freely, and PET_{O2} was held at 55 Torr. The third protocol consisted of an 8 hour exposure to air as a control (Protocol C). During each protocol, hourly measurements were made of ventilation by asking the subject to breathe for 5 min through a mouthpiece connected to a pneumo-tachograph and turbine device.

In addition to the ventilatory measurements, measurements of hypoxic sensitivity were undertaken before the start of the hypoxic exposure, 20 min into the hypoxic exposure, 4 hr into the hypoxic exposure and 8 hr into the hypoxic exposure. These were accomplished using a sequence of 3.5 square wave followed by 4 triangular wave variations in hypoxia (each of period of 90 sec) with the PET_{O2} varying between 45 Torr and 100 Torr. During this assessment of the hypoxic response, PET_{CO2} was held constant at 1-2 Torr above the control air-breathing value throughout.

Apparatus and Techniques

The chamber is 2.3 m in length, 1.7 m in width and 2.2 m in height. The upper panels are clear so that the subject can be observed. Nitrogen can be delivered into the chamber at 1,000 l/min, oxygen at 100 l/min and carbon dioxide at 100 l/min. Carbon dioxide can be absorbed through an absorber circuit containing a soda-lime bed through which the chamber gas can be re-circulated at 2,000 l/min. The chamber is fitted with an air conditioning unit which removes heat and water vapour generated as part of the CO_2 absorbing process. All gas supplies and the absorber circuit are controlled by computer.

The subject is required to wear the finger probe of a pulse oximeter and a fine nasal catheter which is attached to a mass spectrometer outside the chamber. Otherwise the subject is free of encumbrance. The CO_2 and O_2 signals from the mass spectrometer are passed to a computer which picks out the end-tidal values. At 5 min intervals, the computer compares the average end-tidal values with the desired values and adjusts the composition in the chamber to drive the end-tidal values of the subject towards the desired end-tidal values. The chamber contained a pneumotachograph and turbine measuring device (11) so that every hour ventilation could be measured.

The assessments of hypoxic sensitivity were undertaken outside the chamber since they required the use of our dynamic end-tidal forcing system (10,12). Using a model of the respiratory system, a prediction is made of the inspired P_{CO2} and P_{O2} profile that should generate the desired end-tidal functions. Then the experiment is started, and gas of the appropriate composition is generated using a computer-controlled fast gas-mixing system. The subject breathes this gas via a mouthpiece and the end-tidal compositions are monitored by mass spectrometry. The actual end-tidal gas composition is compared with the desired end-tidal composition on a breath-by-breath basis by computer, and the computer then

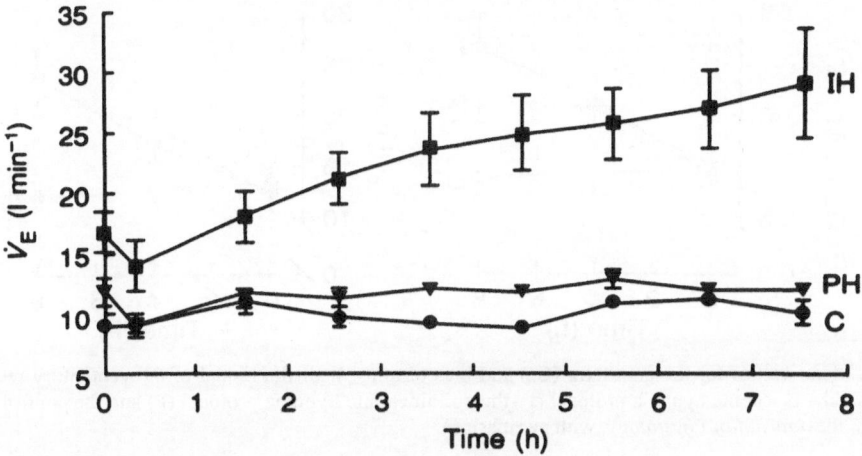

Figure 1. Mean ventilation (± SEM) for the isocapnic protocol (IH), the poikilocapnic protocol (PH) and the control protocol (C). (From the *Journal of Physiology*, with permission.)

adjusts the inspired gas composition in order to minimise the difference between the desired end-tidal values and the actual end-tidal values.

Data Analysis

The average ventilation measured every hour in the chamber were subjected to analysis of variance to determine whether there were any significant differences between the three protocols. A model of the respiratory response to hypoxia consisting of a gain term, time constant, bias term and pure delay (model 3 of (5)) was fitted to the breath-by-breath ventilation from the experiments to determine hypoxic sensitivity. The gain terms and bias terms were subjected to analysis of variance to determine whether there were any significant differences between the protocols.

RESULTS

The mean values for ventilation are shown in Fig.1. The magnitude of the initial ventilatory response to both isocapnic hypoxia and poikilocapnic hypoxia can be seen by comparing the ventilation at time zero for the two hypoxic protocols with the control ventilation. Over the subsequent 20 min a decline in ventilation is observed for both hypoxic protocols consistent with HVD. Ventilation during poikilocapnic hypoxia can be seen to be mildly elevated compared with control for the rest of the experimental period. In contrast, during isocapnic hypoxia there is a progressive rise in ventilation over the remainder of the experimental period. The magnitude of the rise is very large, and the result was highly significant (P<0.001, ANOVA).

The mean values for the gain term (or ventilatory sensitivity to hypoxia, G_p) and for the bias term (component of ventilation which is insensitive to hypoxia, $\dot{V}_c$) are shown in Fig 2. There appears to be no change in either with the control exposure. However, during both the isocapnic and the poikilocapnic hypoxic exposures, there appears to be a progressive increase in G_p and $\dot{V}_c$. The changes in G_p and $\dot{V}_c$ during both hypoxic protocols were significantly different from the changes during control (P<0.001, ANOVA). There were no

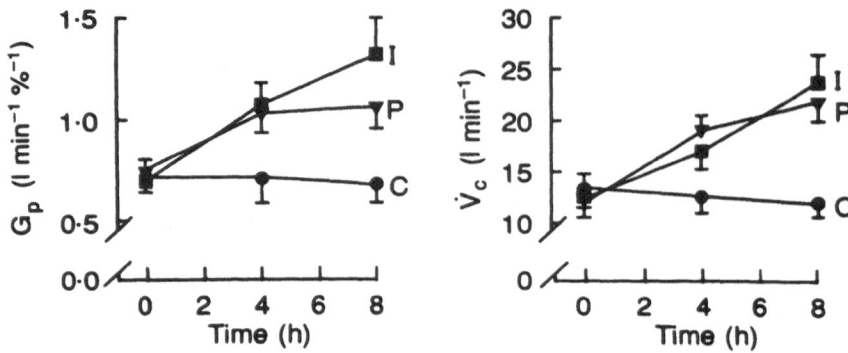

Figure 2. Mean values for the gain term (G_p) and bias term ($\dot{V}_c$) for the model of the ventilatory response to hypoxia for the isocapnic hypoxia protocol (I), the poikilocapnic hypoxia protocol (P) and the control protocol (C). (Form the *Journal of Physiology*, with permission).

significant differences between the effects of the isocapnic and poikilocapnic hypoxic exposures on these parameters.

DISCUSSION

The first result of this study was that ventilation rises considerably over an 8 hour period of isocapnic hypoxia once the initial process of HVD has been completed. This rise in ventilation could be due to a special effect of isocapnic hypoxia as compared with poikilocapnic hypoxia, where no progressive rise was detected. Alternatively, the same underlying changes in the respiratory control system could be developing during poikilo-capnic hypoxia as well as during isocapnic hypoxia, but the expression of these changes (as a rising ventilation) is masked by the concurrent respiratory alkalosis.

The second result of this study is that both the ventilatory sensitivity to hypoxia (at fixed PET_{CO2}) and the component of ventilation that is insensitive to hypoxia (at fixed PET_{CO2}) increase during both poikilocapnic hypoxia and isocapnic hypoxia. No such change was detected with control air breathing conditions. This result suggests that the changes in respiratory control that occur are specifically the result of hypoxia, and do not depend on the isocapnic condition being maintained.

There have been previous studies of more sustained hypoxia in humans in which there has been an attempt to maintain isocapnia. Eger et al. (9) compared the effects of 8 hours of hypoxia with and without the maintenance of isocapnia. They found that the $\dot{V}E/PET_{CO2}$ response line was shifted to the left after both the isocapnic and the poikilocapnic hypoxic exposures, although the effect was more pronounced in the case of the poikilocapnic exposure. They did not assess hypoxic sensitivity. Our results are broadly consistent with these findings. Cruz et al (6) used a hypobaric chamber to compare the effects of 100 hours of hypoxia without and with CO_2 added to the atmosphere. They observed VAH with poikilocapnic hypoxia, but not with isocapnic hypoxia. However, they did not really study the ventilatory changes over the eight hour time course employed in this study. They found no significant changes in hypoxic sensitivity with the isocapnic protocol, but they may have made a type II statistical error as the figure in their paper shows the average hypoxic sensitivity almost doubled compared with control.

A series of experiments studying sustained isocapnic hypoxia has been undertaken in awake goats, using a technique whereby one of the carotid bodies can be perfused independently of the systemic circulation and where the other carotid body has been denervated (2,3,4). These experiments have shown that hypoxia confined to the carotid body is sufficient to induce VAH in the goat (i.e. systemic hypoxia is not required) and that VAH still occurs if the systemic hypocapnic alkalosis is prevented by adding CO_2 to the inspired gas. Our findings are reasonably consistent with these findings in the goat, and they suggest that a direct action of hypoxia on the carotid body may be important in explaining the rise in ventilatory sensitivity to hypoxia that we observe in our experiments. Our finding of a rise in the component of ventilation that is insensitive to hypoxia does not appear to be present in the results from the awake goat when systemic hypocapnia has been prevented, and is more difficult to explain.

REFERENCES

1. Astrand, P. (1953). The respiratory activity in man exposed to prolonged hypoxia. *Acta physiologica scandinavica*, **30**: 342-368.
2. Bisgard, G.E., Busch, M.A. & Forster, H.V. (1986). Ventilatory acclimatization to hypoxia is not dependent on cerebral hypocapnic alkalosis. *Journal of Applied Physiology*, **60**: 1011-1015.
3. Busch, M.A., Bisgard, G.E. & Forster, H.V. (1985). Ventilatory acclimatization to hypoxia is not dependent on arterial hypoxemia. *Journal of Applied Physiology*, **58**: 1874-1880.
4. Busch, M.A., Bisgard, G.E., Mesina, J.E. & Forster, H.V. (1983). The effects of unilateral carotid body excision on ventilatory control in goats. *Respiration Physiology*, **54**: 353-361.
5. Clement, I.D. & Robbins, P.A. (1993). Dynamics of the ventilatory response to hypoxia in humans. *Respiration Physiology*, **92**: 253-275.
6. Cruz, J.C., Reeves, J.T., Grover, R.F., Maher, J.T., McCullough, R.E., Cymerman, A. & Denniston, J.C. (1980). Ventilatory acclimatization to high altitude is prevented by CO_2 breathing. *Respiration*, **39**: 121-130.
7. Douglas, C.G., Haldane, J.S., Henderson, Y. & Schneider, E.C. (1913). Physiological observations made on Pike's Peak, Colorado, with special reference to adaptation to low barometric pressure. *Philosophical Transactions of the Royal Society*, **B203**: 185-318.
8. Easton, P.A., Slykerman, L.J. & Anthonisen, N.R. (1986). Ventilatory response to sustained hypoxia in normal adults. *Journal of Applied Physiology*, **61**: 906-911.
9. Eger, E.I.I., Kellogg, R.H., Mines, A.H., Lima-Ostos, M., Morrill, C.G. & Kent, D.W. (1968). Influence of CO_2 on ventilatory acclimatization to altitude. *Journal of Applied Physiology*, **24**: 607-615.
10. Howson, M.G., Khamnei, S., McIntyre, M.E., O'Connor, D.F. & Robbins, P.A. (1987). A rapid computer-controlled binary gas-mixing system for studies in respiratory control. *Journal of Physiology*, **394**: 7P.
11. Howson, M.G., Khamnei, S., O'Connor, D.F. & Robbins, P.A. (1986). The properties of a turbine device for measuring respiratory volumes in man. *Journal of Physiology*, **382**: 12P.
12. Robbins, P.A., Swanson, G.D. & Howson, M.G. (1982). A prediction-correction scheme for forcing alveolar gases along certain time courses. *Journal of Applied Physiology*, **52**: 1353-1357.
13. Weil, J.V. & Zwillich, C.W. (1976). Assessment of ventilatory response to hypoxia: Methods and interpretation. *Chest*, **70**(Supplement): 124-128.

53

INDIVIDUAL DIFFERENCES IN VENTILATORY AND HR RESPONSES TO PROGRESSIVE HYPOXIA FOLLOWING 100% O$_2$ EXPOSURE IN HUMANS

Y. Honda,[1] A. Masuda,[1] T. Kobayashi,[1] M. Tanaka,[1] S. Masuyama,[2] H. Kimura,[2] and T. Kuriyama[2]

[1] Department of Physiology
[2] Department of Chest Medicine
School of Medicine
Chiba University
260 Japan

INTRODUCTION

Ventilatory response to sustained hypoxia is known to exhibit a biphasic profile: an initial rapid augmentation followed by a gradual decline. The former is induced by an excitation of the peripheral chemoreceptor. Interestingly, regarding the latter secondary depression (defined as hypoxic ventilatory depression, HVD), it has been suggested by several investigators that its magnitude is in some way also determined by a centrally mediated discharge from the peripheral chemoreceptor (4).

The presence of peripheral chemoreceptor activities in ambient air breathing in humans at sea level is well established to be in the range of 10 - 20% of the resting ventilation (2,5). Accordingly, the presenceof ambient air HVD is also speculated to exist. HVD was found to become progressively resolved by increasing O$_2$ concentration in the inspired air (3).

In the present study, we examined the possible role of ambient air HVD in the control of ventilation by prior exposure of healthy humans to 100% O$_2$.

MATERIALS AND METHODS

Fourteen healthy subjects (8 males and 6 females) ranging from 20 to 40yrs old were studied with informed consent. The subjects were connected to a respiratory circuit and initially breathed either 100 % O$_2$ or room air for a 10 min period, and then while keeping P$_{ET_{CO_2}}$ constant, hypoxia was progressively induced by rebreathing 10 to 15 liters of room air contained in a rubber bag. End-tidal P$_{O_2}$ was lowered at a rate of about 5 Torr/min and rebreathing was

Modeling and Control of Ventilation, Edited by S. J. G. Semple, L. Adams, and B. J. Whipp
Plenum Press, New York, 1995

283

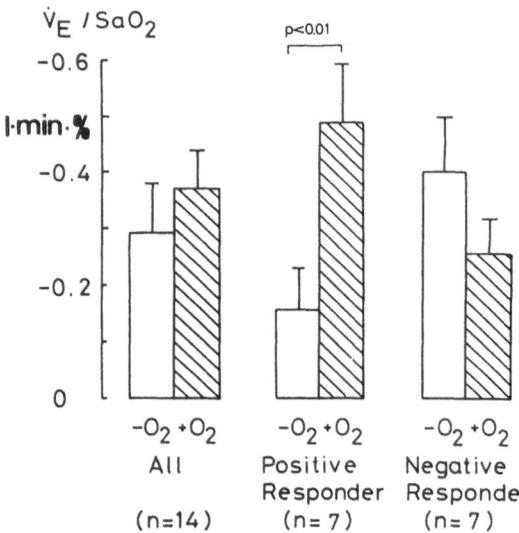

Figure 1. Average slope of hypoxic ventilatory response curve in all subjects as well as positive and negative responders. Open column: $-O_2$ run. Hatched column: $+O_2$ run. In the positive responder group, the slope was significantly augmented following hyperoxic exposure ($p < 0.01$).

terminated when SaO_2 dropped to 80%. This progressive hypoxia test with prior 100% O_2 or room air breathing was defined as $+O_2$ and $-O_2$ runs, respectively. Each subject was examined under a set of + and - O_2 runs 3 times in random order. Breath-by-breath V_T, P_{O_2} and P_{CO_2} were continuously recorded by a hot wire flowmeter and a rapid response O_2 and CO_2 analyzer. SaO_2 and HR were also continuously monitored by a pulse oximeter. A hypoxic ventilatory response (HVR) curve was obtained by linear regression analysis between SaO_2 and minute ventilation($\dot{V}_E$). Subjects consistently exhibiting higher HVR in the $+O_2$ run than in the $-O_2$ run in all three sets of trials were defined as the positive responders, whereas those who did not show such consistent tendency were defined as the negative responders.

RESULTS

Of the 14 subjects, there were 7 positive and 7 negative responders, with the male:female ratio being 3:4 in the former and 5:2 in the latter group. No definite gender difference appeared to exist in terms of distribution in the two groups.

Fig. 1 illustrates the average slope of the hypoxic ventilatory response curves of the + and - O_2 runs in all subjects as well as in positive and negative responders.

The average increment in the HVR slope in the positive responders was also significantly larger than in the negative responders. It was also noted that the average magnitude of HVR without prior O_2 exposure in the positive responders was considerably smaller than that in the negative responders, although the difference was not significant. Fig. 2 shows the average HR-$\dot{V}_E$ response slope obtained by the progressive hypoxia test. Significant depression was observed in the positive responders, whereas no statistical difference was detected in the negative responders as well as in the whole group.

DISCUSSION

We found in half of the subjects examined that their HVR was consistently augmented following 10 min of hyperoxic exposure and termed them positive responders. Since they

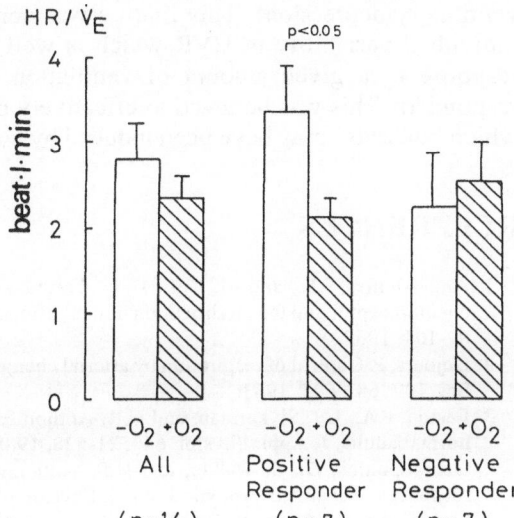

Figure 2. Average HR-VE response slope in the three groups as represented in Fig. 1. In the positive responder group, the slope was significantly depressed following hyperoxic exposure (p < 0.05).

exhibited lower HVR than the negative responders prior to O_2 exposure, we assumed that their HVR was effectively depressed when tested from room air breathing. Therefore, this finding was designated "Ambient air HVD".

Georgopoulos et al. (4) found in humans that augmented peripheral chemoreceptor activities by almitrine administration induced augmented HVD during sustained hypoxia. Thus, the magnitude of afferent peripheral chemoreceptor discharges appears in some way to be related to the degree of centrally mediated HVD. The presence of peripheral chemoreceptor activities during ambient air breathing at sea level is well established in humans. Dejours (2) estimated it to be about 10% and Severinghaus et al.(6) 17% of the resting ventilation by the single breath O_2 test and elevated $PaCO_2$ level after peripheral chemodenervation, respectively. Furthermore, we recently found in cats that the threshold P_{O_2} for carotid sinus nerve discharges was far below the normoxemic P_{O_2} level (Natsui, Kuwana and Honda, unpublished observation). These considerations led us to believe that the existence of "Ambient air HVD" seems physiologically tenable.

It is well known that HVR is individually very different, and even in normal healthy subjects it is not rare that it is severely depressed (6). The present finding may, at least in part, explain the variation in HVR among different individuals.

In the positive responders, the HR response to a given amount of ventilation ($HR/\dot{V}_E$) in the $+O_2$ run was significantly depressed. The possible underlying physiologic mechanism may be an enhanced bradycardic effect due to augmented peripheral chemoreceptor activities (1). This resulted in the prevention of excessive tachycardia which otherwise may have been induced by an elevated pulmonary vagal inflation reflex.

CONCLUSION

Out of 14 subjects examined, 7 consistently exhibited an enhanced hypoxic ventilatory response (HVR) following a 10 min hyperoxic exposure (positive responders). The other 7 showed an inconsistent tendency (negative responders). Prior to the hyperoxic exposure, the positive responders had lower HVR than the negative responders. It was assumed that they had effectively depressed ventilation while breathing room air (Ambient air hypoxic ventilatory depression). This finding was considered to explain, at least in part, the great individual variability in HVR which is well known to exist in healthy humans. The HR response to a given amount of ventilation was significantly depressed in the positive responders. This was believed to effectively prevent excessive tachycardia during hypoxia which otherwise may have been induced by an augmented pulmonary vagal inflation reflex.

REFERENCES

1. Angel-James, J.E., and M. DeB. Daly. Cardiovascular responses in apnoeic asphyxia: role of arterial chemoreceptor and the modification of their effects by a pulmonary vagal inflation reflex, J. Physiol. 201: 87-104, 1969.
2. Dejours, P. Control of respiration by arterial chemoreceptor. In: Regulation of respiration, Ann. NY Acad. Sci. 109: 683-695, 1963.
3. Easton, P.A., L.J. Slykerman, and N.R. Anthonisen. Recovery of the ventilatory response to hypoxia in normal adults. J. Appl. Physiol. 64: 521-528, 1988.
4. Georgopoulos, D., S. Walker, and N.R. Anthonisen. Increased chemoreceptor output and ventilatory response to sustained hypoxia. J. Appl. Physiol. 67: 1157-1163, 1989.
5. Severinghaus, J. W., C.R. Bainton, and A. Carcelen. Respiratory insensitivity to hypoxia in chronically hypoxic man. Respir. Physiol. 1: 308-334, 1966.
6. Severinghaus, J.W. Hypoxic respiratory drive and its loss during chronic hypoxia. Clin. Physiol. (Tokyo) 12: 57-79. 1972.

54

CHANGES IN BLOOD FLOW IN THE MIDDLE CEREBRAL ARTERY IN RESPONSE TO ACUTE ISOCAPNIC HYPOXIA IN HUMANS

M. J. Poulin and P. A. Robbins

University Laboratory of Physiology
Parks Road, Oxford OX1 3PT
United Kingdom

INTRODUCTION

While it is known that cerebral blood flow is increased by hypoxia (5), the dynamics of the cerebral blood flow response to hypoxia have not yet been adequately described. This is important to respiratory physiologists because an increase in cerebral blood flow will affect the difference between the P_{CO_2} in the arterial blood and the local P_{CO_2} of the central chemoreceptors, thus affecting the stimulus at the central chemoreceptors.

One method which has the potential to provide a continuous measurement of blood flow with the required time resolution is transcranial Doppler ultrasound. However, previous studies using transcranial Doppler ultrasound to assess changes in cerebral blood flow in response to hypoxia have measured either maximum velocity or intensity-weighted mean velocity, which are only proportional to flow if the cross-sectional area of the vessel remains constant (7). It has been suggested that a more suitable index of cerebral blood flow is the product of the intensity-weighted mean of the average velocity signal multiplied by the overall intensity (power) signal (8). Since the power signal is proportional to cross-sectional area (2), the proposed index allows for any changes in cross-sectional area of the vessel.

Studies of the dynamic response of the cerebral vasculature to hypoxia also require precise control continuously over both the arterial P_{CO_2} and P_{O_2}. This is because of the marked sensitivity of cerebral blood flow to arterial P_{CO_2}. A dynamic end-tidal forcing apparatus was employed in the current study to obtain the necessary control over end-tidal P_{CO_2} and P_{O_2}.

Thus, the purpose of this study was to assess middle cerebral artery blood flow (MCAF) in response to 20 min of acute isocapnic hypoxia in a group of healthy young men. Transcranial Doppler ultrasound was used to measure the intensity-weighted mean of the velocity signal and the overall intensity (power) from which MCAF was determined.

Modeling and Control of Ventilation, Edited by S. J. G. Semple, L. Adams, and B. J. Whipp
Plenum Press, New York, 1995

METHODS

Subjects

Four healthy young students volunteered for this study. The study requirements were fully explained in written and verbal forms to all participants, with each giving informed consent prior to participation in the study. The research was approved by the Central Oxford Research Ethics Committee. Each participant visited the laboratory, at the same time of day, on five occasions. Subjects were requested not to eat or drink caffeine-containing beverages within four hours prior to their scheduled testing sessions in the laboratory.

Measurement of Cerebral Blood Flow

A 2-MHz pulsed Doppler ultrasound system (PCDop 842, SciMed, Bristol UK) was used to measure backscattered Doppler signals from the right middle cerebral artery. The Doppler signals, including maximum and mean Doppler frequency shifts and power, were collected every 10 msec and saved to a file for later analysis. The heart was monitored continuously via ECG electrodes in a modified V-5 configuration. From this arrangement, the occurrence of QRS complexes was detected and the timing of each was saved to a file to enable beat-by-beat calculation of MCAF from the Doppler signals once the experiment had been completed.

The middle cerebral artery was identified by an insonation pathway through a right temporal window just above the zygomatic arch using standard techniques which have been described previously (1,9). Ultrasound gel was applied to the skin and hair of the temporal window as well as to the probe before the experimenter proceeded to locate and identify the main segment of the middle cerebral artery. Optimization of the Doppler signals from the middle cerebral artery was performed by varying the sample volume depth in incremental steps and, at each depth, varying the angle of insonance to obtain the best quality signals for the Doppler frequency shifts and power. Ultrasound gel was then re-applied sparingly to both the insonation site and the probe before securely positioning the probe in a headband device (Müller and Moll Fixation, Nicolet Instruments Limited, Warwick UK) to ensure optimal insonation position and angle for the duration of the experiment.

Results for MCAF were calculated every 10 msec as the product of the intensity-weighted mean of the velocity signal (MCAV) multiplied by the total power. Beat-by-beat calculations of MCAF and power were then performed and expressed as a percentage of the average value over a 3 min pre-hypoxic period. Group results for all subjects were ensemble-averaged over 15 second periods.

Apparatus and Technique for Dynamic-End-Tidal Forcings

Accurate control of the end-tidal gases was achieved using a computer-controlled fast gas mixing system which has been previously described in more detail (4,10).

Protocol

Prior to each experiment, subjects rested for 20 min. During this time, room air measurements of resting Doppler signals and $P_{ET_{CO_2}}$ were collected. Resting $P_{ET_{CO_2}}$ was measured using a nasal catheter which disturbs quiet breathing less than a mouthpiece and noseclip arrangement.

Table 1. Eucapnic end-tidal P_{CO_2} and distance (depth) from probe to the start of the Doppler sample volume for detecting signals from the middle cerebral artery for each subject

Subject	$P_{ET_{CO_2}}$ (Torr)	MCA depth (cm)
940	41.3	5.03
959	39.6	4.99
966	40.6	4.87
967	39.8	5.06
Mean ($\pm$ SD)	40.3 $\pm$ 0.8	4.99 $\pm$ 0.08

The experimental protocol consisted of a short 6-7 min period when $P_{ET_{O_2}}$ was held at 100 Torr. Then $P_{ET_{O_2}}$ was decreased within one or two breaths to 50 Torr and maintained constant for twenty minutes. Finally, $P_{ET_{O_2}}$ was returned in one step to 100 Torr and maintained constant for a further 10 min. $P_{ET_{CO_2}}$ was elevated by 1-2 Torr at the start of the experiment and maintained at that level for the duration of the protocol. Mean values for $P_{ET_{O_2}}$ and $P_{ET_{CO_2}}$ were determined on a breath-by-breath basis and group results for all subjects were ensemble-averaged over 30 seconds.

Assessment of the statistical significance of results was undertaken using Students paired t-tests. The probability level assumed for statistical significance was $p \leq 0.05$.

RESULTS

The four subjects who undertook the study had an average age of 20.3 $\pm$ 1.7 years (mean $\pm$ SD), an average height of 1.85 $\pm$ 0.06 m and an average weight of 77.8 $\pm$ 10.6 kg. None had a history of cardiovascular or respiratory disease and all had normal systolic (118.5 $\pm$ 6.0 mm Hg) and diastolic (78.5 $\pm$ 1.0) blood pressure.

Table 1 lists the individual values and the group mean for eucapnic $P_{ET_{CO_2}}$ which was held constant throughout. Also shown in Table 1 are the individual values and the group mean for the depth of the Doppler sample volume at which the main segment of middle cerebral artery was insonated. Small variations in depth are attributed to differences in skull size (9).

Figure 1 shows the group responses of MCAF and power to 20 min isocapnic hypoxia. The top panel shows the time-related changes in $P_{ET_{O_2}}$ and $P_{ET_{CO_2}}$, achieved by using a dynamic end-tidal forcing technique. The bottom panel shows the results for MCAF and power. Compared with baseline (100%), MCAF increased by 11.0 ($\pm$ 3.6) % with hypoxia, reaching steady-state responses within approximately 5-10 min and remaining at this level for the duration of the hypoxic challenge. When $P_{ET_{O_2}}$ was returned to 100 Torr, MCAF decreased to values similar to the pre-hypoxic baseline within approximately 3-5 min and remained at this level for the duration of the recovery period (99.2 $\pm$ 2.3%). Power however, did not change as a result of the hypoxic challenge, remaining at its pre-hypoxic value throughout the protocol (Table 2, Figure 1).

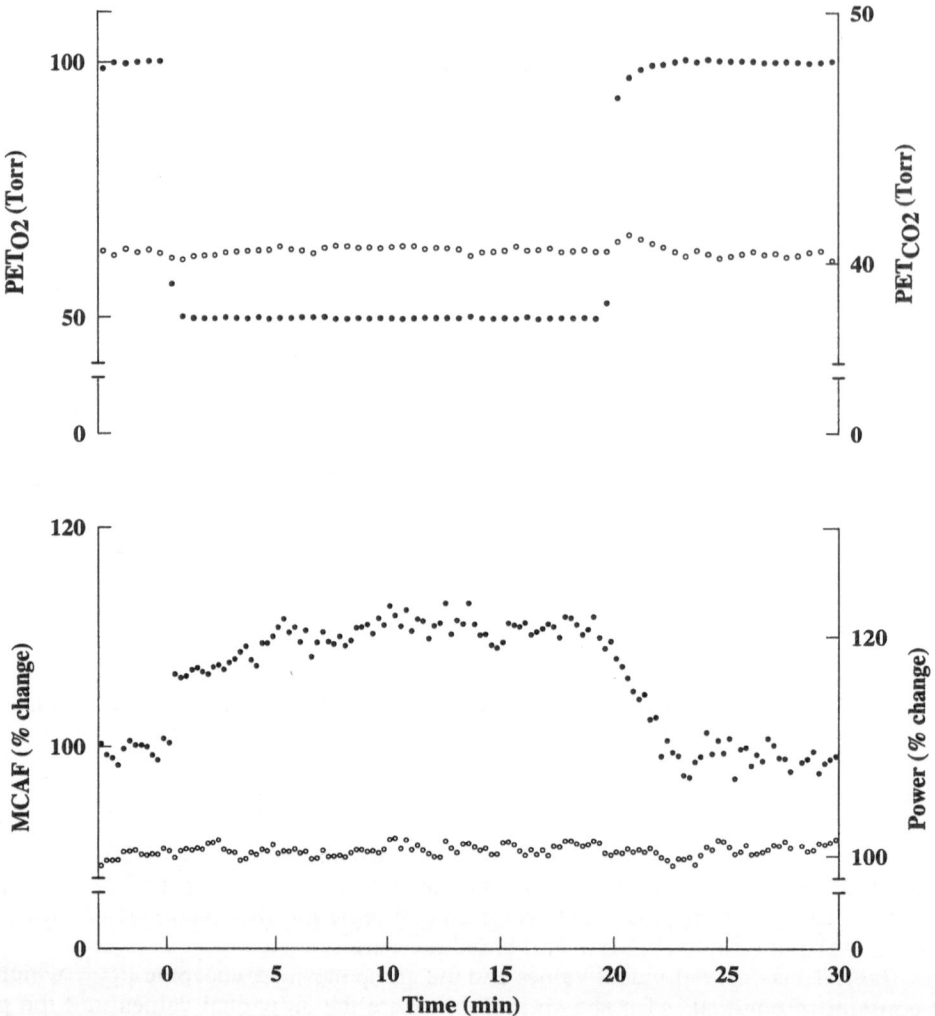

Figure 1. Response of middle cerebral artery blood flow (MCAF) and power to 20 min isocapnic hypoxia. *Top Panel:* Data for $P_{ET_{O_2}}$ (●) and $P_{ET_{CO_2}}$ (○) averaged over 30 s periods. *Bottom Panel:* Data for MCAF (●) and power (○) averaged over 15 s periods.

Table 2. Change in middle cerebral blood flow (MCAF) and power in response to acute isocapnic hypoxia

Variable	Baseline (3 min)	Hypoxia (10 min)	Recovery (3 min)
MCAF (%)	100.0 ± 0.0	111.0 ± 3.6*	99.2 ± 2.3
Power (%)	100.0 ± 0.0	100.8 ± 2.9	100.7 ± 1.2

Values are means $\pm$ SD, normalised from baseline values. Baseline, 3 min pre-hypoxic period; Hypoxia, 10 min period from 10 to 20 min; Recovery, 3 min period from 27 to 30 min.

*Significantly different from baseline ($p < 0.01$).

DISCUSSION

The results from these experiments show a quite modest increase in middle cerebral blood flow of 11% when $P_{ET_{O_2}}$ is lowered from 100 Torr to 50 Torr at a $P_{ET_{CO_2}}$ of 1-2 above resting. Kety and Schmidt (5) along with Shapiro et al. (11) used the N_2O technique to estimate whole brain perfusion, based on the Fick principle, and found increases in cerebral blood flow of 35 and 28%, respectively. Those studies used the steady-state technique to obtain the hypoxic responses. This technique requires several minutes to reach the desired hypoxic levels and the quality of stimulus depend on the ventilatory response itself. Additionally, while in the study by Shapiro et al. (11) CO_2 was added to the inspired gas line in an attempt to restore the $P_{ET_{CO_2}}$ at the control level, no such control was attempted by Kety and Schmidt (5). Ellingsen et al. (3) used transcranial Doppler and reported a 23% increase in carotid artery blood flow velocity in response to steps of moderate hypoxia ($PA_{O_2} = 65$ Torr) with constant $P_{A_{CO_2}}$ (40 Torr). The step changes in end-tidal P_{O_2} and P_{CO_2} however, were also somewhat slow to reach steady state levels. Thus, for the purpose of assessing cerebral blood flow responses to hypoxia, the quality of the stimulus in the above studies was somewhat imperfect and the dynamics of the response remain unclear.

A useful finding of our study is that reducing the $P_{ET_{O_2}}$ to 50 Torr under near eucapnic conditions had little effect on the-cross sectional area of the middle cerebral artery. This suggests that percentage changes in cerebral blood flow velocity reported previously under these conditions (3) are likely to be a reasonably accurate reflection of the underlying changes in blood flow.

Although we did not calculate the time constants of the cerebral blood flow response to hypoxia, our data do suggest an asymmetry in the on- and off- responses, with the on-response appearing to be somewhat slower than the off-response. This temporal profile appears similar to the data of Ellingsen et al. (3).

An increase in cerebral blood flow with hypoxia is of some interest to respiratory physiologists because this will reduce the difference between the local P_{CO_2} of the central chemoreceptors and the P_{CO_2} in the arterial blood, and this effect on its own would be expected to reduce the stimulus at the central chemoreceptors. To date however, the importance of this effect on ventilation remains unclear. Some investigators (12) have shown that, in the anaesthetized cat, the ventilatory depression during brain stem hypoxaemia may be due to a ventral medullary alkalosis associated with the hypoxaemic increase in cerebral blood flow and the washout of CO_2 at the site of the central chemoreceptors. However, in the conscious human, other investigators (6) have presented evidence to suggest that the depression of ventilation by hypoxia may involve instead an alteration of peripheral chemoreflex sensitivity. Our results suggest that, in conscious humans, changes in cerebral blood flow are insufficient to contribute greatly to the depression of ventilation by moderate hypoxia.

We conclude that, in response to moderate hypoxia under conditions of near eucapnia, cerebral blood flow in the middle cerebral artery increases by a modest amount, while the cross-sectional area remains unchanged. The time profile associated with the changes in cerebral blood flow clearly suggests that the response is asymmetric, with the increase in blood flow at the onset of hypoxia being substantially slower than the decrease in blood flow at the relief of hypoxia. The magnitude of the increase in cerebral blood flow however, suggests that changes in cerebral blood flow are unlikely to contribute greatly to the ventilatory decline observed with sustained isocapnic hypoxia.

ACKNOWLEDGMENTS

This study was approved by the Central Oxford Research Ethics Committee and was supported by the Wellcome Trust. M.J. Poulin was supported by an MRC (Canada) postdoctoral research fellowship.

REFERENCES

1. Aaslid, R., T. Markwalder, and H. Nornes. Noninvasive transcranial Doppler ultrasound recording of flow velocity in basal cerebral arteries. *J. Neurosurg.* 57: 769-774, 1982.
2. Doblar, D. D., B. G. Min, R. W. Chapman, E. R. Harbach, W. Welkowitz, and N. H. Edelman. Dynamic characteristics of cerebral blood flow response to sinusoidal hypoxia. *J. Appl. Physiol.* 46: 721-729, 1979.
3. Ellingsen, I., A. Hauge, G. Nicolaysen, M. Thoresen, and L. Walloe. Changes in human cerebral blood flow due to step changes in PAO_2 and $PACO_2$. *Acta Physiol. Scan.* 129: 157-163, 1987.
4. Howson, M. G., S. Khamnei, M. E. McIntyre, D. F. O'Connor, and P. A. Robbins. A rapid computer controlled binary gas mixing system for studies in respiratory control. *J. Physiol. (London)* 403: 103P, 1987.
5. Kety, S. S. and C. F. Schmidt. The effects of altered arterial tensions of carbon dioxide and oxygen on cerebral blood flow and cerebral oxygen consumption of normal young men. *J. Clin. Invest.* 27: 484-492, 1948.
6. Khamnei, S. and P. A. Robbins. Hypoxic depression of ventilation in humans: alternative models for the chemoreflexes. *Respir. Physiol.* 81: 117-134, 1990.
7. Kontos, H. A. Validity of cerebral arterial blood flow calculations from velocity measurements. *Stroke* 20: 1-3, 1989.
8. Lindegaard, K. F., T. Lundar, J. Wiberg, D. Sojberg, R. Aaslid, and H. Nornes. Variations in middle cerebral artery blood flow investigated with non-invasive transcranial blood velocity measurements. *Stroke* 18: 1025-1030, 1987.
9. Padayachee, T. S., F. J. Kirkham, R. R. Lewis, J. Gillard, M. C. E. Hutchinson, and R. G. Gosling. Transcranial measurement of blood velocities in the basal cerebral arteries using pulsed doppler ultrasound: a method of assessing the circle of Willis. *Ultrasound Med. Biol.* 12: 5-14, 1986.
10. Robbins, P. A., G. D. Swanson, and M. G. Howson. A prediction correction scheme for forcing alveolar gases along certain time courses. *J. Appl. Physiol.* 52: 1353-1357, 1982.
11. Shapiro, W., A. J. Wasserman, J. P. Baker, and J. L. Patterson. Cerebrovascular response to acute hypocapnic and eucapnic hypoxia in man. *J. Clin. Invest.* 49: 2362-2368, 1970.
12. Van Beek, J. H. G. M., A. Berkenbosch, J. DeGoede, and C. N. Olievier. Effects of brain stem hypoxaemia on the regulation of breathing. *Respir. Physiol.* 57: 171-188, 1984.

55

MIDDLE CEREBRAL ARTERY BLOOD FLOW VELOCITY STUDIED DURING QUIET BREATHING, REFLEX HYPERCAPNIC BREATHING AND VOLITIONALLY COPIED EUCAPNIC BREATHING IN MAN

Voluntary Control Of Breathing

P. Peters[1] and A. K. Datta[2]

[1] UMDS Department Surgery
St. Thomas' Hospital
London SE1 7EH, United Kingdom
[2] Neuroscience Critical Care Unit, Box 167
Addenbrooke's Hospital
Hills Road, Cambridge CB2 2QQ, United Kingdom

INTRODUCTION

The cerebrovascular circulation can be assessed non-invasively in man using transcranial doppler (TCD) velocimetry (1). The change in flow velocity pulse wave is proportional to the change in blood flow in that vessel, assuming cross sectional area of the artery studied to be constant. Such changes have been shown to correlate well with changes in cerebral perfusion in response to hypercapnia (3). TCD estimates of cerebral blood flow is increased when subjects performed both physical and cognitive activities (6) or when they were presented with visual stimuli (2). This suggests that TCD is capable of detecting changes in cerebral blood flow which accompany cerebral activity. In this study we have tested the hypothesis that volitionally controlled breathing is accompanied by an increase in middle cerebral artery (MCA) blood flow velocity (bfv) independent of changes in PCO_2.

METHODS

14 normal subjects (Age 19-36) were studied with ethical committee approval. Subjects lay supine on a couch in a quiet room with eyes open breathing through a mouthpiece attached to a respiratory valve; end tidal PCO_2 ($P_{ET}CO_2$; infra red; PK

Modeling and Control of Ventilation, Edited by S. J. G. Semple, L. Adams, and B. J. Whipp
Plenum Press, New York, 1995

Morgan), and expiratory flow, integrated to give tidal volume (V_T; pneumotachograph; Fleisch), digital arterial blood pressure (Ohmeda, Finapres) were recorded on FM tape (Racal) and a chartwriter (Devices). Right middle cerebral artery blood flow velocity (MCAbfv) was monitored continuously with transcranial doppler (EME TC2-64); the probe was fixed in place throughout the experiment over the squamous temporal bone using an elastic headband. Each subject was allowed to rest and breathed air until a stable respiratory pattern and $P_{ET}CO_2$ were achieved and then measurements of mean MCAbfv, V_T and $P_{ET}CO_2$ were recorded over a two minute period (REST). The subjects were then asked to voluntarily match their pattern of resting ventilation (VOL COPY); they were given visual feedback via a slowly scrolling digital oscilloscope screen (Gould) of the previously recorded resting V_T trace and their current V_T trace. $P_{ET}CO_2$ was again recorded to ensure matching of ventilation to the REST value for at least one minute. MCAbfv measurements averaged over two further minutes of VOL COPY were then made. Visual feedback was then removed and unknown to subjects the inspirate was changed to 5%CO_2/95% air to determine hypercapnic reactivity index (3); once a stable hypercapnic $P_{ET}CO_2$ was achieved, measurements of mean MCAbfv, expiratory minute volume and $P_{ET}CO_2$ were recorded over a further two minute period (HYPERCAPNIA).

RESULTS

HYPERCAPNIA produced an increase in MCAbfv compared to REST (median hypercapnic reactivity index 31.3 units, range 17.7-91.9; all these values are within the normal range). No change in blood pressure was seen between REST and VOL COPY. Overall there was a mean increase in MCAbfv during VOL COPY compared to REST of 8.8% (95% confidence interval 1.8%-15.8%; $p < 0.05$, Wilcoxon); this increase in MCAbfv occurred despite a small (1 mmHg; $p=0.05$, Wilcoxon) decrease in mean $P_{ET}CO_2$. The results for change in MCAbfv and $P_{ET}CO_2$ are plotted for each subject in Fig.1.

Three subjects were unable to match the REST ventilation during VOL COPY (mean $P_{ET}CO_2 > \pm 1.5$ mmHg compared to REST). In the other 11 subjects the mean increase in MCAbfv during VOLCOPY compared to REST was 12.2% (95% confidence interval 5-19.3%, $p < 0.005$, Wilcoxon).

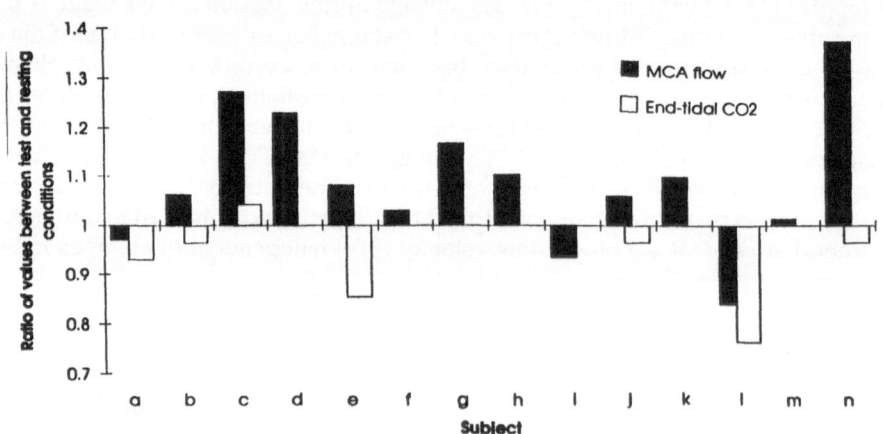

Figure 1. Percentage Change in MCAbfv and $P_{ET}CO_2$ during VOL COPY compared to REST in each subject.

DISCUSSION

The results of this study show that right MCAbfv is increased when respiration is controlled voluntarily compared to resting breathing even though end tidal CO_2 is matched. These values are analogous to changes seen in cortical blood flow seen bilaterally in the primary motor cortex, the right premotor cortex and supplementary motor area by positron emission scanning when voluntary inspiration (4) and expiration (7) is compared with passive mechanical ventilation. Whilst the region of supply of the cerebral arteries is variable (8), the middle cerebral artery is likely to supply blood to most of the lateral cerebral cortical regions. The changes in middle cerebral artery blood flow velocity seen in this study are therefore likely to reflect changes in relevant motor activity. The mechanism of the increased middle cerebral artery blood flow observed with voluntary breathing is probably similar to that seen during other types of cerebral activation (5), namely that accompanying neuronal activity and metabolism is a metabolite related dilatation of resistance arterioles and consequent increased local blood flow. We conclude that transcranial doppler offers an opportunity of non-invasively investigating respiratory task-related changes in relevant cerebral blood flow simply and repeatably in man; there appears to be an increase in mean MCAbfv during voluntary breathing compared to quiet spontaneous breathing.

ACKNOWLEDGMENTS

The expert assistance of Miss Kate Claridge and Mrs. Ann Donald is gratefully acknowledged

REFERENCES

1. Aaslid R., Markwalder T.M. & Nornes H. (1982) Noninvasive transcranial Doppler ultrasound recording of flow velocity in basal cerebral arteries. J.Neurosurg., 47, 769-774
2. Aaslid R. (1987) Visually evoked dynamic blood flow response of the human cerebral circulation. Stroke, 18, 771-775
3. Bishop CCR, Powell S, Rutt D & Browse (1986) Transcranial Doppler measurement of middle cerebral artery blood flow velocity: a validation study. Stroke, 17, 913-915
4. Colebatch J.G., Adams L., Murphy K., Martin A.J., Lammertsma A.A., Tochon-Danguy H.J., Clark J.C., Friston K.J. & Guz A. (1991) J.Physiol., 443, 91-103
5. Ginsberg M.D., Chang J.Y., Kelley R.E., Yoshii F., Barker W.W., Ingenito G. & Boothe T.E. (1988) Increases in both cerebral glucose utilization and blood flow during execution of a somatosensory task. Ann. Neurol., 23, 152-160.
6. Kelley R.E., Chang J.Y., Scheinmann N.J., Levin B.E., Duncan R.S. & Lee S.C. (1992) Transcranial Doppler assessment of cerebral flow velocity during cognitive tasks. Stroke, 23, 9-14
7. Ramsay S.C., Adams L Murphy K., Corfield D.R., Grootonk S., Bailey D.L., Frackowiack R.S. & Guz A. (1993) Regional Cerebral blood flow during volitional expiration in man: a comparison with volitional inspiration. J.Physiol., 461, 85-101
8. van der Zwan A. & Hillen B.(1991) Review of the variability of the territories of the major cerebral arteries. Stroke, 22, 1078-1084

56

THE EFFECTS OF HYPOXIA AND HYPEROXIA ON THE 1/F NATURE OF BREATH-BY-BREATH VENTILATORY VARIABILITY

S. A. Tuck, Y. Yamamoto, and R. L. Hughson

Department of Kinesiology
University of Waterloo
Waterloo, Ontario N2L 3G1, Canada

INTRODUCTION

Fractal characteristics (self-similar, scale invariant fluctuations) have been found in many physiological time series including the firing of single neurons (10), the opening and closing of ion channels (10), and in heart rate (6) and blood pressure variability (7). A simple method of characterizing fractal fluctuations in a time series is by estimation of the spectral exponent or β from the power spectrum; β is the negative slope of a linear regression of the power spectrum plotted in a log-power, log-frequency plane. White noise has a value of $\beta=0$ (same spectral power at all frequencies) whereas fractal "noise" (also referred to as coloured or nonharmonic noise) has values of $\beta>0$ (see Figure 1). Power spectra with $\beta>0$ are also called inverse power law spectra as spectral power is inversely proportional to frequency.

The method of coarse graining spectral analysis (CGSA) (13), which isolates harmonic and fractal components of time series data, has been used to study the nonharmonic component of heart rate variability in humans (14). The same method has been used in a preliminary study that provided evidence for fractal structure in human breath-by-breath ventilatory variability during exercise (5). This study investigates the pattern of breathing using coarse graining spectral analysis. The primary purpose was to determine if the structure of the breath-by-breath variability could be characterized by computation of the β value and if the β was dependent on the fractional inspired oxygen concentration (F_IO_2). In the process of determining the nature of the pattern of breathing, two methodological issues were raised. Therefore, we also report on the computation of the β value as influenced by 1) the uneven interbreath intervals, and 2) "outlying" values due to large breaths or sighs.

Modeling and Control of Ventilation, Edited by S. J. G. Semple, L. Adams, and B. J. Whipp
Plenum Press, New York, 1995

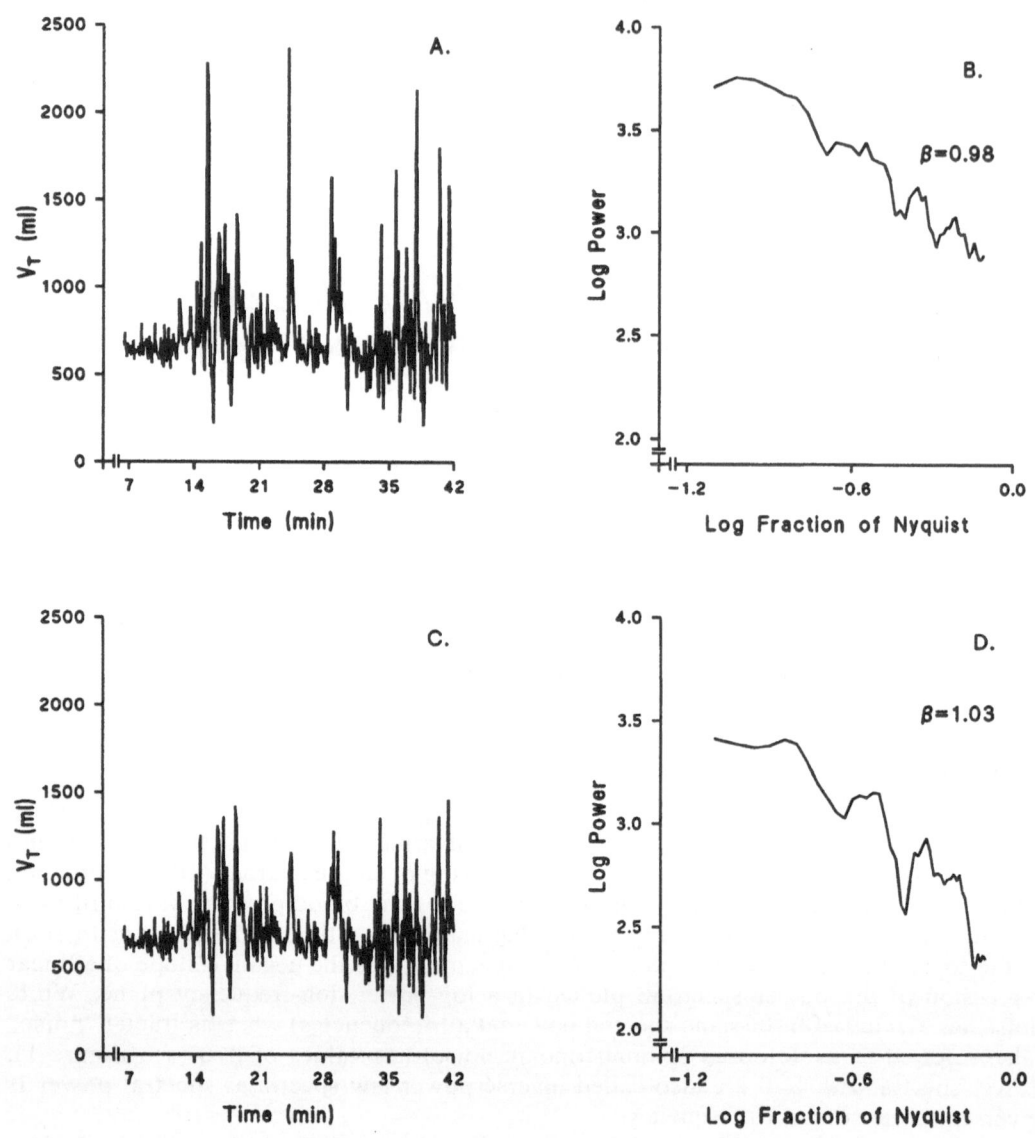

Figure 1.

METHODS

Seven healthy men (age 20-35 years) volunteered for this study after being informed of the nature of the experiments and signing a consent form approved by the Office of Human Research of the university. Subjects were not told that ventilation was the variable of interest; rather, they were told about measurement of heart rate and simply that they would breathe through a face mask. Resting ventilatory and gas exchange measurements were made on seven male subjects while breathing room air (normoxia, $F_IO_2 = 0.2$), hypoxic ($F_IO_2 = 0.14$), and hyperoxic ($F_IO_2 = 0.50$) gas mixtures. The order of the testing was randomized between subjects. In each condition, minute ventilation ($\dot{V}_E$), tidal volume (V_T), breathing frequency

(f_B), inspiratory and expiratory time (T_I and T_E), and estimated mean alveolar PCO_2 (P_ACO_2) (12) were measured every breath for 520 breaths. Ventilation was measured as the subjects breathed through a facemask (Hans Rudolph Inc. Series 7930) modified to allow nasal breathing. Volume was measured with a screen pneumotachograph and corrected to BTPS based on inspired and expired conditions measured near the screen. Calibrations were with a 1 litre syringe pre- and post-test, and with simultaneous collection of expired air in a Tissot spirometer during the test.

Spectral Analysis

As Fourier analysis requires equally spaced samples, the raw time series were realigned with the sampling frequency set to the average time between breaths. A 256 point Fourier transform was performed on these time series, and the technique of coarse-graining spectral analysis was used to separate harmonic from non-harmonic components. The power spectrum of the non-harmonic component was plotted in a log-power, log-frequency plane, and a linear regression was done between 0.08 to 0.8 of Nyquist frequency (where Nyquist = 1/2 mean breathing frequency). The negative slope of the regression within this range was taken as an estimate of the spectral exponent, β.

To investigate the effect of sampling technique on the estimate of β of the f_B time series, the same raw time series were interpolated every second using a zero order "hold" function, and β was estimated as above. To investigate the effects of excluding outlying breaths, values $\pm$ 3 standard deviations (S.D.'s) from the mean in the V_T time series were replaced with the mean of the two neighbouring V_T's for four subjects, and β was calculated as above.

RESULTS

Application of coarse graining spectral analysis to these data indicated that $\geq 81\%$ of the total spectral power was extracted as nonharmonic power with the small remainder extracted as harmonic power. The average estimates of β, the slope of the nonharmonic component, were significantly greater than zero for all variables except $\dot{V}_E$ in the hypoxic tests (Table 1). Thus the variability in the non-harmonic component was not white and could be considered a type of coloured noise.

The β values for $\dot{V}_E$ and V_T were significantly less in the hypoxic, and greater in the hyperoxic tests than in normoxia. While β values for each of T_E and P_ACO_2 were less in hypoxia than normoxia (Table 1).

When the f_B time series sampled at the mean breathing frequency (RLIGN) were compared to the same time series interpolated and sampled at 1 Hz (INTER) (Table 2), the estimate of β was higher for all subjects when the interpolation technique was used.

Four subjects were selected to test the effect of outlier data because of the presence of a number of large breaths or sighs. When outliers $\pm$ 3 S.D's were removed, all four

Table 1. The estimates of β for each ventilatory variable during rest

F_IO_2	$\dot{V}_E$	V_T	f_B	T_I	T_E	P_ACO_2
0.14	0.13±0.16*	0.15±0.12*	0.54±0.25	0.28±0.26	0.33±0.18*	0.39±0.16*
0.21	0.37±0.11	0.27±0.09	0.71±0.20	0.44±0.22	0.57±0.24	0.69±0.18
0.50	0.58±0.15*	0.38±0.06*	0.65±0.15	0.45±0.18	0.51±0.13	0.84±0.17

*$P<0.05$ with respect to normoxia ($F_IO_2 = 0.21$).

Table 2. The effect of sampling technique on β of the f_B time series

	1	2	3	4	5	6	7	Mean±SE
RLIGN	0.61	0.59	0.84	0.61	0.79	0.45	1.05	0.71±0.08
INTER	0.62	0.82	0.89	0.89	0.84	0.80	1.31	0.88±0.08

Values are for individual subjects number 1-7.

Table 3. The effect of including or excluding outliers on the estimate of β of the V_T time series for 4 subjects

Subject	1	5	6	7
Included	0.10	0.11	0.31	0.98
Excluded	0.36	0.34	0.44	1.03

examples showed an increase in the estimate of β compared to the time series with the outliers left in (Table 3).

DISCUSSION

The present study has confirmed that there is a large percentage of spectral power that is not confined to the harmonic component. The observation that the slope (β) of the nonharmonic power when plotted as the log spectral power versus log frequency was significantly greater than zero in all cases tested but one suggested that the noise was not simply white, or random, noise. Changes in inspired O_2 concentration affected the slope of the nonharmonic component, with a generally flatter slope (β value closer to 0) in hypoxia, and steeper slope in hyperoxia.

Methodology

The effectiveness of coarse graining spectral analysis in extracting nonharmonic signals from mixed harmonic and nonharmonic data sets has been confirmed (13). An important consideration when applying any spectral analysis to ventilatory data is sampling technique. As the time between breaths varies from breath to breath, and as Fourier analysis requires equally spaced samples, some manipulation of the raw time series needs to be done. Two common methods are realignment (also called an interval tachogram) and interpolation. An interval tachogram involves realigning the breaths to be separated by the average interbreath interval. Realignment is easily done, however, a drawback is that resolution in terms of time is not uniform (4); any pair or group of breaths at breathing frequencies higher than average breathing frequency get "expanded", whereas those lower than average breathing frequency get "compressed". Interpolation involves use of some function (here, a zero order hold) to add the value of a time series at equal spaced intervals, introducing "new" data, and potentially high frequency noise.

The lower estimation of β with the realignment technique seen in this study could be explained with reference to an inverse power law spectrum which, as shown in Table 1, the ventilatory variables do possess. By definition, spectral power increases as frequency decreases, thus, if breathing frequency has an inverse power law spectrum, greater spectral power would be expected in the lower frequency (< mean f_B) breaths, and less in the higher frequency breaths (> mean f_B). By forcing sampling frequency to be the mean f_B, this has the effect of "shifting" the low frequency/high power breaths up towards the Nyquist frequency, tending to flatten (make less steep) the spectrum below Nyquist. Overall, this bias could be offset by the high frequency/low power breaths shifting down towards Nyquist, but as the regression range used to calculate β is only to 0.8 of Nyquist, only the low frequency/high power breath distortion is included, tending to decrease the values of β. We selected 0.8 of Nyquist as the upper cutoff because the relationship was generally linear for all subjects and all variables to this point.

A component of normal ventilation is occasional large breaths or sighs, manifest primarily in tidal volume. It is important to decide whether such breaths that lie far from the mean value should be excluded from analysis. These outliers contribute large amounts of spectral power, primarily in the higher frequencies, given their very large amplitudes, thus, their exclusion in this study was shown to increase the estimates of β. An issue that remains to be resolved is whether these large breaths are being regulated by a mechanism independent of the factors underlying the normal breath-by-breath pattern of breathing.

Physiological Implications

Our rationale for the experimental design of this study was that we believed there might be differences in the patterns of breath-by-breath variability that were dependent on F_IO_2. With hyperoxia, the carotid chemoreceptors are often "silenced" (11) implying that chemoreceptor control of ventilation will be exclusively via the central chemoreceptors. On the contrary, hypoxia potentiates the output of the carotid chemoreceptors so that ventilatory control can be effected more rapidly in response to changes in arterial blood gas composition. These hypotheses were supported by the outcome of this study. The pattern of breath-by-breath variability exhibited less deviation away from the mean value in hypoxia. The β values were closer to 0. In contrast, the pattern in hyperoxia showed more systematic variation away from the mean and the β was closer to 1 for many variables. This suggests that the central and peripheral chemoreceptors influence the pattern of breathing, and that they do so in different ways.

The current results that supported the concept of differing patterns of breath-by-breath variability contrast with previous conclusions of no change. However, the other studies analyzed distribution histograms with hypercapnia (8), or run length and turning points during hypoxia and hypercapnia (2). Indeed, if the current study had relied on coefficient of variation for which there was no difference between F_IO_2 (data not presented) rather than the spectral exponent, we too would have concluded no effect. The findings are in agreement with experiments on cats in which deviation from the mean was reduced by hypoxia (3).

Of central issue, given the non-zero values of β, is whether the pattern of ventilatory variability represents a fractal process or not. Fractal is, in some literature, taken for $1 \leq \beta \leq 3$, while $0 < \beta < 1$ is also a self-similar process classed as a type of coloured noise with long-range power law autocorrelation (10). Benchetrit and Bertrand (1) used an autoregressive modelling approach to show the existence of "memory" in the breathing pattern. It might be that the underlying respiratory pattern generator and its output are fractal and/or chaotic (9), but this will require further study of the components that make up the respiratory control system.

ACKNOWLEDGEMENTS

This research was supported by Natural Sciences and Engineering Research Council, Canada.

REFERENCES

1 Benchetrit, G. and F. Bertrand. A short term memory in the respiratory centres: statistical analysis. *Resp. Physiol.* 23: 147-158, 1975.

2 Bolton, D. P. G. and J. Marsh. Analysis and interpretation of turning points and run lengths in breath-by-breath ventilatory variables. *J. Physiol. (London)* 351: 451-459, 1984.

3 Eldridge, F. L., D. Paydarfar, P. G. Wagner, and R. T. Dowell. Phase resetting of respiratory rhythm: effect of changing respiratory "drive". *Am. J. Physiol.* 257: R271-R277, 1989.

4 Goodman, L. Oscillatory behavior of ventilation in resting man. *IEEE Trans. Biomed. Eng.* BME-11: 82-93, 1964.

5 Hughson, R. L. and Y. Yamamoto. On the fractal nature of breath-by-breath variation in ventilation during dynamic exercise. In: *Control of Breathing and Its Modeling Perspectives*, edited by Y. Honda, Y. Miyamoto, K. Konno, and J. Widdicombe. New York: Plenum Press, 1992, pp.255-262.

6 Kobayashi, M. and T. Musha. 1/f Fluctuation of heartbeat period. *IEEE Trans. Biomed. Eng.* BME-29: 456-457, 1982.

7 Marsh, D. J., J. L. Osborn, and A. W. Cowley,Jr. $1/f$ Fluctuations in arterial pressure and regulation of renal blood flow in dogs. *Am. J. Physiol. Renal,Fluid Electrolyte Physiol.* 258: F1394-F1400, 1990.

8 Newsome-Davis, J. and D. Stagg. Interrelationships of the volume and time components of individiual breaths in resting man. *J. Physiol. (Lond.)* 245: 481-498, 1975.

9 Sammon, M. Symmetry, bifurcations, and chaos in a distributed respiratory control system. *J. Appl. Physiol.* 77: 2481-2495, 1994.

10 Stone, L. Coloured noise or low-dimensional chaos. *Proc. R. Soc. Lond. [Biol.]* 250: 77-81, 1992.

11 Ward, S. A. Peripheral and central chemoreceptor control of ventilation during exercise in humans. *Can. J. Appl. Physiol.* 19: 305-333, 1994.

12 Whipp, B. J., N. Lamarra, S. A. Ward, J. A. Davis, and K. Wasserman. Estimating arterial PCO2 from flow-weighted and time-average alveolar PCO2 during exercise. In: *Respiratory Control: A Modeling Perspective*, edited by G. D. Swanson, F. S. Grodins, and R. L. Hughson. New York: Plenum Press, 1989, p. 91-100.

13 Yamamoto, Y. and R. L. Hughson. Extracting fractal components from time series. *Physica D* 68: 250-264, 1993.

14 Yamamoto, Y. and R. L. Hughson. On the fractal nature of heart rate variability in humans: effects of data length and β-adrenergic blockade. *Am. J. Physiol.* 266: R40-R49, 1994.

57

CHOLINERGIC DIMENSIONS TO CAROTID BODY CHEMOTRANSDUCTION

Robert S. Fitzgerald, Machiko Shirahata, and Tohru Ide

Departments of Environmental Health Sciences, (Division of Physiology),
    Physiology, Medicine, and Anesthesiology/Critical Care Medicine
The Johns Hopkins Medical Institutions
Baltimore, Maryland 21205

INTRODUCTION

Prevalent current opinion depicts the basic chemoreceptive unit in the carotid body as a glomus or Type I cell being a storehouse of several neurotransmitters; apposed to it is a neuron, properly called a dendrite, whose cell body is in the petrosal ganglion and whose axon terminates in nucleus tractus solitarius. Prevalent current opinion describes hypoxic chemotransduction in the following steps: (a) Hypoxia by some mechanism, perhaps involving a change in the configuration of a membrane protein, depolarizes the glomus cell, perhaps by altering the conductance of an ion channel in the membrane; (b) Extracellular calcium enters the glomus cell; (c) One or more excitatory neurotransmitters is released into the synaptic cleft between the glomus cell and the apposed neuron; (d) The neurotransmitter binds to a postsynaptic receptor; (e) An action potential is initiated in the neuron. OUR HYPOTHESIS: *ACETYLCHOLINE IS AN ESSENTIAL EXCITATORY NEUROTRANSMITTER IN THE PROCESS OF HYPOXIC CHEMOTRANSDUCTION.*

Several opinions were offered beginning with the early workers in the field. Some thought that acetylcholine (ACh) was such a neurotransmitter (12,13,16). Others thought that there was little if any chance that ACh was an essential excitatory neurotransmitter involved in hypoxic chemotransduction (2,3,10,11,14,15).

The model behind our experimental testing of the hypothesis was transmission in the superior cervical ganglion. This was one of the first models used by the early investigators. But we changed it a little to include not only nicotinic receptor involvement but also muscarinic receptor involvement. The nicotinic receptor is responsible for the fast excitatory postsynaptic potential (fEPSP), the M_2 receptor is responsible for the slow inhibitory postsynaptic potential (sIPSP), and the M_1 receptor is responsible for the slow excitatory postsynaptic potential (sEPSP)(1).

Modeling and Control of Ventilation, Edited by S. J. G. Semple, L. Adams, and B. J. Whipp
Plenum Press, New York, 1995

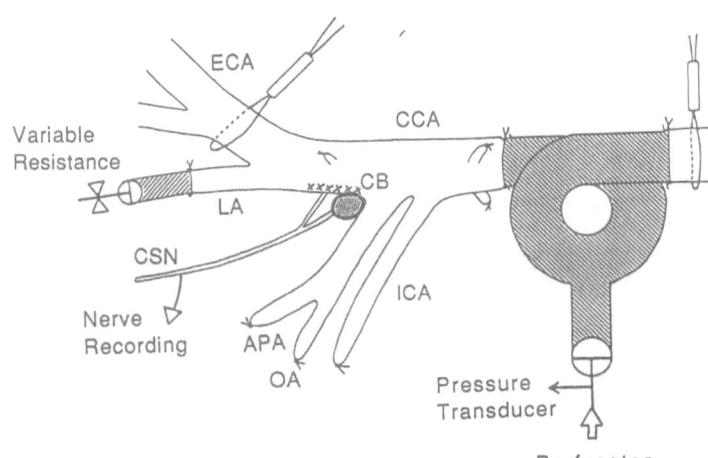

Figure 1. Perfusion Set-Up. For a transient, selective perfusion the snares around the common carotid artery (CCA) and the external carotid artery (ECA) were drawn tight stopping blood flow to the carotid body area, the stopcock was turned and the perfusate pumped into the loop from the water-jacketed bath and tubing at the previous arterial pressure (cf. Figure 2). Outflow of the perfusate from the carotid body area was by the lingual artery into a reservoir and the venous system of the carotid body into the right heart. Pump flow was 5 - 8 ml/min. LA = lingual artery; CSN = carotid sinus nerve; APA = ascending pharyngeal artery; OA = occipital artery; ICA = internal carotid artery From Fitzgerald and Shirahata (8).

METHODS

Our methods have been published recently (8). In brief and shown in Figure 1, we used the anesthetized, paralyzed, artificially ventilated cat; recording from the whole carotid sinus nerve after thermally and mechanically removing the baroreceptors.

In either a hypoxic or a normoxic cat we selectively perfused the carotid body with a hypoxic Krebs Ringer bicarbonate solution (KRB), or with hypoxic KRB which contained various concentrations of various blockers. A nicotinic blocker (mecamylamine) was used together with a muscarinic blocker (atropine).

RESULTS

A sample polygraph record of the results is shown in Figure 2. It should be emphasized that the mecamylamine and atropine are provided *ONLY* to the carotid

Table 1.

Agent	Perfusion #1 (HH KRB) Time (sec)						Perfusion #2 (HH KRB, HH KRB + GALLAM., HH KRB + PIRENZ.) Time (sec)						
	10	20	30	40	50	60	10	20	30	40	50	60	70
HH KRB	65	90	94	96	98	100	85	90	96	100	100	100	100
+GALLAM.	62	80	92	97	98	100	130	140	138	134	130	128	122
+PIRENZ.	55	72	85	96	98	100	85	85	80	78	75	73	70

Units are in percent of the maximum response (in μV to perfusion #1 which occurred at 60 seconds; this was defined as 100%.

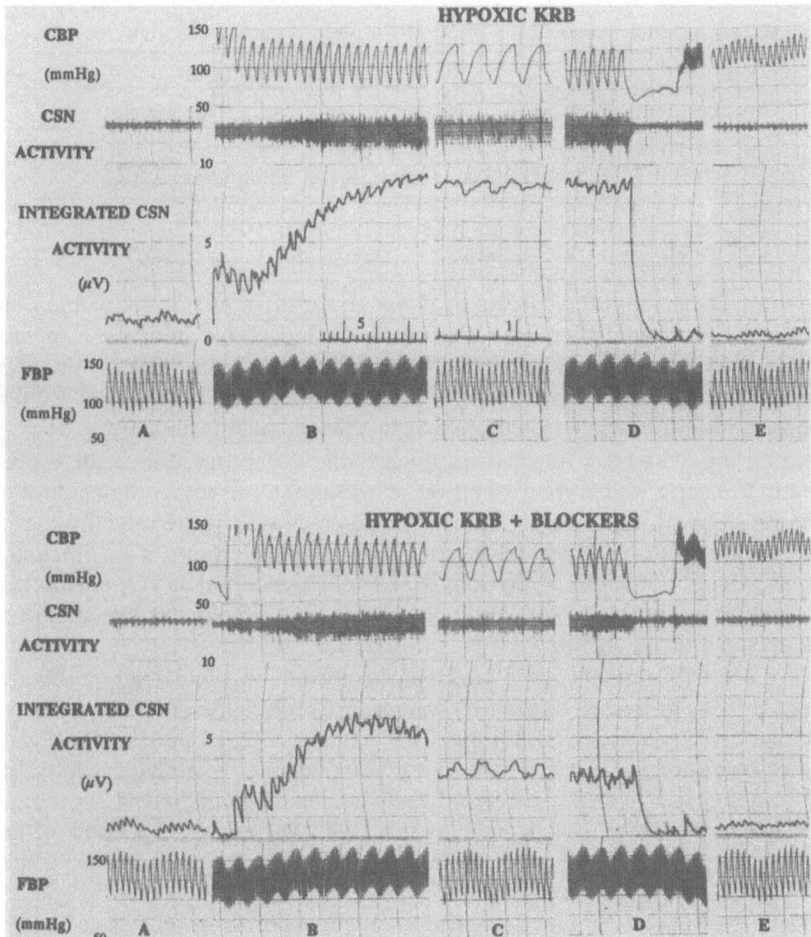

Figure 2. Polygraph trace of hypoxic KRB without blockers (top panel) and with blockers (bottom panel; [in µM] mecamylamine, 402; atropine, 942). In this preparation baroreceptors were left intact. Top panel: A = Control; B = opening seconds of hypoxic perfusion; C = accelerated paper speed; note the oscillation of the raw and integrated nerve activity with the oscillations of the perfusion pressure (CBP = carotid blood pressure) *and* the magnitude (i.e., height) of the trace; D = conclusion of hypoxic perfusion and return to perfusion with normoxic, normocapnic blood; E = 1 min post perfusion . Bottom panel: A, B = as in top panel; C = as in top panel, but note that though the magnitude of the trace is greatly reduced, baroreceptor activity remains unchanged (i.e., for a given magnitude of perfusion pulse pressure there is the same amplitude in integrated neural activity [µV = microvolts] as in the top panel); D, E = as in top panel. CSN = carotid sinus nerve; FBP = femoral arterial blood pressure. Time scale: In panels A, B, D, E (top and bottom) distance between small ticks = 1 second; distance between large ticks = 5 seconds as shown in panel B (top). In panel C (top and bottom) distance between ticks = 1 second as shown in panel C (top). From Fitzgerald and Shirahata (8).

body region, and they are provided *ONLY* in the KRB. The blood is always blocker-free. At the top is the response of the carotid body to a perfusion of hypoxic KRB. At the bottom is the response of the carotid body to a perfusion of equally hypoxic KRB containing mecamylamine and atropine in the 400-500 µM range. Clearly the raw trace and the integrated trace of neural activity were less when the blockers were used.

Table 2.

Perfusate	Time (sec)					
	10	20	30	40	50	60
HHKRB (μV)	2.8	5.8	7.2	7.9	8.0	8.1
HHKRB + AFDX 116 (μV)	3.6	8.1	9.5	9.3	9.2	9.1

Figure 3 presents the results of using three different doses of the combination of nicotinic and muscarinic blockers. The initial period shows the rise in neural activity in response to ventilating the cat on 10% O_2 for 3 minutes, there follows the neural response to either hypoxic KRB or hypoxic KRB + blockers, followed by recovery from the perfusion in which once again blocker-free hypoxic blood is perfusing the carotid body.

Attention was next focused on muscarinic receptors. Since there are M_1 and M_2 receptors in the superior cervical ganglion, we sought a pharmacological tool to see if these receptors functioned in the carotid body. The M_2 receptor in the model was responsible for the sIPSP which tends to hyperpolarize the postsynaptic element. If we blocked it, we should see a greater response to the hypoxic KRB. The M_1 receptor is responsible for the sEPSP, tending to depolarize the postsynaptic element. If we blocked it, we should see a smaller response to the hypoxic KRB.

In these experiments a two perfusion technique in which the second follows immediately upon the termination of the first was used. Table 1 presents the response to Perfusion #1, a blocker-free hypoxic KRB, followed by Perfusion #2, a hypoxic KRB containing either (a) blocker-free hypoxic KRB, or (b) 450 μM gallamine — an M_2 receptor blocker, or (c) 400 μM pirenzepine — an M_1 receptor blocker. In each of the seven experiments the value at 60 sec of Perfusion #1 was defined as the 100% response. The effect of gallamine was maximum at the outset and tended to diminish, whereas the response to pirenzepine took a while to develop and continued to increase.

Finally a reportedly more specific M_2 receptor blocker was used. Dr. Peggy Ganong of Boehringer Ingelheim through Dr. Allison Fryer of this Department graciously provided us with AFDX-116. Hence, again a larger response to hypoxic KRB containing the AFDX-116 than to hypoxic KRB alone was anticipated. Using a single perfusion, we showed in 3 animals that a concentrationof 4 μM had a statistically significant effect (Table 2).

DISCUSSION

The early studies of the many Swedish investigators of the carotid body (12,13,16) plus the later work of Carlos Eyzaguirre and his many colleagues (4,5) and finally our own data support the hypothesis that ACh plays the role of an essential excitatory neurotransmitter in the carotid body's chemotranduction of hypoxia into increased neural activity. The model of excitation, however, is still open to question. Fidone, Gonzalez, Dinger, and their colleagues (6,7,9) integrate the role of acetylcholine with dopamine. They have demonstrated the presence of 3-quinuclidinyl benzilate and alpha-bungarotoxin binding sites on the Type I cells, and failed to demonstrate these sites on the afferent neurons apposed to the Type I cells. They and others have also demonstrated that ACh and other cholinergic agonists stimulate the release of dopamine from the carotid body. This evidence suggests that the role of ACh, released from Type I cells, is to act on autoreceptors for the release of dopamine. And this would then suggest that dopamine is the agent which excites the apposed afferent

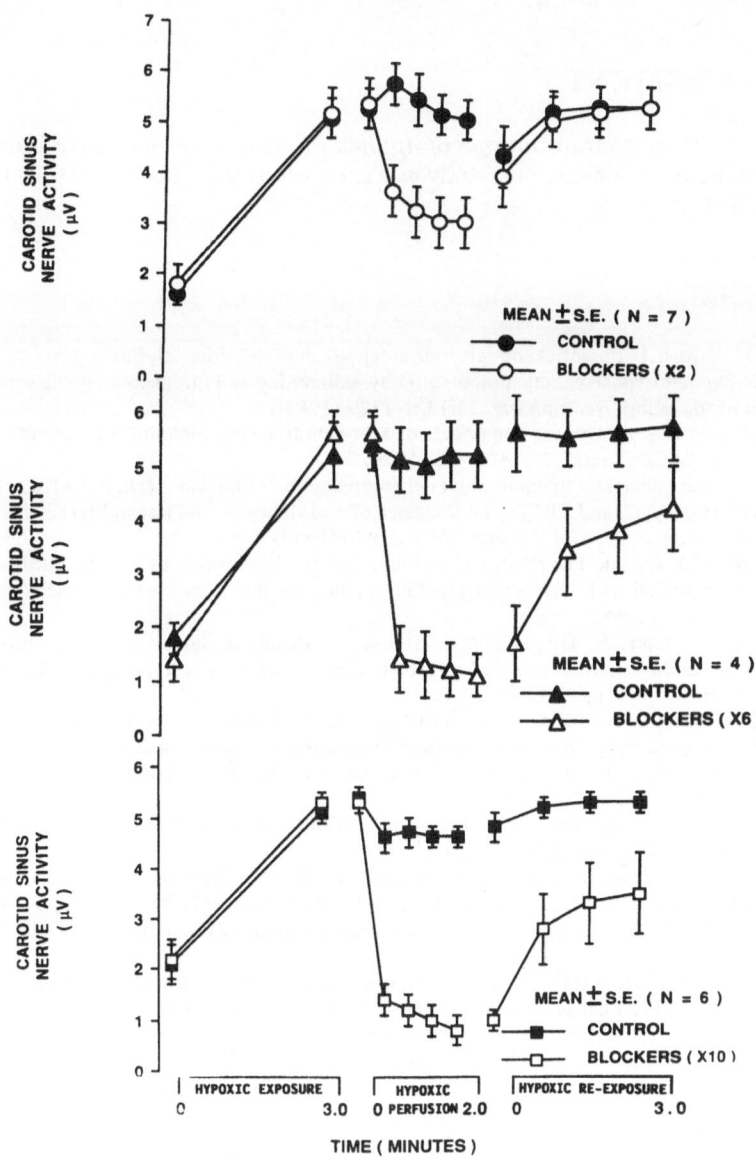

Figure 3. (Top). Time course of the response of the carotid body to selective perfusion with hypoxic KRB. Response is in microvolts (μV). Values are mean ± standard error (n = 7). Closed circles = no blockers; open circles = (in μM) alpha bungarotoxin (α Bgt), 2; mecamylamine, 134; atropine, 314. After the initial exposure to blocker-free hypoxic blood generated by ventilating the cat on 10% O₂, the stopcock was turned and the perfusion made. After 2 min. the stopcock was readjusted and blocker-free hypoxic blood again perfused the carotid body. (Middle). As in 3 (Top). Blocker concentration here was (in μM) α Bgt, 2; mecamylamine, 402; atropine, 942. Reduction in the response to hypoxia at 30 sec when the KRB contained the blockers was complete; i.e., back to the pre-hypoxic control level of activity. (Bottom). As in 3 (Top). Blocker concentration here was (in μM) α Bgt, 2; mecamylamine, 670; atropine, 1570. At 30 sec when the KRB contained the blockers, neural activity was reduced below the initial pre-hypoxic control level. From Fitzgerald and Shirahata (8).

neurons. Whether or not this is the precise sequence of events is controversial at present, and the subject of extensive on-going investigations.

ACKNOWLEDGMENT

Figures reproduced from *Journal of Applied Physiology*(8) with permission from the American Physiological Society. This study was supported by the National Heart Lung Blood Institute, HL 50712.

REFERENCES

1. J. Ashe and C. Yarosh. Differential and selective antagonism of the slow-inhibitory postsynaptic potential and the slow-excitatory postsynaptic potpotential by galloamine and pirenzepine in the superior cervical ganganglion of the rabbit. *Neuropharm.* 23:1321-1329 (1984).
2. R.J. Docherty and D.S. McQueen. The effects of acetylcholine and dopamine on carotid chemosensory activity in the rabbit. *J. Physiol.* 288:411-423 (1979).
3. W. Douglas. Is there chemical transmission at chemoreceptors? *Pharmacol. Rev.* 6:81-83 (1954).
4. C. Eyzaguirre, H. Koyano and J.R. Taylor. Presence of acetylcholine and transmitter release from carotid body chemoreceptors. *J. Physiol. London* 178:463-476 (1965).
5. C. Eyzaguirre and P. Zapata. The release of acetylcholine from carotid body tissues. Further study on the effects of acetylcholine and cholinergic blocking agents on the chemosensory discharge. *J. Physiol. London* 195:589-607 (1968).
6. S. Fidone, C. Gonzalez, B. Dinger and L. Stensas. Transmitter dynamics in the carotid body. In: *Chemoreceptors and Chemoreceptor Reflexes*, edited by H. Acker, A. Trzebski and R.G. O'Regan. New York, Plenum Press, 1990, pp. 3-14.
7. S. Fidone, C. Gonzalez, A. Obeso, A. Gomez-Nino and B. Dinger. Biogenic amine and neuropeptide transmitters in carotid body chemotransmission: experimental findings and perspectives. In: *Hypoxia: The Adaptations*, edited by J.R. Sutton, G. Coates and J.E. Remmers. Toronto and Philadelphia, B.C. Decker, Inc., 1990, pp. 116-126.
8. R.S. Fitzgerald and M. Shirahata. Acetylcholine and carotid body excitation during hypoxia in the cat. *J. Appl. Physiol.* 76:1566-1574 (1994).
9. C. Gonzalez, B. Dinger and S. Fidone. Mechanisms of carotid body chemoreception. In: *Regulation of Breathing*, edited by J.A. Dempsey and A.I. Pack. New York, Marcel Dekker, 1995, pp. 391-471.
10. C. Heymans and E. Neil. *Reflexogenic Areas of the Cardiovascular System*. London: Churchill, 1958, pp. 191.
11. N. Joels and E. Neil. The idea of a sensory transmitter. In: *Arterial Chemoreceptors*, edited by R.W. Torrance. Oxford and Edinburgh, Blackwell Scientific Publications, 1968, pp. 153-178.
12. S. Landgren, G. Liljestrand and Y. Zotterman. The effect of certain autonomic drugs on the action potentials of the sinus nerve. *Acta Physiol. Scand.* 26:264-290 (1952).
13. G. Liljestrand. The problem of transmission at chemoreceptors. *Pharmacol. Rev.* 6:73-76 (1954).
14. D. MCQueen. A quantitative study of the effects of cholinergic drugs on carotid chemoreceptors in the cat. *J. Physiol. London* 273:515-532 (1977).
15. G. Moe, L. Capo and B. Peralta. Action of tetraethyl-ammonium on chemoreceptor and stretch receptor mechanisms. *Am. J. Physiol.* 153:601-605 (1948).
16. U. von Euler, G. Liljestrand and Y. Zotterman. The excitation mechanism of the chemoreceptors of the carotid body. *Skand. Arch. Physiol.* 83:132-152 (1939).

58

GASES AS CHEMICAL MESSENGERS IN THE CAROTID BODY

Role of Nitric Oxide and Carbon Monoxide in Chemoreception

Nanduri R. Prabhakar

Department of Physiology and Biophysics
Case Western Reserve University, School of Medicine
Cleveland, Ohio 44106-4915

INTRODUCTION

Until recently, nitric oxide (NO) and carbon monoxide (CO) were thought to be environmental pollutants that are harmful to the body. It is now well established that mammalian cells not only synthesize NO and CO, but also use them as chemical messengers in various physiological systems (9). Recent studies (7,8,12,13) suggest that NO and CO are produced within the carotid body and they may have important roles in oxygen chemoreception. The purpose of this article is to review briefly the significance of NO and CO as chemical messengers in the carotid body.

NITRIC OXIDE (NO)

In mammalian cells, the enzyme nitric oxide synthase (NOS) catalyzes the formation of NO, from the amino acid L-arginine. Two types of NOS have been purified: a constitutive calcium-calmodulin-dependent type; and an inducible Ca^{2+}-independent form. NOS is a heme containing enzyme and requires NADPH, tetrahydrobiopterin and molecular oxygen for its activity. The distribution of NOS suggests that both neuronal and non-neuronal cells synthesize NO (4,9). Once formed, NO rapidly diffuses into neighboring cells where it is inactivated to nitrites in presence of oxygen.

Methods and Results

The distribution of NOS in the cat carotid body was examined by NADPH-histo-chemistry (8). NOS was found to be localized primarily in the nerve plexuses innervating the chemoreceptor tissue, whereas there was no evidence for NOS in either glomus or type II cells. NOS activity in the carotid body determined by [^{3}H]Citrulline assay averaged 2 ±

Modeling and Control of Ventilation, Edited by S. J. G. Semple, L. Adams, and B. J. Whipp
Plenum Press, New York, 1995

0.1 pmol/min/mg of tissue. Chronic ablation of the sinus nerve abolished NOS positive nerve fibers, whereas sympathectomy or removal of nodose ganglion had no effect (12). These observations suggest that NOS containing nerve fibers are of sinus nerve origin whose cell bodies lie in the petrosal ganglion. This idea is consistent with the finding that NOS positive cells are indeed present in the petrosal ganglion (12).

Various analogues of arginine function as inhibitors of NOS (4) and provide valuable tools to examine the significance of NO in physiological systems. We examined the significance of endogenous NO in the carotid body using L-nitro-ω-arginine (L-NNA), a potent inhibitor of NOS (8). We found that: 1) L-NNA increased baseline sensory activity of the carotid body in a dose-dependent manner; 2) maximum excitation was seen between 100-300μM; and 3) the effect of L-NNA was stereospecific because it was blocked by L-, but not by D-arginine. These findings suggest that endogenous NO is inhibitory to carotid body activity. This notion is further supported by the observations that: 1) sensory discharge of the carotid body could be inhibited in a dose-dependent manner by L-arginine, the substrate for NOS, whereas comparable doses of D-arginine, an enantiomer, had no effect; and 2) sodium nitro prusside (SNP), a nitrosyl compound that liberates NO, also inhibited the chemosensory discharge. Taken together, these findings are consistent with the idea that NO is an inhibitory chemical messenger in the carotid body.

It is well known that one effect of NO is to activate the enzyme soluble guanylate cyclase by binding to its heme moiety. The resulting increase in cGMP mediates most of the known biological actions of NO. The following observations suggest that cGMP is involved in NO-induced chemosensory inhibition. First, inhibition of the sensory discharge by NO donors was associated with a 10 fold increase in cGMP content of the carotid bodies. Second, the chemosensory inhibition by sodium nitro prusside, a NO-donor, could be partially reversed by methylene blue, an inhibitor of guanylate cyclase. Moreover, the sensory excitation by an NOS -inhibitor was associated with a marked reduction in cGMP content of the chemoreceptor tissue (8). These observations support the hypothesis that the actions of NO in the carotid body are mediated in part by a cGMP pathway.

CARBON MONOXIDE (CO)

CO is formed by hemeoxygenase (HO), which oxidatively cleaves the heme ring releasing CO and biliveridin (3). Two forms of HO have been isolated. HO I is present in peripheral tissues such as liver and spleen, and is induced by numerous oxidative stresses (3). On the other hand, HO II is constitutive and present predominantly in neuronal tissues (11). Recent studies suggest that CO may influence certain forms of synaptic plasticity, as hippocampal long-term potentiation is inhibited by inhibitors of HO-II and enhanced by exogenous administration of CO (10,14). These studies suggest that CO may function as a chemical messenger in the nervous system.

Methods and Results

We examined the distribution of HO II in the rat and cat carotid bodies, using an antibody specific for HO-II (7). Many glomus cells were found to be positive for HO-II, whereas no reaction product was evident in nerve fibers. HO II activity can be inhibited by certain metalloporphyrins such as Zn protoporphyrin-9 (ZnPP-9; 11). We tested the effects of ZnPP-9 on sensory discharge of the isolated carotid body. Doses as little as 0.3μM significantly augmented the sensory discharge and the maximum response was seen at 3μM. However, comparable doses of CuPP-9, which is not an inhibitor of HO-II, failed to excite the carotid body sensory activity. Further, the stimulatory actions of ZnPP-9 could be reversed by exogenous administration of CO. Recently, Lahiri et al (2) examined the effects of CO on sensory activity of the isolated carotid body. These authors reported that low doses of CO inhibited carotid body activity. Taken together, these results provide clear evidence that CO is another potent inhibitory chemical messenger in the carotid body.

DISCUSSION

Where might NO and CO be acting in the carotid body? Type I cells are one potential target of NO and CO actions. Our recent results which show that NO-donors markedly alter cytosolic calcium in type I cells are consistent with this notion. Further, this observation suggests that NO might affect the release of neurotransmitter (s) from type I cells, which could cause chemosensory inhibition. In addition, it is well known that NO and CO dilates blood vessels (4), which could result in improved oxygenation of the chemoreceptor tissue and lead to a decrease in sensory discharge. The fact that NOS is distributed predominantly in nerve fibers of sinus nerve origin prompted us to suggest that NO might mediate efferent inhibition of the afferent activity in the carotid body (8). Our hypothesis is supported by a preliminary study by Wang et al. (13) that suggests the involvement of NO in efferent inhibition.

Carotid bodies are endowed with several classes of neurotransmitters including biogenic amines and neuropeptides, which seem necessary for the initiation and maintenance of the hypoxic stimulus at the carotid body (5,6). The findings described above add NO and CO to the growing list of chemical messengers in the carotid body. How might NO and CO contribute to carotid body O_2 chemoreception? It has been postulated that augmentation of carotid body sensory activity is due to decreased availability of an inhibitory neurotransmitter in the chemoreceptor tissue (1) whose identity remains elusive. The facts that synthesis of NO and CO depends absolutely on oxygen availability, and that both are inhibitory to carotid body activity are relevant to this idea. It is possible that, during normoxia, constant production of NO and CO maintain sensory discharge at low levels. It follows that increased sensory discharge during hypoxia is due to "disinhibition" resulting from decreased production of NO and CO. Such a notion is supported by our recent finding that NOS activity in the carotid body could be inhibited by low levels of pO_2 (8). Most importantly, these levels of hypoxia also stimulate chemoreceptor activity, and occur under many pathophysiological situations.

ACKNOWLEDGMENTS

This work was supported by grants from National Institutes of Health, Heart, Lung and Blood Institute, HL-45780; HL- 52038 and a Research Career Development Award HL-02599. The work reported in this article was done in collaboration with Drs. S.H. Snyder, J.L.Dinerman, G.K.Kumar, F.H.Agani, G. Bright, and J.L. Overholt.

REFERENCES

1. Kramer, E. Carotid body chemoreceptor function: hypothesis based on a new circuit model. *Proc. Natl. Acad. Sci. USA* 75: 2507-2511, 1978.
2. Lahiri, S., R. Iturriaga., A. Mokashi., D.K. Ray., and D.Chugh. CO reveals dual mechanisms of O2 chemoreception in the cat carotid body. *Resp. Physiol.* 94: 227-240, 1993.

3. Maines, R.D. Heme oxygenase : function, multiplicity, regulatory mechanisms, and clinical applications. *FASEB J.* 2: 2557-2568, 1988.

4. Moncada, S., R. M. J.Palmer., and E. A. Higgs, E. A. Nitric Oxide: Physiology, pathophysiology, and pharmacology. *Pharmacol. Rev.* 43: 109-142, 1991.

5. Prabhakar, N.R. Significance of excitatory and inhibitory neurochemicals in hypoxic chemo-transmission of the carotid body. In: *Control of Breathing: Modelling and Perspecticves.* Eds: Honda Y et al . Plenum, New York . pp 141-148, 1993.

6. Prabhakar, N.R. Neurotransmitters in the carotid body. In: *Arterial Chemoreceptors: Cell to System.* Eds: R.O' Regan et al, Plenum Press, New York. 1994. pp 57-69.

7. Prabhakar, N.R., J.L. Dinerman., F.H. Agani ., and S.H. Snyder. Carbon monoxide: A role in carotid body chemoreception. *Proc. Natl. Acad. Sci. USA.* 92: 1994-1997, 1995.

8. Prabhakar, N.R., G.K. Kumar., C.H. Chang., F.H. Agani., and M.A. Haxhiu. Nitric oxide in the sensory function of the carotid body. *Brain Res.* 625: 16-22. 1993.

9. Snyder, S.H. Nitric Oxide : First in a new class of neurotransmitters. *Science.* 257: 494-496, 1992.

10. Stevens, C.F., and Y. Wang, Y. Reversal of long term potentiation by inhibitors of heme oxygenase. *Nature* . 364: 147-148, 1993.

11. Verma, A., D.J. Hirsch., C.E. Glatt., G.V. Ronnett., and S.H. Snyder. Carbon monoxide, a putative neural messenger. *Science.* 259: 381-384. 1992.

12. .Wang , Z.Z., D.S. Bredt., S.J. Fidone., and L.J. Stensas. Neurons synthesizing nitric oxide innervating the carotid body. *J.Comp. Neurol.* 336: 419-432, 1993.

13. Wang, Z.Z., L. J. Stensas., D.S. Berdt., B.G. Dinger., and S.J. Fidone, S.J. Mechanisms of carotid body inhibition. In : Arterial chemoreceptors : Cell to System. R.O'Regan. etal. plenum Press, New York., 1994. pp 229-235.

14. Zhuo, M., S. A. Small., E.R. Kandel., and R. D. Hawkins. Nitric Oxide and Carbon Monoxide produce activity-dependent long-term synaptic enhancement in hippocampus. *Science.* 260: 1946-1950, 1993.

59

INTERACTIVE VENTILATORY EFFECTS OF CAROTID BODY HYPOXIA AND HYPOCAPNIA IN THE UNANESTHETIZED DOG

C. A. Smith, K. S. Henderson, and J. A. Dempsey

The John Rankin Laboratory of Pulmonary Medicine
Department of Preventive Medicine
University of Wisconsin, Madison
504 North Walnut Street, Madison, Wisconsin 53705-2368

INTRODUCTION

In a previous study (5) we demonstrated that moderate normoxic hypocapnia, when limited to the carotid sinus region (carotid body), could markedly inhibit ventilation and VT in the awake or sleeping (Non-REM and REM) dog. TI and TE were not changed significantly, thus there was no tendency for carotid body hypocapnia to produce apnoeas. The ventilatory inhibition was graded over the range of carotid body hypocapnia ($\Delta PCBCO_2$ = -7 to -15 Torr relative to eupnoea) but the stimulus: response slope was most sensitive over the eupnoea to -7 Torr range at least during wakefulness and Non-REM sleep. In other words, in these conditions there was clearly no "dog-leg" present in the ventilatory response to carotid body hypocapnia (2). During REM sleep there was no consistent ventilatory response until $\Delta PCBCO_2$ was > -7 Torr (Fig. 1). During wakefulness and Non-REM sleep the inhibitory effect was probably maximal between about -10 and -15 Torr as perfusion of the carotid body with normocapnic and hyperoxic blood ($PCBO_2$>500 Torr) produced ventilatory inhibition that was essentially identical in time course, magnitude and ventilatory pattern to that produced by -10 to -15 Torr carotid body hypocapnia. Given the well known inhibitory effects of hyperoxia on carotid sinus nerve output (3,4) we concluded that carotid body hypocapnia in the -10 to -15 range must virtually silence chemoreceptor output in the carotid body.

The inhibitory effects of carotid body hypocapnia were unopposed by excitatory peripheral stimuli in the studies discussed above. The purpose of the present study was to further characterize the inhibitory effects of carotid body hypocapnia by examining the interaction between the inhibitory stimulus of hypocapnia and the excitatory stimulus of hypoxia in the unanesthetized, neurally-intact dog.

Modeling and Control of Ventilation, Edited by S. J. G. Semple, L. Adams, and B. J. Whipp
Plenum Press, New York, 1995

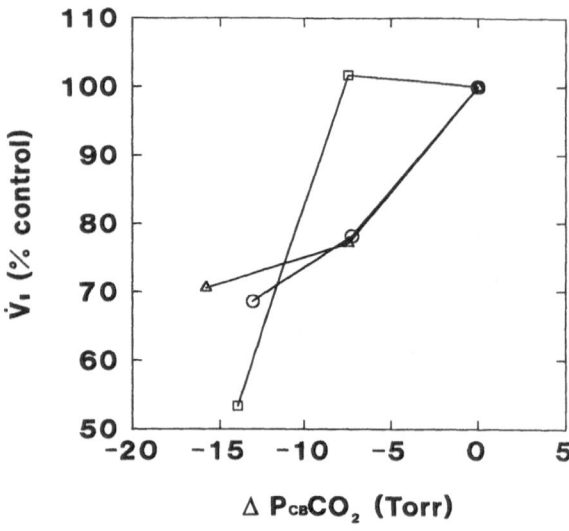

Figure 1. Mean stimulus: response slopes to two levels of carotid body hypocapnia during wakefulness (O), Non-REM (∆), and REM sleep(□).

METHODS

3 female dogs (20 to 25 Kg), trained to wear respiratory masks and lie quietly in the laboratory, were studied . Each was surgically prepared for extracorporeal perfusion of the vascularly isolated carotid sinus region. This technique is based on the preparation developed by Busch *et al.* (1) in the goat. Briefly, on the dog's right side the lingual artery was cannulated, an occluder was placed on the external carotid artery cranial to the lingual artery, and the internal carotid, occipital, ascending pharyngeal, and cranial laryngeal arteries were ligated. The contralateral carotid body was denervated but the arterial supply to the brain was left intact. Four to seven days of recovery elapsed prior to experiments. To perfuse the intact carotid sinus region, the occluder was inflated and blood from an extracorporeal gas exchanger (Capiox 350, Terumo) was pumped at a low flow rate (<100 ml/min) retrograde through the carotid sinus region at a pressure <10 Torr higher than systemic arterial pressure. Switching from endogenous perfusion of the carotid sinus region to extracorporeal perfusion or *vice versa* was achieved in <2 seconds.

Also installed at the time of surgery were bipolar multi-strand EMG electrodes in the costal diaphragm and a small tracheal cannula for measurement of sub-laryngeal airway pressure.

Ventilation, EMG, and airway pressures were recorded on a chart recorder and on a computer-based data acquisition system for subsequent off-line analysis.

Our surgical techniques and experimental protocol were approved by the Animal Care and Use Committee of the University of Wisconsin, Madison.

PROTOCOL

Dogs were exposed to two levels of $PCBO_2$ (53 Torr and 45 Torr) plus carotid body normocapnia, for 5 minutes while breathing room air followed by an additional 5 minutes during which $FICO_2$ was supplemented to maintain arterial normocapnia. This protocol was then repeated except now the carotid body was exposed to the same two levels of hypoxia plus hypocapnia (-10 Torr relative to eupnoea).

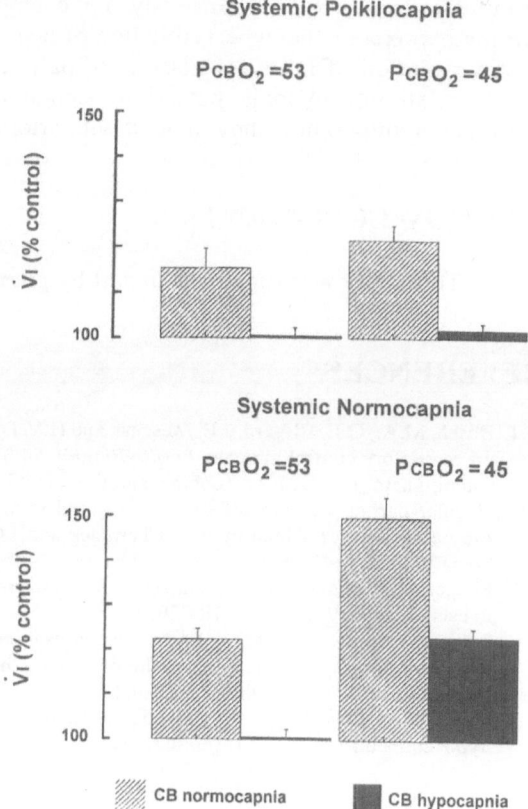

Figure 2. Mean data from one dog illustrating the effects of carotid body hypoxia alone and carotid body hypoxia plus hypocapnia (-10 Torr). Data are presented for both systemic poikilocapnia and normocapnia. Note the marked inhibitory effects of carotid body hypocapnia on the ventilatory response to carotid body hypoxia.

RESULTS

The two levels of normocapnic carotid body hypoxia ($PCBO_2$ 53 Torr and 45 Torr) increased ventilation 15% and 20% respectively when systemic poikilocapnia was permitted. When systemic isocapnia was maintained at the two levels of carotid body hypoxia ventilation increased 23% at $PCBO_2 = 53$ Torr and 47% at $PCBO_2 = 45$ Torr.

When -10 Torr carotid body hypocapnia was presented simultaneously with the two levels of carotid body hypoxia, ventilation was essentially unchanged from control when poikilocapnia was permitted to occur. When systemic normocapnia was maintained ventilation was unchanged at the milder level of carotid body hypoxia and only increased 20% at the more severe level of carotid body hypoxia (Fig. 2).

DISCUSSION

We conclude that moderate hypocapnia at the vascularly isolated but *in situ* carotid body can markedly attenuate the ventilatory response to simultaneous carotid body hypoxia in the unanesthetized dog. Brain hypocapnia is not required for this carotid body-mediated inhibition.

The present preliminary results, if confirmed, will have relevance to ventilatory control in physiological conditions where hyperventilation and an excitatory carotid body stimulus are present simultaneously, for example, the ventilatory response to acute and chronic hypoxia or the hyperventilation of heavy exercise. Stimulus interaction during the hyperventilation of heavy exercise is of particular interest because a number of putative excitatory stimuli (hypoxia, potassium, lactate, norepinephrine) are present simultaneously and it is not known how they interact with arterial hypocapnia.

ACKNOWLEDGMENTS

This study was supported in part by grants from NHLBI.

REFERENCES

1. Busch, M.A., G.E. Bisgard, J.E. Messina, and H.V. Forster. The effects of unilateral carotid body excision on ventilatory control in goats. *Respir. Physiol.* 54:353-361, 1983.
2. Cunningham, D.J.C., P.A. Robbins, and C.B. Wolfe. Integration of ventilatory responses to changes in alveolar partial pressures of CO_2 and O_2 and in arterial pH. In: *Handbook of Physiology, Control of Breathing, Part II*, edited by N.S. Cherniack and J.G. Widdicombe. Bethesda, American Physiological Society, 1986, p.475-528.
3. Fitzgerald, R.S. and D.C. Parks. Effect of hypoxia on carotid chemoreceptor response to carbon dioxide in cats. *Respir. Physiol.* 12:218-229, 1971.
4. Lahiri, S. and R.S. Fitzgerald. Reflex responses to chemoreceptor stimulation. In: *Handbook of Physiology, Control of Breathing, Part 1*, edited by N.S. Cherniack, and J.G. Widdicombe, Bethesda, American Physiological Society, 1986, p. 313-362.
5. Smith, C.A., K.W. Saupe, K.S. Henderson, and J.A. Dempsey. Inhibition of ventilation by carotid body hypocapnia during sleep. *Respir. Crit. Care Med.* 149:A922, 1994.

60

THE POSTNATAL POTENTIATION OF CHEMORECEPTOR SENSITIVITY TO O$_2$ AND CO$_2$ IN THE *IN VITRO* RAT CAROTID BODY IS BLUNTED BY CHRONIC HYPOXAEMIA

Development of Chemosensitivity

P. Kumar, R. C. Landauer, and D. R. Pepper

Department of Physiology
The Medical School
University of Birmingham
Birmingham. B15 2TT, United Kingdom

INTRODUCTION

Peripheral chemoreceptors have long been known as transducers of arterial blood gas tensions and pH. More recently, they have also been described as sensors of arterial potassium concentration (2) and temperature (1) and a role for these receptors in contributing to both resting ventilation and to exercise hyperpnoea has been forwarded on numerous occasions (e.g. see 17). However, a role in the immediate postnatal period, a time when arterial blood gas tensions are changing, has been less well documented. With increasing postnatal age, an increase occurs in the magnitude of the initial ventilatory response to an acute hypoxic challenge (9) which has been attributed to a maturation, or resetting, of peripheral chemoreceptor hypoxic sensitivity (4). That this development could be 'blunted' by preventing the normal postnatal increase in PaO$_2$ (10, 11) indicated that the process of resetting was, in some way, oxygen dependent. However, the mechanisms which may underlie resetting are not known. In addition, such knowledge may also shed some light upon the mechanisms that underlie chemotransduction itself. Ventilatory reflex studies have indicated that a change in CO$_2$ sensitivity may also be occurring during this postnatal period although the results are equivocal (19, 21). Such discrepancies may arise from the effects of CO$_2$ upon central chemoreceptors. The direct effect of CO$_2$ upon peripheral chemoreceptor discharge indicates that an increase in CO$_2$ sensitivity may occur with increasing postnatal age (4, 7, 15, 16). However, to date there have been no reports of the developmental and interactive effects of hypoxia and CO$_2$, nor of the effect of chronic hypoxaemia, upon peripheral chemoreceptor response curves recorded *in vitro*.

Modeling and Control of Ventilation, Edited by S. J. G. Semple, L. Adams, and B. J. Whipp
Plenum Press, New York, 1995

METHODS

Animals and Hypoxic Chamber

Three groups of Wistar rats were studied; neonatal (5-7 days old, n = 10) and adult (49 ± 7 days old, n = 8) normoxic animals and adult rats which had been made chronically hypoxic from birth (46 ± 1 day old, n = 8). The chronically hypoxic group had been born into and subsequently reared in a normobaric, hypoxic chamber from time-mated females. The chamber air was circulated and the O_2 concentration set to 12% (range 11.75-12.25%) by a system whereby deviations from the set concentration were opposed by addition of nitrogen or air, through feedback-controlled solenoid valves. Ambient CO_2 in the chamber was measured and maintained at normal air levels (ca. 0.03%) by circulating the gas through soda lime. Ambient temperature was measured and maintained at 22-24 °C and humidity measured and maintained between 40-50% by circulating the chamber gas through a freezer unit and silica gel.

Carotid Body Preparation

Anaesthesia was induced and maintained with 1.5-2.5% halothane in 100% O_2 or 12% O_2 (balance N_2) for the normoxic and chronically hypoxic groups respectively. The right carotid bifurcation, including the carotid body, sinus nerve and glossopharyngeal nerve, was isolated, the common carotid artery ligated and the bifurcation excised within 15 seconds of ligation of the blood supply. The bifurcation was pinned out onto Sylgard in a small volume tissue bath and superfused at 3ml/min. with warmed (36.7-37 °C), gassed (95% O_2/ 5% CO_2) bicarbonate-buffered saline solution. Excess tissue was removed and the preparation was partially digested in a gassed enzyme solution (0.06% collagenase (Sigma Type II), 0.02% protease (Sigma Type IX)) for 25 minutes at 37°C.

Electrophysiological Recording

Extracellular recordings of afferent single- or few-fibre activity were made from the cut end of the carotid sinus nerve using glass suction electrodes. The P_{O_2} and temperature of the superfusate were continually monitored in the superfusion line immediately before the bath by a membrane oxygen electrode and meter (ISO_2, W.P.I.) and thermocouple. Superfusate P_{CO_2} was determined from samples collected and measured by a blood gas analyser (NovaStat 3). Afferent spike activity and superfusate P_{O_2} were recorded on-line.

Experimental Protocol

Baseline chemoreceptor discharge was obtained by bubbling the superfusate with 95% O_2/ 5% CO_2 which resulted in a dissolved P_{O_2} of ca. 450 mmHg and P_{CO_2} of 35 mmHg. Superfusate P_{O_2} was decreased between ca. 450 and 75 mmHg over 3-4 minutes at each of three fixed levels of P_{CO_2} (35, 46 and 61 mmHg) by precision mixing of the gases which were bubbling through the superfusate reservoir. The superfusate was returned to 95% O_2/ 5% CO_2 levels for 4 minutes between every hypoxic challenge and then maintained at each new level of P_{CO_2} for 3 minutes before the subsequent hypoxic challenge. The entire protocol lasted no longer than 40 minutes.

Data and Statistical Analysis

Few-fibre chemoreceptor discharge was discriminated by amplitude. Action potentials were counted and binned into 10 sec. periods, expressed as the percentage of the maximum binned frequency found during the experiment and correlated against the mean Po_2 during the 10 sec. period using a Macintosh IIci with National Instruments DA and DMA cards running customised LabVIEW 2 software. A single exponential with offset was fitted to each response to hypoxia using an iterative routine with robust weighting (Ultrafit, Biosoft), for every animal at each level of CO_2 according to the following equation;

$$\text{discharge} = \text{offset} + \text{discharge at 70 mmHg } Po_2\ e^{-(PO_2 - 70)/\text{rate constant}}$$

where the 'rate constant' is the increase in mmHg Po_2 required to decrease discharge to $1/e$ of its initial value. Data points which corresponded to a consistent fall in discharge as Po_2 was decreased (failure) were excluded from the analysis. All data are expressed as mean ± SEM.

RESULTS

In all experiments decreasing perfusate Po_2 led to increases in chemoreceptor discharge that were reversible upon return to hyperoxia and gave response curves that were well fit by single exponential functions with offset (Fig. 1). Perfusate Po_2 needed to be reduced to lower levels in the neonatal animals to generate increases in chemoreceptor discharge from which exponential functions could be fitted. In all groups increasing perfusate Pco_2 increased chemoreceptor discharge in both hyperoxia and hypoxia. Failure in hypercapnic hypoxia was common only in adult normoxic animals.

The effect of CO_2 upon the baseline discharge and the shape of the hypoxic response curves was assessed by linear regression analysis of the values found for the horizontal asymptote and the rate constant. This showed a significant increase in baseline discharge, with increasing Pco_2, in the neonatal (0.25 ± 0.07 % discharge/mmHg Pco_2 ; P<0.005), normoxic adult (0.15 ± 0.03 % discharge/mmHg Pco_2; P<0.001) and chronically hypoxic

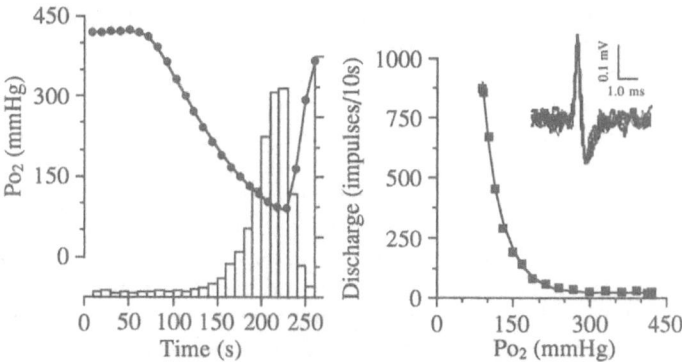

Figure 1. Chemoreceptor response to hypoxia (Pco_2 = 46 mmHg) in a normoxic adult carotid body preparation. Superimposed graphs (left) show perfusate Po_2 (filled circles) and chemoreceptor discharge (open histogram) averaged into 10 sec. intervals throughout a single drop of perfusate Po_2. On the right is shown the discharge plotted against the Po_2 (filled squares) fitted by a single exponential with offset (discharge/ 10 sec = 44.5 + 2860 $\cdot\ e^{-(PO_2 - 70)/37.5}$). Inset shows 8 superimposed traces of the extracellularly recorded action potential.

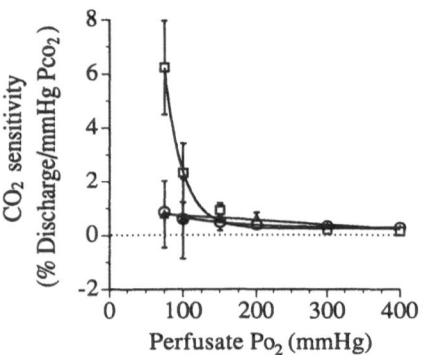

Figure 2. Mean effect of Po_2 upon CO_2 chemosensitivity. The effect of superfusate Po_2 upon the mean ± SEM of CO_2 sensitivity in all neonatal (open circles), normoxic adult (open squares) and chronically hypoxic adult (filled triangles) rats. Values were derived from the slopes of CO_2 response curves, at 6 levels of Po_2, generated from the exponential fits to the Po_2 response curves. Solid lines through each set of data are the best-fit single exponentials with offset.

adult (0.13 ± 0.04 % discharge/mmHg Pco_2; $P<0.005$) animals. The rate constant was however unaffected by Pco_2 in the neonatal (0.12 ± 0.13 mmHg Po_2/mmHg Pco_2; $P>0.300$), normoxic adult (-0.04 ± 0.18 mmHg Po_2/mmHg Pco_2; $P>0.500$) and the chronically hypoxic (0.63 ± 0.48 mmHg Po_2/mmHg Pco_2; $P>0.200$) group. However, the absolute levels of the rate constant were significantly greater in the chronically hypoxic group when compared to the age-matched normoxic adults at each level of Pco_2 ($P<0.050$, unpaired t-test).

From the fitted exponential functions it was possible to derive values for chemoreceptor discharge at six levels of Po_2 between 400 and 75 mmHg at each of the three levels of Pco_2. From these values CO_2 response curves could be generated from which CO_2 sensitivity (% discharge/ mmHg Pco_2) could be calculated at each level of Po_2 (Fig 2). The effect of hypoxia upon CO_2 chemosensitivity was assessed by one factor ANOVA. This revealed that in the normoxic adult group decreasing Po_2 significantly increased CO_2 sensitivity ($P<0.001$), but was without effect upon CO_2 sensitivity in the neonatal and chronically hypoxic preparations ($P>0.500$ and $P>0.999$, respectively).

DISCUSSION

These results are the first to demonstrate that in an *in vitro* carotid body preparation a significant increase in the degree of CO_2 -O_2 interaction occurs at the level of the carotid chemoreceptor between approximately 5 days and 5 weeks after birth and that this development can be prevented by chronic hypoxaemia from birth. Our results therefore show that the adult multiplicative interaction known to exist between CO_2 and hypoxia in the adult (13) is not observed at birth and also add to the idea that chronic hypoxaemic episodes during early postnatal life can lead to a failure to reset chemosensitivity which may lead to a compounding of the chronic hypoxaemia further, perhaps resulting ultimately in handicap or death (12). As it is not possible to ascertain the actual level of tissue Po_2 during our experimental protocol we did not make comparisons between groups with regard to Po_2-dependent components of the Po_2 response curves and have instead only made such comparisons within single animals or within a single group of animals. However, a comparison between groups of the effect of Pco_2 upon Po_2-independent components of each response curve can be made. Thus, Pco_2 increases the horizontal asymptote in all groups suggesting that normal development and chronic hypoxaemia is without any qualitative effect upon CO_2 chemotransduction *per se*. On the other hand, whilst Pco_2 was without effect upon the rate constants of the exponential Po_2 response curves in all 3 groups, the actual level of the rate constant was significantly greater in the chronically hypoxic animals when compared to the normoxic adults, giving rise to 'flatter' response curves in the former group. Thus, at birth and in chronically hypoxic adults, whilst a chemoreceptor response to both hypercapnia and hypoxia can be observed, the response to the two stimuli given together asphyxially is considerably less than that observed in the normoxic adult. Interestingly, this difference becomes greater with increasing hypoxia which might explain some of the ambiguity in previously reported studies of the development of CO_2 chemosensitivity. We wish to suggest therefore that the firmly established increase that occurs in postnatal hypoxic ventilatory reflex and chemoreceptor afferent sensitivity may in fact be due to the increased interaction between the imposed hypoxic stimulus and the steady level of Pco_2 at which experiments are performed. Unfortunately, it is not possible to distinguish between these two possibilities at present. It is also possible to do more than speculate upon the mechanisms that might underlie such a development and its impairment by chronic hypoxaemia. It is generally accepted that an elevation in intracellular Ca^{2+} in the type I cells of the carotid body is a necessary prerequisite of chemotransduction although controversy remains centred around the source of this elevation(3, 5, 6, 20). The most recent studies imply entry of Ca^{2+} through voltage sensitive calcium channels following stimulus-induced cell depolarisation by inhibition of a class of K^+-channels (8, 14, 18). It is conceivable that the development of postnatal CO_2 -O_2 sensitivity requires an O_2 -dependent maturation in the sensitivity of these channels. Unfortunately, no developmental studies have been performed in a single species although one study (22), using chronically hypoxic rats aged 9-13 days showed that these animals exhibited a significant reduction in K^+-current density that was in part due to a reduction of a charybdotoxin-sensitive K^+-channel component which had been previously implicated in normal hypoxic chemotransduction.

In conclusion we have shown that a development of CO_2 -O_2 interaction at the carotid body occurs with increasing postnatal age and that this process in intrinsic to the chemoreceptor and appears to be dependent upon arterial oxygenation.

ACKNOWLEDGMENTS

We are grateful for the financial support of the MRC and the Wellcome Trust.

REFERENCES

1. Alcayaga, J., Y. Sanhueza and P. Zapata. Thermal-dependence of chemosensory activity in the carotid-body superfused invitro. *Brain Res.* 600; 103-111, 1993.
2. Band, D.M. and R.A.F. Linton. The effect of potassium on carotid body chemoreceptor discharge in the anaesthetized cat. *J. Physiol.* 381: 39-47, 1986.
3. Biscoe, T.J. and M.R. Duchen. Responses of type I cells dissociated from the rabbit carotid body to hypoxia. *J. Physiol.* 428: 39-59, 1990.
4. Blanco, C.E., G.S. Dawes, M.A. Hanson and H.B. McCooke. The response to hypoxia of arterial chemoreceptors in fetal sheep and new-born lambs. *J. Physiol.* 351: 25-37, 1984.
5. Buckler, K.J. and R.D. Vaughan-Jones. Effects of hypoxia on membrane potential and intracellular calcium in rat neonatal carotid body type I cells. *J. Physiol.* 476.3: 423-428, 1994.
6. Buckler, K.J. and R.D. Vaughan-Jones. Effects of hypercapnia on membrane potential and intracellular calcium in rat carotid body type I cells. *J. Physiol.* 478.1: 157-171, 1994.
7. Carroll, J.L., O.S. Bamford and R.S. Fitzgerald. Postnatal maturation of carotid chemoreceptor responses to O_2 and CO_2 in the cat. *J. Appl. Physiol.* 75: 2383-2391, 1993.
8. Delpiano, M.A and J. Hescheler. Evidence for a Po_2-sensitive K^+ channel in the type-I cell of the rabbit carotid body. *FEBS Lett.* 249: 195-198, 1989.
9. Eden, G.J. and M.A. Hanson. Maturation of the response to acute hypoxia in the newborn rat. *J. Physiol.* 392: 1-9, 1987.
10. Eden, G.J. and M.A. Hanson. Effects of chronic hypoxia from birth on the ventilatory responses to acute hypoxia in the newborn rat. *J. Physiol.* 392: 11-19, 1987.

11. Hanson, M.A., P. Kumar and B.A. Williams. The effect of chronic hypoxia upon the development of respiratory chemoreflexes in the newborn kitten. *J. Physiol.* 411: 563-574, 1989.

12. Hunt, C.E., K. McCulloch and R.T. Brouillette. Diminished hypoxic ventilatory responses in near-miss sudden infant death syndrome. *J. Appl. Physiol.* 50: 1313-1317, 1981.

13. Lahiri, S. and R.G. Delaney. Stimulus interaction in the responses of carotid body chemoreceptor single afferent fibres. *Respir. Physiol.* 24: 249-266, 1975.

14. Lopez-Lopez, J., C. Gonzalez, J. Urena and J. Lopez-Barneo. Low Po_2 selectively inhibits K^+-channel activity in chemoreceptor cells of the mammalian carotid body. *J. Gen. Physiol.* 93: 1001-1015, 1989.

15. Marchal, F., A. Bairam, P. Haouzi, J.P. Crance, C. Digiulio, P. Vert and S. Lahiri. Carotid chemoreceptor response to natural stimuli in the newborn kitten. *Respir. Physiol.* 87: 183-193, 1992.

16. Mulligan, E.M. Discharge properties of carotid bodies: developmental aspects. In: *Developmental Neurobiology of Breathing*, edited by. G.G. Haddad and J.P. Farber. New York: Dekker, 1991, p. 321-340.

17. Paterson, D.J. Potassium and ventilation in exercise. *J. Appl. Physiol.* 72: 811-820, 1992.

18. Peers, C. Hypoxic suppression of K^+-channels in type I carotid body cells: Selective effect on the Ca^{2+}-activated K^+-current. *Neuroscience Lett.* 119: 253-256, 1990.

19. Rigatto, H.R., J.P. Brady, B. Chir and R. Delatorre Verduzco. Chemoreceptor reflexes in preterm infants. II. The effects of gestational and postnatal age on the ventilatory response to inhaled carbon dioxide. *Pediatrics* 55: 614-620, 1975.

20. Rocher, A., A. Obeso, C. Gonzalez and B. Herreros. Ionic mechanisms for the transduction of acidic stimuli in rabbit carotid body glomus cells. *J. Physiol.* 433: 533-548, 1991.

21. Wolsink, J.G., A. Berkenbosch, J. DeGoede, and C.N. Olievier. The effects of hypoxia on the ventilatory response to sudden changes in CO_2 in newborn piglets. *J. Physiol.* 456: 39-48, 1992.

22. Wyatt, C.N., C. Peers, C. Wright and D. Bee. K^+ currents in isolated type I carotid body cells of neonatal rats born and reared in hypoxia and normoxia. *J. Physiol.* 475.P: 128P, 1994.

61

THE EXCITATION OF CAROTID BODY CHEMORECEPTORS OF THE CAT BY POTASSIUM AND NORADRENALINE

G. Heinert, D. J. Paterson, G. E. Bisgard, N. Xia, R. Painter, and P. C. G. Nye

University Laboratory of Physiology
Parks Road, Oxford OX1 3PT
United Kingdom

INTRODUCTION

This chapter describes the combined effects of noradrenaline (NA) and potassium (K^+) on carotid body discharge at different levels of P_{O_2}. We used an equation derived from that of Lloyd and Cunningham[4,9] for the description of the ventilatory responses of man to hypercapnia and hypoxia. In 1963 Lloyd et al. demonstrated that noradrenaline (NA) increases ventilation in man and described its effect mathematically[5]. Later it was shown that K^+ increases carotid body discharge [2,3] and that the effect of K^+ is more pronounced in hypoxia [3]. In exercise the hypoxic sensitivity of the carotid body is increased [1,11]. Is this due to modulation by sympathetic efferent nerves, by circulating compounds or by central potentiation?

In severe exercise the arterial concentrations of many compounds that can excite the carotid body increase. These compounds include lactic acid, H^+-ions, K^+ and catecholamines. To date only their individual effects on carotid body discharge have been studied. However, all these compounds converge on the carotid body and their combined effect is unknown. Do the effects of these compounds potentiate each other, are they additive or do they cancel each other out? The objective of this study was to analyse the effects of NA and K^+ on the CO_2/O_2-discharge dose-response curve, both alone and in combination. In order to understand better the potential role of these stimuli during the hyperpnoea and exercise.

METHODS

Six cats (yielding 9 fibre preparations) were anaesthetised with sodium pentobarbitone (40mg/kg i.p.). After tracheal cannulation, the left femoral artery was cannulated for blood pressure measurement, and the left and right femoral veins were cannulated for infusion of NA and K^+ and supplements of anaesthetic (Figure 1). Infusions were made by

Modeling and Control of Ventilation, Edited by S. J. G. Semple, L. Adams, and B. J. Whipp
Plenum Press, New York, 1995

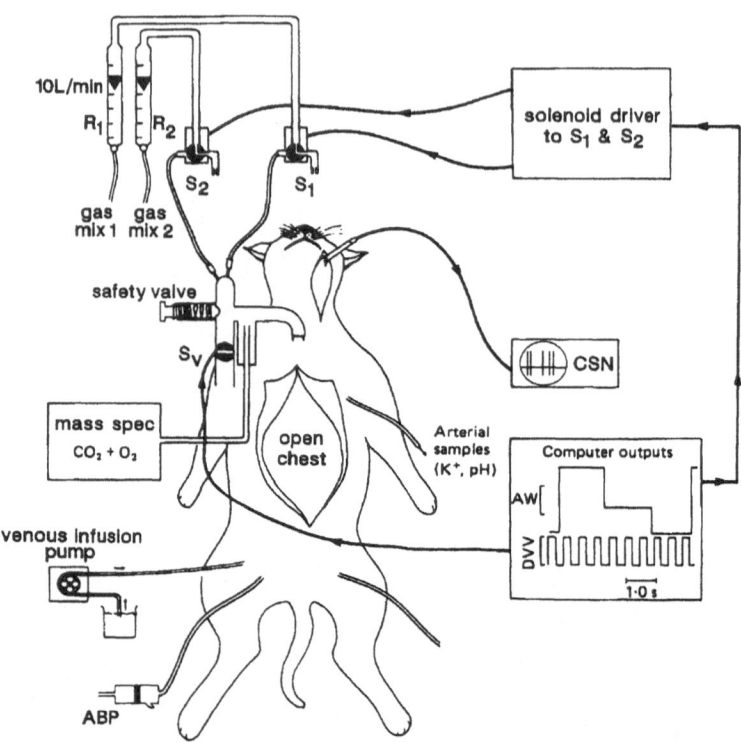

Figure 1. Experimental preparation - a high frequency ventilator for generating waveforms of arterial gas tension in the anaesthetized cat (with permission[8]).

calibrated peristaltic pumps. The cats were thoracotomised and artificially hyperventilated (~40ml tidal volume at 2 Hz) using a computer-controlled ventilator [8]. CO_2 was added to the inspiratory air so that end tidal P_{CO_2} was about 5% (38 Torr). Tracheal gas tensions were continuously monitored by mass spectrometry (VG Micromass model 201). Arterial blood pressure, tracheal gas tensions and sinus nerve discharge were recorded on magnetic tape (Racal Store 7). Blood K^+ level and blood gas tensions were frequently measured by a Radiometer ABL505 blood gas and electrolyte analyser. The discharge from the nine single or few-fibre preparations from the left carotid sinus nerve was recorded under mineral oil. We used stainless steel hook electrodes and the lateral approach described by Goodman [6]. Discharge was amplified (Tektronix 122 and Tektronix 502A oscilloscope).

Experimental Protocol

The inspired P_{O_2} composition was varied in ramps from about 30% to 5% (230 to 38Torr). The CO_2 concentration was kept constant at 5% (38 Torr) by changing the N_2 concentration. Each pair of falling and rising ramps had a cycle time of 2min. Intravenous infusions of KCl and/or NA (150mM KCl 1ml/kg/min and 1µg/kg/min NA) were started at the beginning of a ramp and lasted for 2 ramp cycles. During each infusion arterial blood samples were taken for analysis of pH, gas tensions and K^+ concentration. $[K^+]_a$ increased from 3.7±0.2mM in control to 6.2±0.2mM during the K^+ and NA+K^+ infusion. For curve fitting only the data from the second falling ramp (increasing hypoxia) were used.

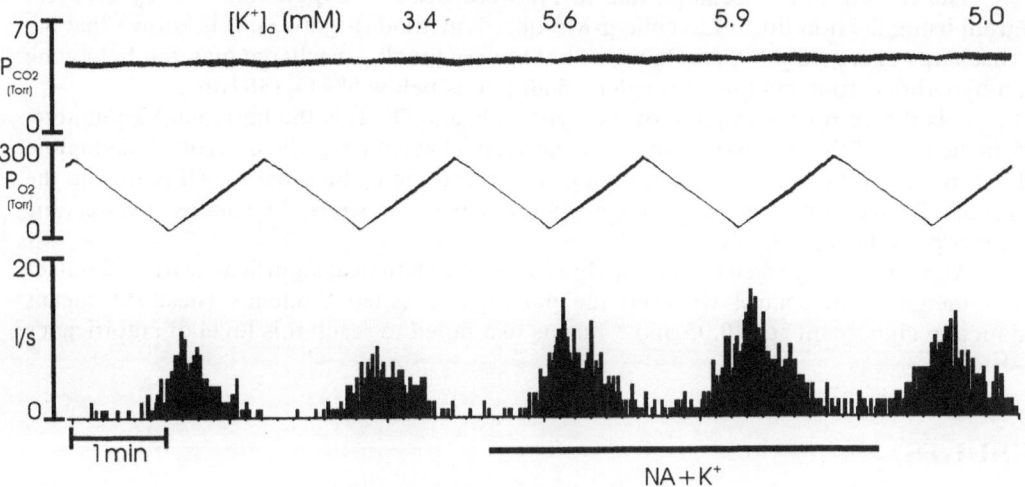

Figure 2. Raw data trace.

Curve Fitting

The number of chemoreceptor fibres present in each preparation was estimated from the maximal discharge frequency. The discharge was counted in one second bins. Each bin was allocated the average P_{O_2} value of the tracheal gases as an estimate of the arterial P_{O_2} (Figure 2). This estimate is valid, since the lung was hyperventilated, so that the alveolar gas tension was almost the same as the inspired gas tension. This is reflected in the small changes of P_{O_2} and P_{CO_2} during the respiratory cycle. The alveolar-arterial difference is likely to be affected only by direct shunt, since the detrimental effect of diffusion/perfusion inequilibrium is reduced by hyperventilation. Diffusion limitation causing an alveolar-arterial P_{O_2} difference is also unlikely under the experimental conditions. An estimated four second circulation lung to carotid body delay was taken into account.

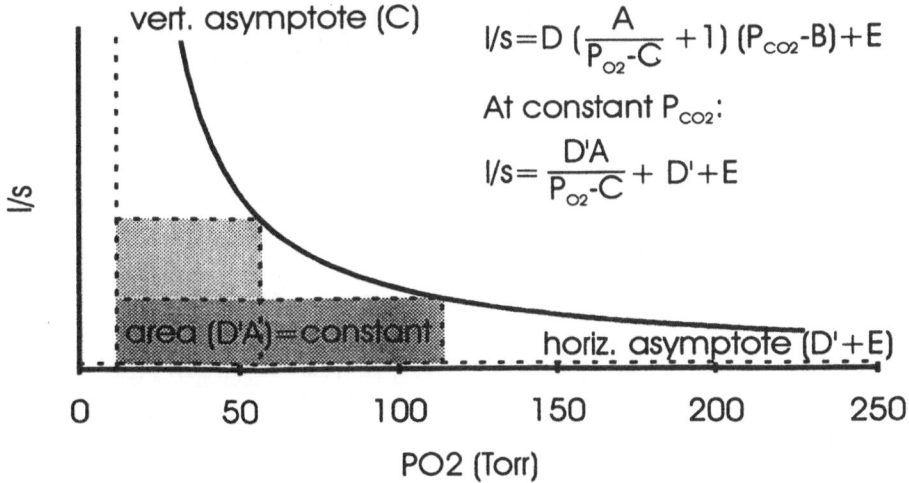

Figure 3. Equation used for modelling data.

The responses of discharge rate to P_{O_2} were fitted to hyperbolae [4,9,10] by a Pascal program using the non-linear Levenberg-Marquardt method (Figure 3). It is known that the P_{O_2}/discharge curve fits a hyperbola well[7], but at very low P_{O_2} discharge may fall below the fitted hyperbola. To avoid this we neglected all points below 5% O_2 (38 Torr).

C is the vertical asymptote of the hyperbola and D'+E is the horizontal asymptote. D'A is the area of the rectangle formed by the vertical asymptote, the horizontal asymptote and any point on the hyperbola. The area of this rectangle is the same for all points on the hyperbola. It gives a measure of how sharply the hyperbola bends: the smaller the area, the more sharply it bends.

All results are given as means ±1SE. To test the statistical significance of differences of parameters from control we used the paired, two tailed Student's t-test. ** means statistically significant at p<0.05 and * means just failed to reach this level of significance (p<0.1).

RESULTS

Hyperbolae fitted the response of chemoreceptors to hypotia well. An example of a fit for one preparation is shown in Figure 4.

The gain D'A increased significantly during infusions of NA, K$^+$ and NA+K$^+$. The vertical asymptote C and the horizontal asymptote D'+E tended to be decreased by all stimuli (Table 1).

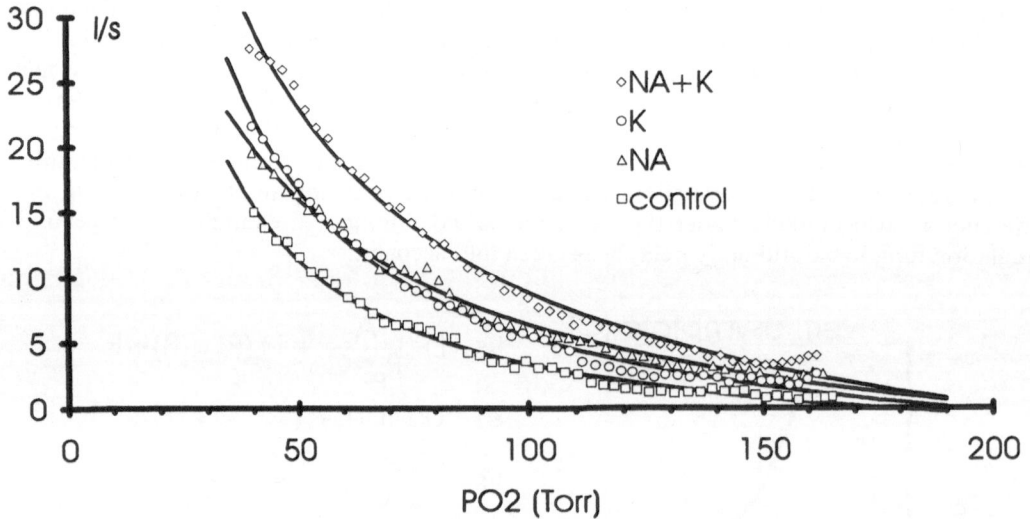

Figure 4. Hyperbolae fitted to 5s running averaged data in one preparation.

Table 1. Mean parameters ±1SE

	C (Torr)	D'A (I s^{-1} Torr)	D'+E (I s^{-1})	chi^2/n ((I s^{-1})^2)
Control	1 ± 3	754 ± 125	-2.9 ± 0.5	0.27 ± 0.04
NA	-39 ± 13**	2381 ± 570**	-6.8 ± 1.7**	0.87 ± 0.22
K$^+$	-36 ± 21*	2523 ± 714**	-8.3 ± 1.7**	0.97 ± 0.21
NA+K$^+$	-30 ± 13*	2994 ± 917**	-9.5 ± 2.5**	0.80 ± 0.21

Table 2. Parameters of the hyperbolae fitted to the interpolated, averaged data

	C (Torr)	D'A (I s⁻¹ Torr)	D'+E (I s⁻¹)	chi²/n ((I s⁻¹)²)
Control	4	697	-2.7	0.0008
NA	-32	1973	-6.0	0.0009
K⁺	-19	1884	-7.1	0.0012
NA+K⁺	-24	2368	-8.1	0.0025

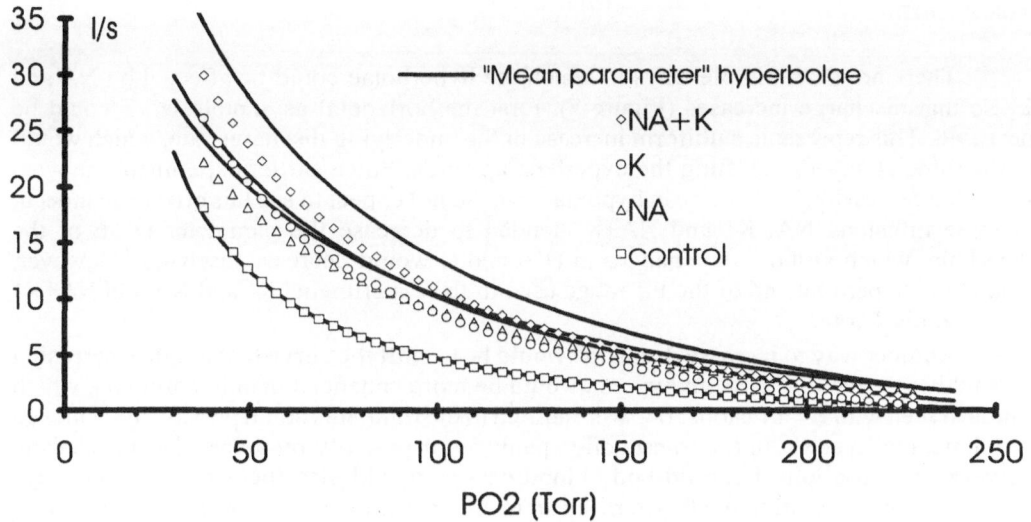

Figure 5. Interpolated, averaged data and "mean parameter" hyperbolae

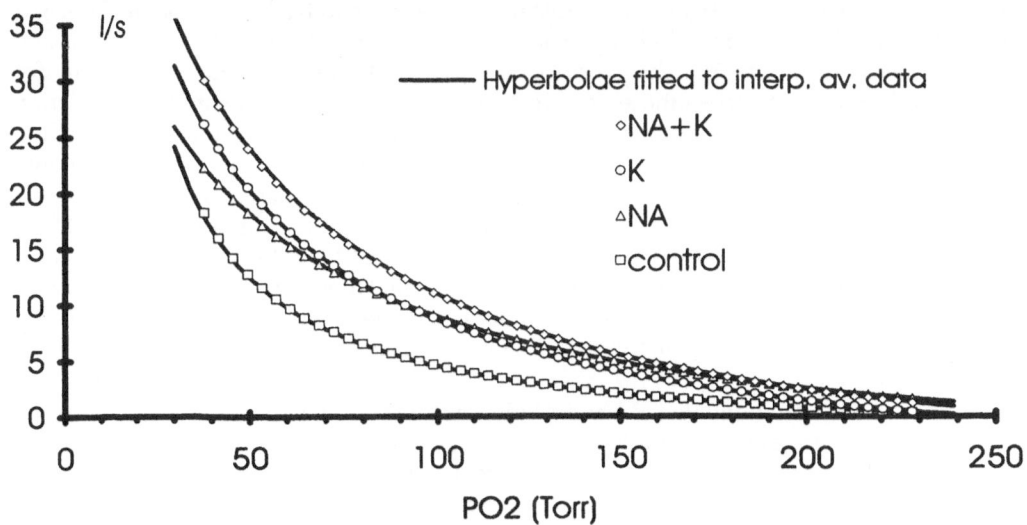

Figure 6. Interpolated, averaged data and hyperbolae fitted to it

A hyperbola constructed from the means of the parameters obtained from each preparation is shown in Figure 5. Using the fitted hyperbolae for interpolation between the points of each preparation's response, the mean discharges for the nine fibres were calculated in the range from 30% to 5% (230 to 38 Torr) P_{O_2}. Despite the fact that these data points do not lie on true hyperbolae, hyperbolae fitted the interpolated, averaged data points well (Table 2, Figure 6).

The gain was highest with NA and K+ combined, but the effects of these two stimuli were less than additive at all levels of hypotia.

Discussion

There are three simple ways in which the hyperbolae could be affected by NA and K+ so that discharge increases (Figure 7). First the horizontal asymptote D'+E could be increased. This represents a uniform increase in the underlying discharge rate, which would be the same at any P_{O_2}, shifting the hyperbola upwards. Since earlier experiments showed that K^+ increases discharge more in hypoxia [3], we did not expect to see this effect and indeed, all three infusions NA, K^+ and NA+K^+ tended to decrease the parameter D'+E of the hyperbola, which without the changes in D'A and C would decrease discharge. However, even at the hyperoxic end of the P_{O_2} range used in the experiments K^+ and NA and NA+K^+ increased discharge.

Another way to increase discharge would be to shift the curve to the right (increasing C). In this case the discharge in hypoxia would be more enhanced than in normoxia, which is indeed seen with K^+ infusions. If C were shifted to the right, this could possibly be achieved by mimicking hypoxia in the transduction pathway or possibly by a vascular mechanism. However, constriction of carotid body blood vessels would also increase P_{CO_2} and hence increase D' (incorporating the P_{CO_2} term). But in all three infusions C was shifted to the left, which tends to decrease the discharge. Because C was shifted to the left, NA and K^+ do not appear to act by mimicking hypoxia in the transduction pathway.

The third possibility is to increase D'A. In this case the absolute increase in discharge in hypoxia would be bigger than in normoxia. An increase in D'A reflects an increased gain of the transduction pathway. In all three cases of the NA, K^+ and NA+K^+ infusions D'A was increased, illustrating how the discharge was increased. Although the effects on D'+E and C tend to decrease discharge, the increase in D'A predominates. The mechanism for increasing D'A may be to affect the sensitivity of the transduction mechanism. Constriction of the carotid body vessels resulting in a lower local carotid body P_{O_2} and higher P_{CO_2} could also lead to an increase in D'A (D' incorporates the P_{CO_2} term). But a vascular effect should also increase C and D'+E, which was not observed. The effect of increased levels of CO_2 is to increase hypoxic sensitivity as well. Is it possible that NA and K^+ mimic the effect of CO_2 to increase the gain of the transduction mechanism? This seems unlikely, since in this case D' (incorporating the P_{CO_2} term) would be increased as well. Hence D'+E should be increased as well, but on the contrary D'+E was decreased. It seems that the factor A (hypoxic sensitivity) was predominantly increased, which is consistent with results published for ventilation in man [5].

One must bear in mind that NA and K^+ also shifted C to the left and D'+E downward. This implies that a crossing over of the NA, K^+ and NA+K^+ curves is possible in hyperoxia and in very severe hypoxia. In this case the effects of NA and K^+ on D'+E and C may predominate and they may be inhibitory. However, these points lie *outside* the data points collected during experiments that were used for curve fitting. Within the P_{O_2} range used in the experiments NA and K^+ increased carotid body discharge.

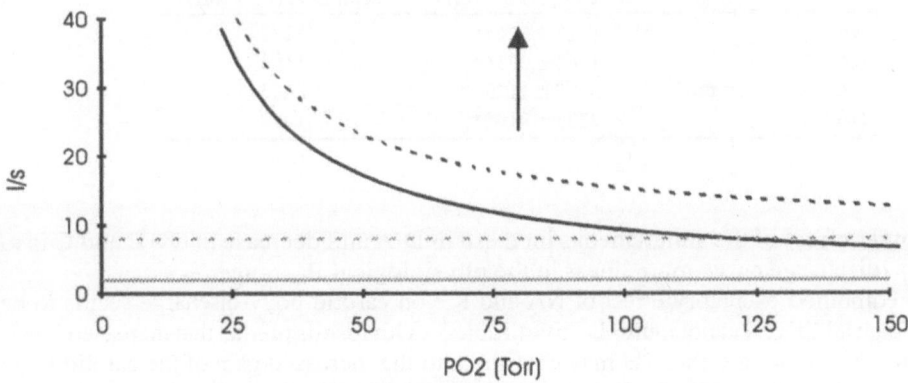

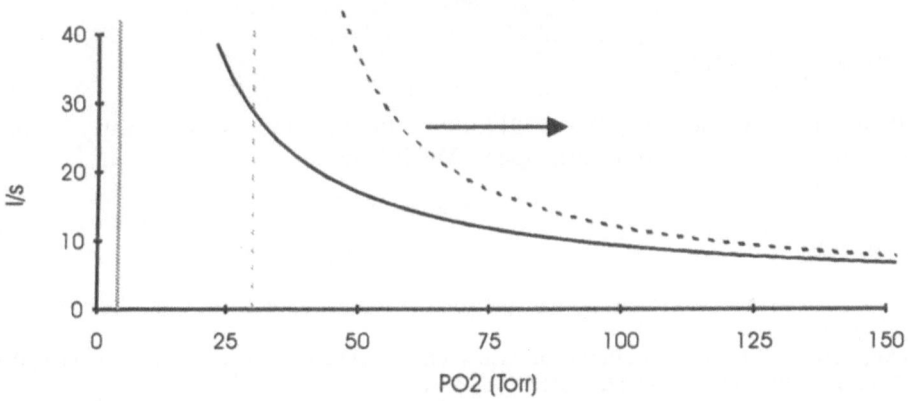

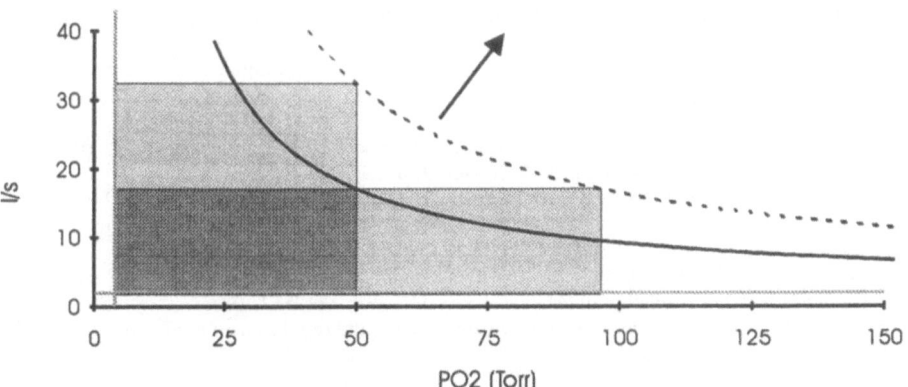

Figure 7. Different ways in which discharge might be increased.

Table 3. Effect on the parameter D'A

	Hyperbola	
	Mean parameter delta D'A (I s^{-1} Torr)	Interpolated, averaged delta D'A (I s^{-1} Torr)
NA	1627 ± 526**	1277
K^+	1768 ± 673**	1186
NA and K^+ summed	3395 ± 1179**	2463
NA+K^+	2238 ± 910**	1672

Another effect of the simultaneous increase in D'A and decrease in D'+E and C is to make the P_{O_2}/discharge curve more linear in the physiological P_{O_2} range.

The combined excitatory effect of NA and K+ on carotid body discharge seems to be slightly less than additive and not multiplicative (Table 3). Our results predict that increased levels of NA and K+ during severe exercise may contribute to the increased gain of the carotid body mainly by increasing D'A (probably A, the hypoxic sensitivity). However, the effects are modest and may not account for the whole increase in peripheral gain reported by Asmussen and Nielsen1.

ACKNOWLEDGMENT

We thank the MRC and NIH [HL15473] for their funding. GEB was a Fogarty Senior International Fellow [1970]. GH is a Christopher Welch Scholar.

REFERENCES

1. Asmussen, E., and M. Nielsen. Studies on the regulation of respiration in heavy work. Acta Physiol. Scand. 12: 171-188, 1946.
2. Band, D.M., and R.A.F. Linton. The effect of potassium on carotid body chemoreceptor discharge in the anaesthetized cat. J Physiol. 381: 39-47, 1986.
3. Burger, R.E., J.A. Estavillo, P. Kumar, P.C.G. Nye, and D.J. Paterson. Effects of potassium, oxygen and carbon dioxide on the steady-state discharge of cat carotid body chemoreceptors. J. Physiol. 401: 519-531, 1988
4. Cunningham, D.J.C. Some quantitative aspects of the regulation of human respiration in exercise. Brit.med.Bull. 19: 25-30, 1963.
5. Cunningham, D.J.C., E.N. Hey, J.M. Patrick, and B.B. Lloyd. The effect of noradrenaline infusion on the relation between pulmonary ventilation and the alveolar PO2 and PCO2 in man. Ann NY. Acad Sci. 109: 756-771, 1963.
6. Goodman, N.W. Efferent control of arterial chemoreceptors mediated by glossopharyngeal fibres and artefacts introduced by stimulation techniques. J Physiol. 230: 295-311, 1973.
7. Heinert, G., R. Painter, and P.C.G. Nye. Hyperbolic fitting of carotid body discharge excited by ramps of CO2 and O2 in the anaesthetized cat. J Physiol. Proceedings of Birmingham Meeting of the Physiol. Society December 1994.
8. Kumar, P., P.C.G. Nye, and R.W. Torrance. Do oxygen tension variations contribute to the respiratory oscillations of chemoreceptor discharge in the cat?. J Physiol. 395: 531-552, 1988.
9. Lloyd, B.B., and D.J.C. Cunningham. A quantitative approach to the regulation of human respiration. In: The Regulation of Human Respiration. ed DJC Cunningham and BB Lloyd. Blackwell:Oxford : 331-349, 1963.
10. Nye, P.C.G., and R. Painter. Quantifying the steady-state discharge of the carotid body of the anaesthetized cat. J Physiol. 417: 172P, 1989.
11. Weil, J.V., Byrne-Quinn, E., Sodal, I.E., Kline,J.S., Mccullough, R.E., and Filley, G.F. Augmentation of chemosensitivity during mild exercise in normal man. J. Appl. Physiol. 33: 813-819, 1972.

62

ACTIVATION OF LIMBIC STRUCTURES DURING CO$_2$-STIMULATED BREATHING IN AWAKE MAN

D. R. Corfield,[1] G. R. Fink,[2,3] S. C. Ramsay,[2] K. Murphy,[1] H. R. Harty,[1]
J. D. G. Watson,[2] L. Adams,[1] R. S. J. Frackowiak,[4] and A. Guz[1]

[1] Department of Medicine, Charing Cross and Westminster Medical School
London W6 8RF, United Kingdom
[2] MRC Cyclotron Unit
London W12 0HS, United Kingdom
[3] Max-Planck Institute for Neurological Research
Cologne, Germany
[4] Wellcome Department of Cognitive Neurology, Institute of Neurology
London WC1N 3BG, United Kingdom

INTRODUCTION

In awake man, the ventilatory response to CO$_2$ is highly variable. In addition, breathing CO$_2$ is associated with sensations of dyspnoea. Such observations suggest that, in man, supra-brainstem structures might modulate the ventilatory response to CO$_2$. This hypothesis is supported by observations in animals that the respiratory response to CO$_2$ is influenced by the hypothalamus [11] and that respiratory-related activity is present during CO$_2$-stimulated breathing in neurones located in the thalamus [1].

The present study, using positron emission tomography (PET), was designed to test the hypothesis that supra-brainstem structures, in particular the primary motor cortex [7], would also be involved in the ventilatory response to inhaled CO$_2$ in awake man. Subjects were scanned during periods of CO$_2$-stimulated breathing (CO$_2$-stim) and during periods of eucapnic positive pressure ventilation (Control).

METHODS

Five healthy male subjects (aged 23 - 49 years) were studied with ethical approval; all had a normal ventilatory response to inhaled CO$_2$. Each subject either breathed sponta-neously or was ventilated through a tightly-fitting nasal mask whilst lying supine. Airflow, tidal PCO$_2$ and arterial oxygen saturation (SaO$_2$) were measured. Pressure within the nasal mask (Pnasal) was used as an index of upper airway pressure. Prior to the study day, subjects

Modeling and Control of Ventilation, Edited by S. J. G. Semple, L. Adams, and B. J. Whipp
Plenum Press, New York, 1995

331

experienced CO_2-stim and were trained to relax their breathing during positive pressure ventilation.

CO_2-stim was performed for periods of 6 minutes; for this, subjects breathed via a dead-space attached to the breathing circuit. The fraction of inspired CO_2 (F_ICO_2), and therefore $P_{ET}CO_2$, could be controlled by trickling 100% CO_2 directly into the dead-space. Passive eucapnic ventilation was performed, as Control, for periods of 6 minutes; this was achieved by subjects relaxing their respiratory muscles during intermittent positive pressure ventilation (Servo 900B, Siemens-Elma, UK). The adequacy of the relaxation was judged from the recordings of Pnasal [2]. During Control, tidal volume (V_T) and respiratory frequency (f_R)were similar to that produced during CO_2-stim; $P_{ET}CO_2$ was maintained around eucapnia by increasing F_ICO_2.

PET scans were performed alternately during six CO_2-stim and six Control runs. Scans were timed so that the data collection was coincident with the last two minutes of either CO_2-stim or Control. Scans were performed, using a Siemens-CTI 953B scanner in 3D mode, to determine relative regional cerebral blood flow (rCBF) by measuring the regional distribution of cerebral radioactivity following the intravenous infusion of radiolabelled water (each scan $\approx$ 15 mCi $H_2^{15}O$)[4,6]. To analyse the grouped data, individuals' PET images were transformed into standard stereotactic space [10] using information from magnetic resonance image data (obtained on a separate occasion).

The results presented, for the relative rCBF measurements, are based on a categorical (subtraction) analysis between the activation (CO_2-stim) and the control task (passive ventilation). A pixel-based ANCOVA was performed using 'global' activity (reflecting blood flow) as the covariate to control for differences in global cerebral blood flow (gCBF) associated with the different conditions. The condition-specific (activation or control) mean values of relative rCBF and the associated error variances were then calculated for each pixel across all subjects. Comparison of the two conditions was then carried out for each pixel using t-statistics.

RESULTS

The group mean ($\pm$ s.e.m.) $P_{ET}CO_2$ was significantly higher during CO_2-stim than during Control (50.3 $\pm$ 1.7 vs 38.4 $\pm$ 1.0 mmHg respectively; $p < 0.05$ ANOVA). During CO_2-stim, V_T, and consequently minute ventilation, was slightly but significantly greater than during Control (V_T: 1.79 $\pm$ 0.20 vs 1.62 $\pm$ 0.20 l; minute ventilation: 27.4 $\pm$ 2.1 versus 23.7 $\pm$ 1.7 l.min^{-1}) There were no statistically significant differences in the group mean values for respiratory timing (f_R: 16.4 $\pm$ 2.7 versus 15.5 $\pm$ 2.2 breaths.min^{-1}). Examination of the Pnasal trace, during Control, indicated that all subjects were able to relax their respiratory muscles and that breathing was passive during this condition.

Sensations associated with the two conditions were assessed in a supplementary study. Descriptors associated with a changed pattern of breathing were associated with both CO_2-stim (n = 4) and Control (n = 5). All subjects reported symptoms of discomfort associated with their breathing during CO_2-stim.

Figure 1 illustrates the areas of significantly increased relative rCBF associated with CO_2-stim in the data averaged across all individuals and all runs.

DISCUSSION

To determine increases in neuronal activation that are associated with CO_2 stimulation, cerebral blood flow measured during a control condition must be subtracted from that

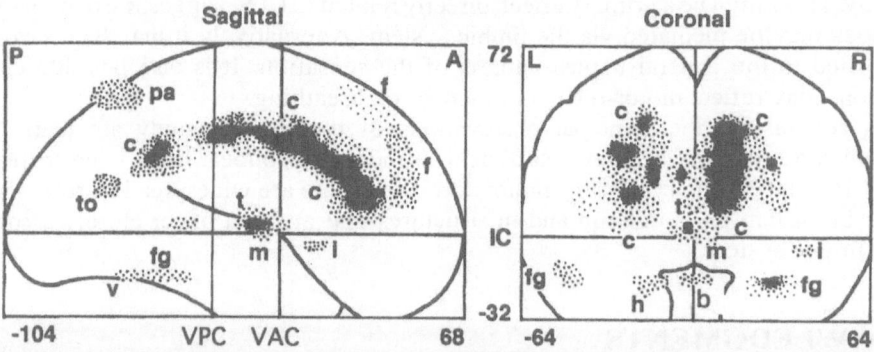

Figure 1. Projections of significant increases in rCBF associated with CO_2-stim. Areas of significant ($p < 0.05$; corrected for multiple comparisons; $Z > 3.68$) relative rCBF increases during CO_2-stim, averaged from the group of 5 subjects, are shown as projections onto representations of sterotactic space [10]. The left (sagittal) image views the brain from the side, the right (coronal) image from the back. A = anterior, P = posterior, L = left, R = right, VAC = vertical plane through the anterior commissure, VPC = vertical plane through the posterior commissure. Numbers by axes refer to co-ordinates of stereotactic space and indicate the distance (mm) to right or left of the mid-sagittal line, the distance anterior or posterior to VAC and the distance above or below the intercommissural line (IC). Significance increases with shading density. The anatomical locations of the areas of activation are indicated by lettering placed next to the corresponding region of interest (b: upper brainstem, m: midbrain/hypothalamus, t: thalamus, h: hippocampus/parahippocampus, fg: fusiform gyrus, c: cingulate area, i: insula, f: frontal cortex, to: temporo-occipital cortex, pa: parietal cortex, v: cerebellar vermis). No changes in relative rCBF were seen within the primary sensory-motor cortex.

determined during CO_2 stimulation. In the present study, and in common with previous studies from our group using PET [2,8], we have used passive mechanical positive pressure ventilation at an elevated level of minute ventilation that matches that produced during the 'activation' condition. The intention of using this control, rather than the alternative of spontaneous breathing at rest, is that sensory differences between the two conditions related to differences in the level of ventilation are minimised and therefore increases in relative rCBF should reflect neuronal activation associated specifically with CO_2 stimulation.

It is generally assumed that an increase in rCBF is due to an increase in local metabolic demand that is produced by a change in neuronal activity. Our principal concern in this study is that CO_2 itself has a direct action on cerebral blood flow, due to vasodilatation, that is independent of metabolism. The analysis assumes that increases in rCBF are independent of gCBF; however, it is possible that increased neuronal activation may be associated with a change in the response of rCBF when gCBF is elevated. Previously Ramsay *et al.* [9] increased steady-state $P_{ET}CO_2$ and showed that visual stimulation was associated with an increased rCBF in the visual cortex that was independent of the changes in gCBF induced by the increase in $P_{ET}CO_2$ (when such changes in gCBF were allowed for using an analysis of covariance). Assuming that other cortical areas behave in a similar manner to the visual cortex, the results indicate that activation-dependent increases in rCBF will occur and will be quantitatively independent of gCBF, even when gCBF is elevated by CO_2.

The present results suggest that a region of neuronal activation extends from the upper brainstem, up through the midbrain and hypothalamus, to the thalamus during CO_2 stimulation. Activation of the limbic system [5] is consistent with the changes in relative rCBF seen here within the cingulate area, parahippocampus, hippocampus, fusiform gyrus and insula. Such activation would also be consistent with associated activation occurring within the parietal and frontal cortices. Activation of the limbic system might be explained by the uncomfortable sensations associated with CO_2 breathing that were reported by all subjects in the study. This might be a primary effect, directly related to CO_2, suggesting that sensations of dyspnoea may be mediated via the limbic system. Alternatively it may be a secondary effect, related to the general unpleasantness of the sensation. It is also possible that the observations may reflect motor-related influences on breathing.

In conclusion, the principal observations of the present study are that, during CO_2-stimulated breathing at a level sufficient to induce dyspnoea, there is no increase in relative rCBF in the primary sensory-motor cortex but there are widespread increases within the upper brainstem and midbrain and in structures that are part of, or closely associated with, the limbic system.

ACKNOWLEDGMENTS

This work formed part of a Wellcome Trust Programme Grant to A.G. We thank the staff at the MRC Cyclotron Unit and at the NMR Unit, Royal Postgraduate Medical School who made the studies possible. This work were first presented to the Physiological Society [3].

REFERENCES

1. Chen, Z., Eldridge, F. L. & Wagner, P. G. (1992). Respiratory-associated thalamic activity is related to level of respiratory drive. *Respiration Physiology* 90, 99-113.
2. Colebatch, J. G., Adams, L., Murphy, K., Martin, A. J., Lammertsma, A. A., Tochon–Danguy H.J., Clark, J. C., Friston, K. J. & Guz, A. (1991). Regional cerebral blood flow during volitional breathing in man. *Journal of Physiology* 443, 91-103.
3. Corfield, D. R., Fink, G. R., Ramsay, S. C., Murphy, K., Watson, J. D. G., Adams, L., Frackowiak, R. S. J. & Guz, A. (1994). Cortical and subcortical activations during CO_2-stimulated breathing in conscious man. *Journal of Physiology* 476, 80P.
4. Fox, P. T. & Mintun, M. A. (1989). Non-invasive functional brain mapping by change distribution analysis of average PET images of $H_2{}^{15}O$ tissue activity. *Journal of Nuclear Medicine* 30, 141-149.
5. Maclean, P. D. (1992). The limbic system concept. In *The temporal Lobes and the Limbic System.* ed. Trimble, M. R. & Bolwig, T. G., pp. 1-14. Wrightson Biomedical Publishing Ltd, Petersfield, Hampshire, U.K.
6. Mazziotta, J. C., Huang, S. C., Phelps, M. E., Carson, R. E., MacDonald, N. S. & Mahoney, K. (1985). A non-invasive positron computed tomography technique using oxygen-15 labelled water for evaluation of a neural behavioural task battery. *Journal of Cerebral Blood Flow and Metabolism* 5, 70-78.
7. Murphy, K., Mier, A., Adams, L. & Guz, A. (1990). Putative cerebral cortical involvement in the ventilatory response to inhaled CO_2 in conscious man. *Journal of Physiology* 420, 1-18.
8. Ramsay, S. C., Adams, L., Murphy, K., Corfield, D. R., Grootoonk, S., Bailey, D. L., Frackowiak, R. S. J. & Guz, A. (1993a). Regional cerebral blood-flow during volitional expiration in man - a comparison with volitional inspiration. *Journal of Physiology* 461, 85-101.
9. Ramsay, S. C., Murphy, K., Shea, S. A., Friston, K. J., Lammertsma, A. A., Clark, J. C., Adams, L., Guz, A. & Frackowiack, R. S. J. (1993b). Changes in global cerebral blood flow in humans: effect on regional cerebral blood flow during a neural activation task. *Journal of Physiology* 471, 521-534.
10. Talairach, J. & Tournoux, P. (1988). Coplanar Stereotaxic Atlas of the Human Brain. Thieme, Stuttguard, Germany.

11. Waldrop, T. G. (1991). Posterior hypothalamic modulation of the respiratory response to CO_2 in cats. *Pflügers Archive* 418, 7-13.

63

IMPROVEMENTS TO THE PRBS METHOD FOR MEASURING VENTILATORY RESPONSE TO CO$_2$

S. D. Mottram

University Department of Anaesthesia
Leicester Royal Infirmary
Leicester LE1 5WW, United Kingdom

INTRODUCTION

In recent years the most widely used dynamic method for measuring the ventilatory response to CO$_2$ has been based on a model by Swanson and Bellville (7). This technique, known as dynamic end-tidal forcing (DEF), uses apparatus capable of rapid feedback control of F$_i$CO$_2$ to force step changes in end-tidal PCO$_2$. A mathematical model is then fitted to the ventilatory response and the parameters of the model are taken to be estimates of the gain of the central and peripheral chemoreceptors, the "apnoea point", etc. This method has been used in recent clinical investigations to estimate the effects of drugs on the central and peripheral chemoreceptor systems (1, 2, 3). Whilst clearly being suited to this kind of study, the technique has features that may make it less useful for basic physiological studies in respiratory control.

Because DEF involves directly controlling P$_{et}$CO$_2$, the response being measured may appear to be open-loop, with the P$_{et}$CO$_2$ (and hopefully the P$_a$CO$_2$) being independent of ventilation. However, an alteration in P$_{et}$CO$_2$ brought about by a change in V_1 may only be corrected by altering F$_i$CO$_2$ on the following breath. Thus the system is actually open-loop with respect to slow changes in the subjects physiology, but closed loop for changes lasting less than T$_{tot}$. Because the behaviour of dynamic control systems may be significantly different in open-loop conditions compared to closed-loop, there must be some doubt as to the relationship between the model parameters measured with the DEF technique, and the parameters under normal physiological conditions. The significance of this possible flaw in the technique is not known at present.

To provide an alternative method for comparison with DEF, I have attempted to develop a closed-loop method, based on the existing PRBS technique, capable of estimating the central and peripheral components of the ventilatory response.

Modeling and Control of Ventilation, Edited by S. J. G. Semple, L. Adams, and B. J. Whipp
Plenum Press, New York, 1995

The Pseudo-Random Binary Sequence (PRBS) Method

The PRBS method has been widely used in the engineering literature for determining the properties of control systems (5). The technique produces an estimate of the system's impulse response; that is the response to an infinitely short, infinitely large change in the input, where the area under the impulse is one unit. Because these methods are used with sampled data, the closest approximation to an impulse that can be obtained is an input with the value 1 in the first position in the data set, and zero at all other positions. The PRBS measurement process involves providing an input signal consisting of high and low values according to a pattern with special statistical properties. This input sequence (which is actually the PRBS from which the technique gets its name) consists of the numbers 0 and 1 arranged in an order such that the auto-correlation of the sequence approximates an impulse. The output of the system is recorded at time points corresponding to each value in the PRBS sequence. The cross-correlation between the input PRBS and the recorded output approximates the impulse response of the system being measured. In practice there are statistical problems with using the cross-correlation directly, and a more complex method of calculation is used.

The PRBS was first applied to measurement of the ventilatory response to CO_2 by Sohrab and Yamashiro (6). In this case, because respiratory data are only measured once per breath, the impulse response is equivalent to the response to a single breath with an F_ICO_2 of 1%. Although it is possible to measure the ventilatory response to a single breath containing CO_2 (as in the DeJours test (4)), this method has the disadvantage that many repetitions are required to remove the effects of noise, and a delay is needed between each repeat to allow the subject to regain equilibrium. The measurements required to gain an accurate estimate of the ventilatory response are therefore very time-consuming. The PRBS method has the advantage that it is capable of estimating the impulse response even in the presence of a considerable amount of noise, and so fewer repetitions are required. However the PRBS technique assumes that the system being measured is linear and that the properties of the system do not change with time. These assumptions may be violated in making respiratory measurements.

In order to obtain a PRBS response, a subject is connected to a face mask which can be switched breath-by-breath between two different inspiratory gas mixtures - either air or a gas mixture containing CO_2. The subject inspires air during breaths where the PRBS digit is 0, and from the CO_2 mixture when the digit is 1. V_I is recorded using a pneumotachograph attached to the mask, and F_ICO_2 and $P_{ET}CO_2$ are measured using a fast CO_2 analyser aspirating gas through a cannula placed just infront of the lips.

MATHEMATICAL MODEL

To develop a method capable of estimating central and peripheral chemoreceptor gains, a model of the control system is needed. Because the PRBS method assumes that the system being measured is linear, the Swanson and Bellville model (where the time "constant" of the central chemoreceptor varies with P_aCO_2) cannot be used. However, the fluctuations in P_aCO_2 produced by the PRBS are relatively small, so it may be possible to use an approximation with a true time constant. A simplified version can be proposed as follows:

$$V_I = V_I^c + V_I^p \tag{1}$$

$$\frac{dV_I^c}{d} \cdot \frac{1}{\alpha_c} = q_c[C(t - \tau_c) - k] - V_I^c(t) \tag{2}$$

$$\frac{dV_1^P}{d} \cdot \frac{1}{\alpha_p} = q_p[C(t - \tau_p) - k] - V_1^P(t)$$

$$(3)$$

where V_1 is minute ventilation, V_1^c is the component of the ventilatory response attributed to the central chemoreceptor and V_1^P is the peripheral component, α_c and α_p are the central and peripheral rate terms, $C(t)$ is the P_aCO_2 at time t, τ_c and τ_p are the central and peripheral circulatory time delays and k is the apnoea point. q_c and q_p are the central and peripheral chemoreceptor gains, respectively.

To develop the method of measurement, we need to make a series of assumptions. Firstly, we must assume that the relationship between F_1CO_2 and $P_{ET}CO_2$ can be described by a transfer function H_{xc} - that is we can obtain an estimate of the impulse response of $P_{ET}CO_2$ to F_1CO_2, denoted h_{xc}. We must also assume that we can obtain an impulse response of ventilation to F_1CO_2, written h_{xv}. If the transfer function H_{xv} (and hence the impulse response) exists, it follows that a third transfer function, H_{cv}, also exists, such that $H_{xv} = H_{xc} \cdot H_{cv}$. In the time domain this can be represented as

$$h_{xv} = h_{xc} * h_{cv}$$

$$(4)$$

where the symbol "*" represents convolution, and h_{cv} is the response of ventilation to an impulse in $P_{et}CO_2$.

If a PRBS recording is made, and F_1CO_2, $P_{et}CO_2$ and V_1 data are available, both h_{xv} and h_{xc} may be calculated. Because we are attempting to find the relationship between $P_{et}CO_2$ and V_1, we need to obtain a value of h_{cv} that satisfies equation 4.

If the model described by equations 1 to 3 is solved for an impulse response in $C(t)$ we may obtain the general form

$$h_{cv}(t) = H(t - \tau_c)A_c e^{\alpha_c[t - \tau_c]} + H(t - \tau_p)A_p e^{\alpha_p[t - \tau_p]}$$

$$(5)$$

where $H(t)$ is a step function equal to 0 where $t <= 0$ and 1 when $t > 0$. In this form, A_c/α_c is equal to the central chemoreceptor gain, q_c and A_p/α_p equals the peripheral gain q_p. In a practical situation, estimating from a PRBS recording, the values of h_{xv} and h_{xc} are sampled, and errors are present in the original data set and the estimates. We can therefore introduce a discrete-time version of the problem:

$$h_{xv}[n] = (h_{xc}[n] * h_{cv}[m]) + e[n]$$

$$(6)$$

where n and m represent the sample number, and $e[n]$ represents sampled noise. The parameters of the respiratory control model are therefore obtained by choosing the parameters of a sampled version of equation 5 to obtain an estimate of $h_{cv}[m]$ which minimises $e[n]$ in equation 6.

TEST METHODS

In order to begin testing the validity of this analysis, a series of 14 PRBS recording were made in 6 healthy naive unmedicated volunteers, following approval by the regional Ethical Committee. Subjects sat at rest in a comfortable chair and were connected to the measurement apparatus. Recordings of V_1, $P_{ET}CO_2$ and F_1CO_2 were commenced after 10 minutes rest, and the subjects listened to tape recorded books via headphones throughout the measurements. The test stimulus consisted of a 127 breath sequence with F_1CO_2 switching between 0 and a value between 0.05 and 0.07. In each recording the sequence was repeated

Vi l/min

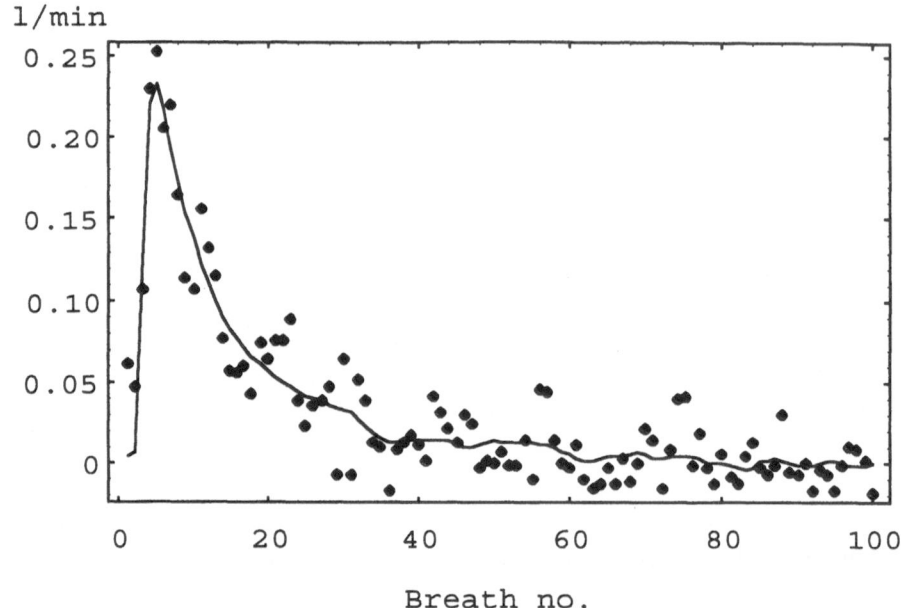

Breath no.

Figure 1. Ventilation response to an impulse in F_iCO_2.

at least 6 times, and the first sequence was eliminated before analysis, to allow time for the subject to reach a stable operating point. The data were analysed using a standard cross-correlation method, and the impulse responses for each repeat of the sequence were averaged to give a single estimate of h_{xc} and h_{xv}. Parameters of the respiratory controller model were estimated by a purpose-written computer program, using the Nelder-Mead simplex method to choose values of A_c, α_c, A_p and α_p which minimised the sum of the squares of the errors $e[n]$ in Equation 6. The process was repeated using a grid-search method to find the best values of τ_c and τ_p, expressed as multiples of the mean breath duration in the recording.

RESULTS

Figure 1 shows an impulse response of V_1 to F_1CO_2, with the best fit curve generated using the test program. The parameters obtained in all the measurements are given in table 1. Of the 14 recordings used, 13 produced clear impulse responses with varying amounts of "noise" present. In one case the h_{xv} response was of particularly poor quality. It is believed that this was due to the presence of an irregular respiratory rate and large numbers of sighs

Table 1. Summary data from 14 PRBS studies

	Mean	Min	Max	Units
Total gain	8.87	1.19	23.06	$1 \cdot min^{-1} \cdot kPa^{-1}$
Central gain	6.70	0.62	14.54	$1 \cdot min^{-1} \cdot kPa^{-1}$
Peripheral gain	2.17	0.57	8.81	$1 \cdot min^{-1} \cdot kPa^{-1}$
Peripheral/total	0.24	0.09	0.54	
Apnoea point	3.36	-1.90	4.8	kPa

in the recording. This data set produced very small values for chemoreceptor gains and a negative value for the apnoea point. This set of data was not excluded from the analysis, so figures in the table reflect the poorer recording as well as the majority of better quality studies.

DISCUSSION

As can be seen from the data above, the method described appears to produce estimates of the respiratory controller parameters similar to those in other dynamic studies, particularly in the case of the central and peripheral gains, and the ratio of peripheral to total gain. This suggests that despite the assumptions that are inherent in the analysis, the method may be of some use in physiological studies. In order to investigate the accuracy and repeatability of the technique, simulation studies where the true values of the parameters are known, will be necessary. Only when these studies have been carried out will it be possible to directly compare this method with the DEF and other techniques.

Acknowledgement

I would like to thank Dr C.D. Hanning and Dr J. Fothergill for advice and encouragement in this work.

REFERENCES

1. Berkenbosch A, Olievier CN, Wolsink JG, DeGoede J, Rupreht J. Effects of morphine and physostigmine on the ventilatory response to carbon dioxide. *Anesthesiology.* **80:** 1303-1310, 1994.
2. Berkenbosch A., Rupreht J, DeGoede J, Olievier CN, Wolsink JG. Effects of Eseroline on the ventilatory response to CO_2. *European J. Pharmacology.* **232:** 21-27, 1993
3. Dahan A, Maarten JL, van den Enden J, Berkenbosch A, DeGoede J, Olievier ICW, van Kleef JW, Bovill JG. Effects of subanesthetic Halothane on the ventilatory responses to hypercapnia and acute hypoxia in healthy volunteers. *Anesthesiology* **8(4)**, 727-738.
4. DeJours P. Chemoreflexes in breathing. *Physiol. Reviews.* **42(3):** 335-358, 1962.
5. Hill JD, McMurtry GJ. An application of digital computers to linear system identification. *IEEE Trans. Automatic Control.* 536-538, October 1964.
6. Sohrab S and Yamashiro SM. Pseudo-random testing of ventilatory response to inspired carbon dioxide in man. *Journal of Applied Physiology.* **49(6)**, 1000-1009, 1980
7. Swanson GD and Bellville JW. Step changes in end-tidal CO_2: methods and implications. *Journal of Applied Physiology.* **39: 377-385, 1975.**

64

INTRALARYNGEAL CO₂ REDUCES THE INSPIRATORY DRIVE IN CATS BY SENSORY FEEDBACK FROM THE LARYNX

Małgorzata Szereda-Przestaszewska and Beata Wypych

Department of Neurophysiology
P.A.S. Medical Research Centre
3, Dworkowa str., 00-784 Warsaw, Poland

INTRODUCTION

Larynx is a powerful reflexogenic area in regulation of breathing. Majority of the respiratory responses from the laryngeal airway is defensive, hypoventilatory in function. As a part of the dead space larynx is exposed to carbon dioxide (CO_2) during each expiration. It has been shown that intralaryngeal CO_2 in animal preparation elicits a decrease in ventilation (6). This effect has a sensory pathway in the superior laryngeal nerve. Most fruitful recent search for the CO_2 - responsive receptors within the laryngeal passage did not reveal any specialized endings. Few receptors are stimulated by intralaryngeal CO_2 (1, 7), while activity of the remaining ones is attenuated (4). This decreased afferent input from the laryngeal airway might well contribute to changes in the respiratory pattern induced by excess of CO_2. We reasoned that overall increased ventilation evoked by the inhaled CO_2 is attenuated by the laryngeal inhibitory input. The present experiments were designed to study the effects of increased concentration of carbon dioxide in the functionally separated laryngeal airway on respiratory pattern and to assess the role of vagal and sensory laryngeal feedback in this respiratory response.

METHODS

Eighteen cats anaesthetized with sodium pentobarbitone (30 mg · kg⁻¹, i.p.), later supplemented with chloralose (16 mg · kg⁻¹, i.v.), were used. They breathed spontaneously. The trachea was divided below the larynx and both ends of the tracheal cannula were put to the lower and upper trachea. The latter was inserted under cricoid cartilage. The cannula had two side-arms and in its mid-portion was divided by septum into two separate parts. Tidal volume was measured at the distal side arm. Humidified and warmed air or 10% CO_2 in air at a constant rate of 1.5 l · min⁻¹ was flowed through the proximal side arm and the oral cannula, positioned just above the vocal folds (Fig. 1).

Modeling and Control of Ventilation, Edited by S. J. G. Semple, L. Adams, and B. J. Whipp
Plenum Press, New York, 1995

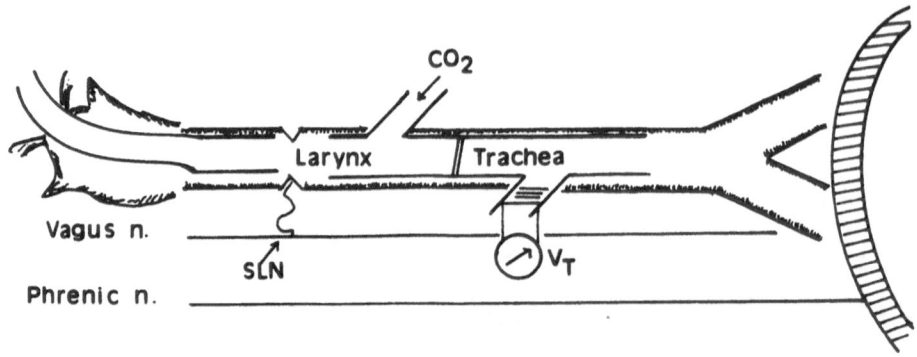

Figure 1. Schematic representation of the experimental setup. Gases were given to the larynx in the expiratory direction and were not breathed by the cats. SLN, superior laryngeal nerve.

Air and CO_2 were supplied alternatively in two minutes' periods. Air was used as control value. The two recurrent laryngeal nerves were identified and spared. The C_4 - C_5 root of the right phrenic nerve was cleared, cut, desheathed and prepared for recording. The cervical vagus nerves as well as the superior laryngeal nerves (SLNs) were separated, isolated and prepared for division later in the experiment.

A femoral vein and femoral artery were catheterized for further injections and to monitor blood pressure, respectively.

Arterial pressure was measured with pressure manometer (C.K. 01, Mera - Tronik) and blood pressure monitor (MCK 4011S). Tidal volume was recorded from pneumotachograph (Electrospirometer CS6, Mercury). End-tidal CO_2 was measured with a capnograph (Engström Eliza plus). Action potentials of the phrenic nerve were amplified and integrated (NL 104A and NL 703, NeuroLog TM, Digitimer). Time constant of the integrator was 100 msec. All recordings were registered with Honeywell Omnilight Recorder 8M 36. Rectal temperature was monitored and maintained at 37-39°C throughout the experiment.

The effects of increased fractional CO_2 in the larynx were recorded in the intact animals, following midcervical vagotomy and after subsequent division of the superior laryngeal nerves (SLNs).

Inspiratory time (T_I) and expiratory time (T_E) were determined from the start and the peak of the phrenic neurogram and breathing frequency was computed.

The ventilatory responses were assessed by comparing the averaged data of ten respiratory cycles when 10% CO_2 was flown through the larynx to the data obtained with identical flow of air used as control value.

The results were calculated as original values (mean ± S.E.M.). Statistical significance was evaluated with Student's t-test for paired and independent means, as appropriate. $P<0.05$ was deemed significant.

RESULTS

The ventilatory responses to continuous intralaryngeal administration of 10% CO_2 enriched gas mixture were recorded in eighteen cats. In the intact and subsequently vagotomized animals intralaryngeal CO_2 significantly reduced tidal volume. The response was more apparent soon after introduction of CO_2 into laryngeal airstream than at the end of two

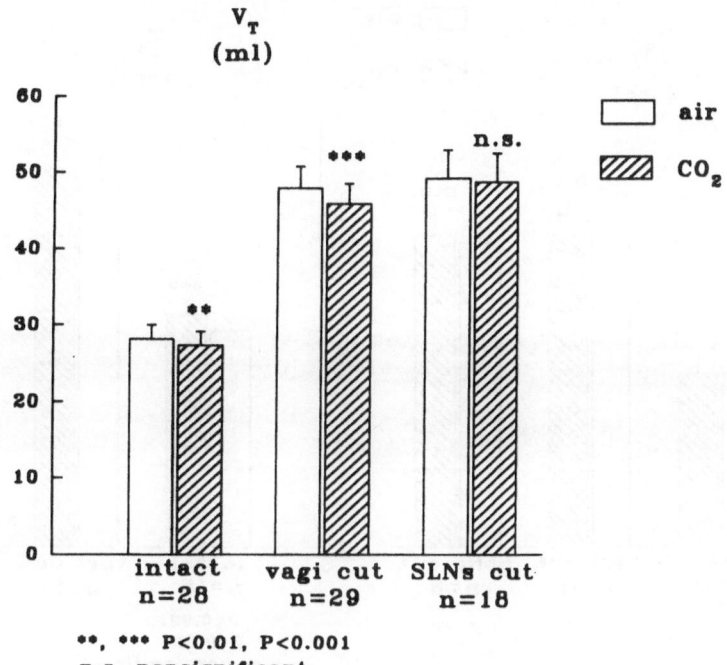

Figure 2. Mean changes in tidal volume induced by intralaryngeal CO_2 in eighteen cats. Data are presented as means ± S.E.M. n denotes the number of trials.

minute exposure, which is in general agreement with previous reports (3, 6). The reduction in tidal volume was small but consistent in all animals studied (Fig. 2).

In subsequently vagotomized cats the decrease in tidal volume was more apparent but not significantly different from the intact state (P>0.05). As found previously by Boushey and Richardson (6) and Bartlett and associates, (3) division of the superior laryngeal nerves abolished the ventilatory response to intralaryngeal CO_2.

The volume effects were accompanied by the prolongation of the respiratory cycle. As illustrated in Fig. 3 the inspiratory time (T_I) showed no appreciable response. This result was attributed entirely to significant prolongation of the expiratory time (T_E) in the intact and vagotomized cats, with no significant difference between the two conditions.

Due to the reduction in tidal volume and the decrease in respiratory rate minute ventilation was significantly depressed. Fig. 4. presents a similar effect of intralaryngeal CO_2 on minute ventilation in the intact and vagotomized cats.

DISCUSSION

The raised CO_2 in the laryngeal airstream reduced the respiratory drive. It was due to the decrease in tidal volume with no change in the inspiratory time. The expiratory time was prolonged, resulting in reduction in the respiratory rate. In consequence minute ventilation fell. This ventilatory inhibition was modest, but highly consistent and reproducible in all animals prior to the division of the superior laryngeal nerves. The extent of this respiratory response to intralaryngeal CO_2 was similar in the neurally intact and vagotomized cats. As shown in Fig. 2 the inflow of intralaryngeal CO_2 resulted in reduction of tidal volume. To

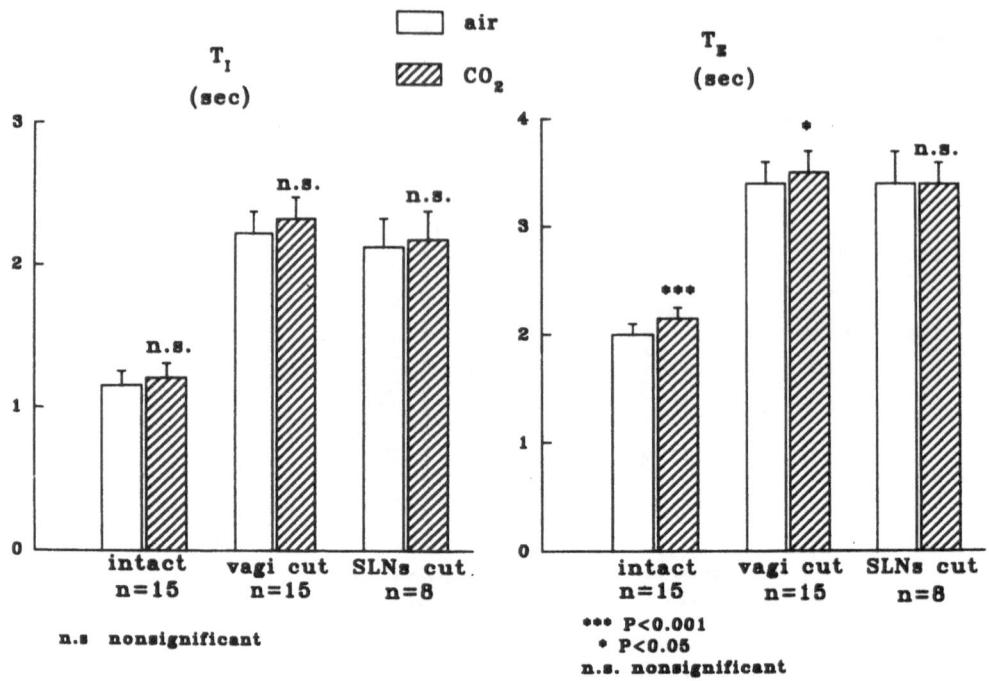

Figure 3. Summary of mean values ± S.E.M. of the inspiratory and expiratory time responses to intralaryngeal

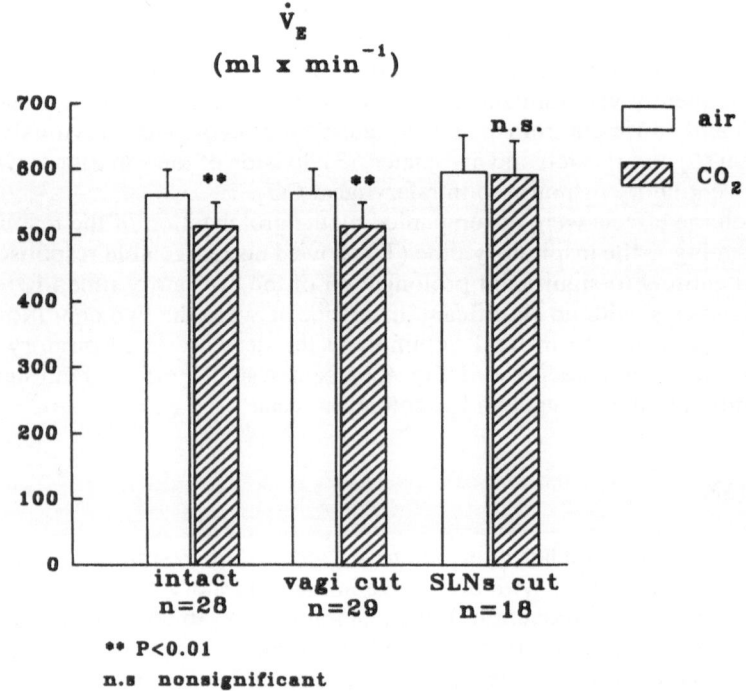

Figure 4. The effects (mean ± S.E.M.) of the excess of CO_2 in the laryngeal airway on minute ventilation.

our knowledge, the present study is the first to show the decrease in tidal volume effected by excess of CO_2 in the laryngeal airstream. Our finding is in general agreement with the result obtained by Bartlett and associates, (1992) in decerebrate, paralysed cats, showing decrease in peak phrenic activity in response to raised CO_2 within the larynx. In this report phrenic inhibition, however, was suggested to be larger after ablating vagi. The difference might occur in the animal preparation. Our cats were anaesthetized and spontaneously breathing. The expiratory airstream enriched with 10% CO_2 arriving at the larynx left the respiratory system through the oral cannula, thus omitting the upper airway. This procedure excludes the extra inhibitory effects of carbon dioxide in the upper airway (9).

Under identical conditions to those in the present study Boushey and Richardson (6) reported inconsistent changes in tidal volume. The reduction in breathing rate and the depression of minute ventilation, however, were quantitatively similar to our current findings. These effects of CO_2 on the latter variables were greater in the decerebrate cats reported in the same study (6), which displays the inhibitory effects of general anaesthesia on laryngeal reflexes.

The timing component of the respiratory pattern - an increase in the expiratory time, which we have observed (Fig. 3) is entirely consistent with the results of Bartlett and associates, (5), described in decerebrate, vagotomized cats. In their study the same percent-·age of intralaryngeal CO_2, as used by us induced 10-13% prolongation of T_E. Our cats presented a smaller degree of prolongation of the expiratory time, but they were anaesthetized and breathed with the omission of the respiratory passages above the larynx.

The depression of ventilation reported in this study did not differ significantly between the initially intact cats and then subsequently vagotomized at the midcervical level. This is in contrast to the aforementioned observation of Bartlett and associates, (3) but reinforced by recent preliminary report of Sahin and associates (11), showing that vagotomy fails to affect the ventilatory depression evoked by intralaryngeal CO_2. The transient reduction in ventilation is carried out beyond the vagal pathway. Reflex inhibition of the central respiratory drive is mediated by the laryngeal afferent input.

Considering laryngeal mechanoreceptors responding to pressure and flow (12) in experiment of the present research we evaluated the respiratory effects of carbon dioxide as compared to plain air. This approach gave us considerable information on modest but significant hypoventilatory influence of increased carbon dioxide within the laryngeal airway. As noted in the introduction, recent research has not delineated the clear-cut CO_2 sensitive receptors of the superior laryngeal nerve. Besides, the activity of most laryngeal receptors is not stimulated or occurs to be inhibited by excess of carbon dioxide (1, 4, 8). Whether this hypoventilatory laryngeal chemoreflex is due to the inhibition of slowly adapting laryngeal mechanoreceptors with qualitative similarity to the inhibition of pulmonary ones (2) remains, as yet, an open question. In line with previous reports (3, 5, 6, 10) the present experiment brought the corroborative evidence that sensory denervation of the larynx eliminates the ventilatory depression evoked by excess of CO_2 in the laryngeal airstream. The results of our study support our hypothesis that increased CO_2 within the larynx has an inhibitory effect on the pattern of breathing. Modest respiratory depression is modulated by the afferents in the superior laryngeal nerve and the exact definition of the responsible receptors requires further studies.

ACKNOWLEDGMENT

This work was supported by grant 663619201 from State Committee for Scientific Research.

REFERENCES

1. Anderson, J.W., F.B. Sant'Ambrogio, G.P. Orani, G. Sant'Ambrogio, and O.P.Mathew. Carbon dioxide - responsive laryngeal receptors in the dog. *Respir. Physiol.* 82: 217-226, 1990.

2. Bartlett, D.Jr., and G. Sant'Ambrogio. Effects of local and systemic hypercapnia on the discharge of stretch receptors in the airways of the dog. *Respir. Physiol.* 26: 91-99, 1976.

3. Bartlett, D.Jr., S.L. Knuth, and J.C. Leiter. Alternation of ventilatory activity by intralaryngeal CO_2 in the cat. *J. Physiol. Lond.* 457: 177-185, 1992.

4. Bartlett, D.Jr., and S.L. Knuth. Responses of laryngeal receptors to intralaryngeal CO_2 in the cat. *J. Physiol. Lond.* 457: 187-193, 1992.

5. Bartlett, D.Jr., S.L. Knuth, and M.J. Godwin. Influence of laryngeal CO_2 on respiratory activities of motor nerves to accessory muscles. *Respir. Physiol.* 90: 289-297, 1992.

6. Boushey, H.A., and P.S. Richardson. The reflex effects of intralaryngeal carbon dioxide on the pattern of breathing. *J. Physiol. Lond.* 228: 181-191, 1973.

7. Bradford, A., P. Nolan, D. McKeogh, C. Bannon, and R.G. O'Regan. The responses of superior laryngeal nerve afferent fibres to laryngeal airway CO_2 concentration in the anaesthetized cat. *Exp. Physiol.* 75: 267-270, 1990.

8. Bradford, A., P. Nolan, R.G. O'Regan, and D. McKeogh. Carbon dioxide - sensitive superior laryngeal nerve afferents in the anaesthetized cat. *Exp. Physiol.* 78: 787-798, 1993.

9. Lee, L.Y., R.F. Morton, M.J. McIntosh, and J.A. Turbek. An isolated upper airway preparation in conscious dogs. *J. Appl. Physiol.* 60: 2123-2127, 1986.

10. Nolan, P., A. Bradford, R.G. O'Regan, and D. McKeogh. The effects of changes in laryngeal airway CO_2 concentration on genioglossus muscle activity in the anaesthetized cat. *Exp. Physiol.* 75: 271-274, 1990.

11. Sahin, G., N. Karaturan, T. Oruç, I. Güner, L.C. Çakar, and M. Terzioglu. The effects of air flow rate and CO_2 on the isolated laryngeal airway and the resultant changes of ventilation. *Eur. Resp. J. Suppl.* 18: p.199s, 1994.

12. Sant'Ambrogio, G., O.P. Mathew, J.T. Fisher, and F.B. Sant'Ambrogio. Laryngeal receptors responding to transmural pressure, airflow and local muscle activity. *Respir. Physiol.* 54: 317-330, 1983.

65

INVESTIGATION OF CENTRAL CO$_2$-SENSITIVITY AROUND EUCAPNIA IN AWAKE HUMANS USING A BRIEF HYPOXIC STIMULUS

C. A. Roberts, D. R. Corfield, K. Murphy, S. De Cort, L. Adams, and A. Guz

Department of Medicine
Charing Cross and Westminster Medical School
London W6 8RP, United Kingdom

INTRODUCTION

In awake man, ventilation appears little influenced by changes in PCO$_2$ below a threshold that is at, or around, eucapnia (see [2]). However animal studies suggest that both peripheral and central chemoreceptor sensitivity may persist at low PCO$_2$ [5,6]. Studies in man of respiratory control during hypocapnia are difficult to interpret for they are generally performed post-hyperventilation, when PCO$_2$ is not in a steady-state and when the neural effects of the active hyperventilation may still be present [4], or when hypoxia is being used as the stimulus for hyperventilation [8].

Here we describe a new method to test human chemosensitivity during steady-state eucapnia and hypocapnia. Experiments were performed during 'passive' mechanical ventilation when rhythmic respiratory activity was absent. The respiratory response to a transient hypoxic stimulus was determined by quantifying the loss of 'passivity'. The stimulus was designed to excite the peripheral chemoreceptors in a way that was independent of the steady-state PCO$_2$; the modulation of the response to the stimulus, by the steady-state PCO$_2$, will reflect changes in chemosensitivity that are related to the level of CO$_2$ acting centrally.

METHODS

Healthy subjects (aged 23-43 years) were mechanically ventilated with air using positive inspiratory pressure through a close-fitting nasal mask attached to a volume-cycled ventilator (Servo 900, Siemens). Airway pressure was measured via a catheter from a port in the nasal mask (P$_{mask}$). Electromyographic activity of the diaphragm (EMG$_D$) was measured from surface electrodes placed on the chest wall [7], positioned to optimise the

Modeling and Control of Ventilation, Edited by S. J. G. Semple, L. Adams, and B. J. Whipp
Plenum Press, New York, 1995

recording of inspiratory-related activity. EMG_D was filtered to remove the electocardiograph artefact and then integrated.

The minute volume of the ventilator was adjusted, by increasing tidal volume (V_T) whilst keeping respiratory timing (i.e. T_I and T_E) constant, to reduce end-tidal PCO_2 ($P_{ET}CO_2$) to approximately 25 mmHg ($V_T \approx 1.5$ l; $f_R \approx 15$ breaths.min^{-1}). Once this was achieved the ventilator settings remained constant for the remainder of the experiment. Subjects were encouraged to relax all respiratory muscle activity by verbal instruction from the investigator. Passivity was determined from the P_{mask} and EMG_D activity; such passivity was characterised by: 1) the smooth reproducible profile of the inflation pressure, 2) consistency in the peak inflation pressure and 3) the absence of rhythmic respiratory-related activity in EMG_D. Breaths assisted by intrinsic respiratory activity can be detected from changes in the profile of the inflation pressure, reduction of the peak inflation pressure and activity in EMG_D [3]. Once ventilation was 'passive', feedback from the investigator ceased.

The fraction of inspired CO_2 (F_ICO_2) was then increased to return $P_{ET}CO_2$ to the subject's eucapnic level (determined from resting breathing assessed before the period of mechanical ventilation). After 5 minutes at eucapnia, the subject was given a hypoxic stimulus by switching the inspired gas to a mixture of CO_2 in N_2 for 3 or 4 breaths, such that the F_ICO_2, and therefore $P_{ET}CO_2$, remained fixed at the level it had been immediately prior to the stimulus. The number of breaths in the stimulus was adjusted until the hypoxia produced a fall in S_aO_2 to approximately 85% and a clear respiratory response, indicated by a transient loss of passivity. The number of breaths in the stimulus then remained constant (range 3-7 breaths).

The hypoxic stimulus was then repeated at different steady-state levels of $P_{ET}CO_2$. Each steady-state was achieved by adjusting the F_ICO_2 whilst the ventilator frequency and volume remained unchanged and then holding $P_{ET}CO_2$ constant for 5 minutes prior to each stimulus. When the hypoxic stimulus was delivered at a steady-state $P_{ET}CO_2$ below eucapnia it was coupled with an increase in F_ICO_2 to return $P_{ET}CO_2$ transiently to the subject's eucapnic level. When the hypoxic stimulus was delivered during hypercapnia F_ICO_2 was decreased to return $P_{ET}CO_2$ transiently to eucapnia. In this way, it was intended that an equivalent stimulus of 'hypoxia at eucapnia' would stimulate the peripheral chemoreceptors each time whilst the central chemoreceptors remained at a CO_2 level related to the steady-state $P_{ET}CO_2$ level. The respiratory responses to the hypoxic stimuli at the different steady-state $P_{ET}CO_2$ levels were quantified by measuring the changes in the peak inflation pressure ($\Delta Pmax$) and the change in the amplitude of the integrated EMG_D (ΔEMG_D). These changes were assessed as the difference between a control period immediately prior to the hypoxic stimulus and a period containing the entire respiratory response to the stimulus.

RESULTS

During eucapnia and before the hypoxic stimulus, each subject was being 'passively ventilated' as indicated by the smooth reproducible profile of the inflation pressure, the consistency of the peak inflation pressure and the absence of rhythmic EMG_D activity. The hypoxic stimuli produced falls in S_aO_2 to approximately 85%; $P_{ET}CO_2$ remained unchanged i.e. at eucapnia. As the S_aO_2 fell, the subjects transiently lost passivity, as seen by the change in the pressure profile of P_{mask}, decreases in peak inflation pressure and the appearance of bursts of EMG_D activity. Initial studies, in six subjects, indicated that such responses to the hypoxic stimuli were absent when the steady-state $P_{ET}CO_2$ was reduced to 25 mmHg.

Further studies, in four individuals, demonstrated that the respiratory responses were graded. At a steady-state $P_{ET}CO_2$ 7.5 mmHg below eucapnia (eucapnia - 7.5) the responses were virtually absent. As the steady-state $P_{ET}CO_2$ increased, the response to the hypoxic

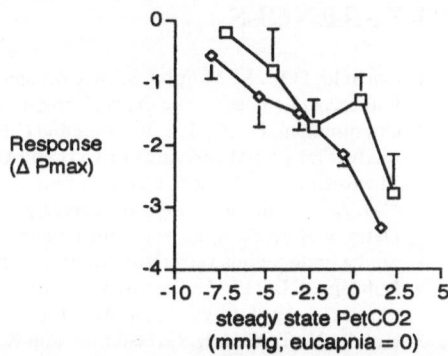

Figure 1. Modulation of the response to the hypoxic stimulus by changes in steady-state $P_{ET}CO_2$. Mean values ($\pm$ sem) for 4 subjects of ΔPmax at each of 5 steady-state $P_{ET}CO_2$ levels. Each subject was studied on two seperate days (Day 1: square; Day 2: diamond). The steady-state $P_{ET}CO_2$ levels were normalised between subjects by expressing them as the change in $P_{ET}CO_2$ from each subject's eucapnic value. The responses to the stimuli reduced significantly ($p < 0.05$) as steady state $P_{ET}CO_2$ decreased.

stimulus increased significantly ($p < 0.05$; ANOVA) i.e. the stimulus produced a greater fall in Pmax and a greater increase in EMG_D activity. The largest response to the stimulus was seen at eucapnia and eucapnia+2.5 mmHg. The mean results for the group (Fig. 1), clearly show the modulation of the respiratory response to hypoxia with changes in steady-state $P_{ET}CO_2$, the response of ΔPmax increasing as steady-state $P_{ET}CO_2$ is increased. A similar modulation of ΔEMG_D by increasing steady-state $P_{ET}CO_2$ was also seen.

DISCUSSION

The respiratory response produced by each hypoxic stimulus is the result of an interaction between the peripheral and central chemoreceptor inputs to the respiratory centre. Throughout each study, it was intended that a constant hypoxic stimulus would be delivered to the peripheral chemoreceptors, the stimulus being independent of the prevailing steady-state PCO_2; in addition, it was intended that the central chemoreceptors would remain stimulated by the steady-state PCO_2. In this way, for each stimulus, the peripheral chemoreceptor input to the respiratory centre would be constant and the changes in respiratory output, with steady-state PCO_2, would therefore reflect changes in chemosensitivity that are related to the level of CO_2 acting centrally. To ensure this, the hypoxic stimuli delivered to the peripheral chemoreceptors during the steady-states of hypercapnia or hypocapnia were coupled with changes in F_ICO_2 to transiently return $P_{ET}CO_2$ to the eucapnic level. In this way the stimulus at the peripheral chemoreceptor was always 'hypoxia at eucapnia' despite the steady-state $P_{ET}CO_2$ level.

The principal finding of this study is that, in awake man, the respiratory response to a fixed transient peripheral chemoreceptor stimulus is modulated by changes in steady-state $P_{ET}CO_2$. The size of the respiratory response decreased gradually with steady-state PCO_2, from 2.5 mmHg above eucapnia down to about 7.5 mmHg below eucapnia, at which point the respiratory response was virtually absent. There is no evidence for a 'threshold' within this range. We cannot draw any conclusions about chemosensitivity at, or less than, 7.5 mmHg below eucapnia.

ACKNOWLEDGMENTS

Supported by a Wellcome Trust Programme Grant to A.G. Elements of this work were first presented to the Physiological Society [1].

REFERENCES

1. Corfield, D.R., K. Murphy, S. De Cort and A. Guz (1993). Hypoxic challenge during hypocapnia in man: Is there a chemosensitive drive to breathe at low PCO_2? J. Physiol., 467: 117P.
2. Cunningham, D.J.C., P.A. Robbins and C.B. Wolff (1986). Integration of respiratory responses to changes in alveolar partial pressures of CO_2 and O_2 and in arterial pH. In: Handbook of Physiology: Section 3, The Respiratory System, Control of breathing, Part 1, ed N.S. Cherniack and J.G. Widdicombe. Bethesda, MD: American Physiological Society, pp. 475-528.,
3. Datta, A.K., S.A. Shea, R.L. Horner and A. Guz (1991). The influence of induced hypocapnia and sleep on the endogenous respiratory rhythm in humans. J. Physiol., 440: 17-33.
4. Eldridge, F.L. (1973). Post-hyperventilation breathing: different effects of active and passive hyperventilation. J. Appl. Physiol., 34: 422-430.
5. Honda, Y., T. Natsui, N. Hasumura and K. Nakamura (1963). Threshold PCO_2 for respiratory system in acute hypoxia of dogs. J. Appl. Physiol., 18: 1053-1056.
6. Lahiri, S. and R.G. Delaney (1975). Relationship between carotid chemoreceptor activity and ventilation in the cat. Respir. Physiol., 24: 267-286.
7. Lansing, R. and J. Savelle (1989). Chest surface recording of diaphragm potentials in man. Electroencephalog. and Clin. Neurophysiol., 72: 59-68.
8. Nielsen, M. and H. Smith (1952). Studies on the regulation of respiration in acute hypoxia. Acta Physiol. Scand., 24: 293.

66

CENTRAL-PERIPHERAL VENTILATORY CHEMOREFLEX INTERACTION IN HUMANS

Claudette M. St. Croix, D. A. Cunningham, and D. H. Paterson

Centre for Activity and Ageing[*]
University of Western Ontario
London, Ontario, Canada, N6A 3K7

INTRODUCTION

It has been recognized since the work done by Nielsen and Smith (8) that hypoxia and hypercapnia interact multiplicatively in their effects on breathing. While this interaction has been convincingly demonstrated to occur at the level of the peripheral chemoreceptors in man (4,7) the possibility that there is some degree of central interaction as well can not be excluded (3,4).

The most widely accepted model characterizing the interactions between chemical respiratory feedback stimuli describes hypoxia and the CO_2-H^+ complex interacting at the level of the peripheral chemoreceptor, with the drives from the periphery and from the central chemosensitive area adding together in their effects on ventilation (3). The independence of the peripheral and central contributions to ventilation has been assumed in many studies involving parameter estimation in humans (1, 4).

The main problem in determining the nature of the peripheral-central chemoreflex interaction in humans is the difficulty in changing the peripheral CO_2-H^+ drive without affecting the drive at the central chemoreceptor. Robbins (9) used the differing speeds of response of the central and peripheral chemoreceptors to enable a temporal separation of their chemical stimulation. A comparison was made between the ventilatory response to hypoxia when both the central and peripheral chemoreceptor environments were eucapnic, and the hypoxic response when the central chemoreceptor was exposed to diminishing hypercapnia. In two of the three subjects studied, the ventilatory response to hypoxia was significantly augmented when central PCO_2 was high, but returned to baseline as central eucapnia was restored. These results provided some evidence for a degree of multiplicative interaction between the central and peripheral chemoreceptors in humans. Further studies by Clement et al. (2), however, in the same laboratory, but employing a different protocol, failed to demonstrate a significant interaction between peripheral hypoxic sensitivity and central chemoreflex stimulation.

---

[*] Affiliated with the Faculties of Kinesiology and Medicine at the University of Western Ontario and The Lawson Research Institute at the St. Joseph's Health Centre.

Modeling and Control of Ventilation, Edited by S. J. G. Semple, L. Adams, and B. J. Whipp
Plenum Press, New York, 1995

The purpose of this study was to investigate the interaction between the central and peripheral ventilatory chemoreflex loops. The study was modelled after that of Robbins (9) with modifications to the administration of the experimental protocols. It was hypothesized that the results of this study would strengthen the evidence for interaction between the central and peripheral chemoreflex loops.

METHODS

Experimental Protocols

Subjects breathed through a mouthpiece with the nose occluded. Inspired and expired ventilation flow rates were measured using a low resistance bi-directional turbine (Alpha Technologies, VMM 110) and volume transducer (Sensor Medics VMM-2A). Respiratory flows and timing information were measured using a pneumotachograph (Hans Rudolph, Inc. Model 3800) and differential pressure transducer (Validyne MP45-871). Inspired and expired gases were sampled continuously (20 ml/s) at the mouth and analyzed by a mass spectrometer (AIRSPEC MGA 2000).

A data acquisition computer sampled the variables every 20 ms. Accurate control of end-tidal gases was achieved using a computer controlled fast gas-mixing system similar to that described in more detail by Howson et al. (5) and Robbins et al. (10). The control computer compared the measured end-tidal gas tensions with the target end-tidal tensions (entered into the control computer before the experiment according to the protocol). The variables used for feedback control were $P_{ET}CO_2$ and $P_{ET}O_2$. The inspired PCO_2 and PO_2 required were converted by an algorithm into appropriate values for flows of CO_2, O_2, and N_2. The sensing process for $P_{ET}CO_2$ and $P_{ET}O_2$ was repeated at the end of each breath and the control computer adjusted the gas mixture to force the end-tidal PCO_2 and PO_2 towards the desired values.

Three experimental protocols were modelled after Robbins (9). In protocol A, subjects were exposed to an end-tidal PCO_2 of 8-10 torr above resting $P_{ET}CO_2$, with $P_{ET}O_2$ = 100 torr, for 8 min. 30 s after the hypercapnic stimulus was withdrawn, a 5 min hypoxic stimulus ($P_{ET}O_2$ = 50 torr) was introduced. The 30 s interval should be sufficient for the peripheral chemoreceptor to have adapted to the new level of CO_2, however, the central chemoreceptor environment changes more slowly. Over the subsequent 5 min of hypoxia, therefore, the central chemoreceptor was exposed to diminishing hypercapnia. Protocol B was similar to A, but without the hypoxic step ($P_{ET}O_2$ was maintained at 100 torr). In protocol C, the step down in $P_{ET}O_2$ was administered without the prior hypercapnia ($P_{ET}CO_2$ maintained at resting levels).

Each subject was studied on a minimum of six different occasions. Three periods of breathing on the apparatus were planned for each visit, corresponding to one of the three protocols. Breathing sessions were separated by at least 30 min. This study differed from that of Robbins (9) in the administration of the experimental protocols. Robbins (9) administered protocols A and B in one breathing period and two type C protocols in a second breathing period.

Data Analysis

Five subjects contributed six sets of data to each of the three protocols. For each protocol, a mean of the respiratory variables for the 2 min steady-state period prior to the hypoxic step was calculated along with means for each 30 s period following the step. These results were combined to yield an average response for each subject to each protocol. The effect of hypoxia on ventilation in protocol A was examined by subtracting the response in protocol B from the hypoxic response in A. The effect of hypoxia in protocol C was measured by subtracting each 30 s data point from the 2 min control point. The results of these calculations gave the ventilatory response to hypoxia under the two conditions. The responses were compared with a one-sided two sample t-test. If the effect of hypoxia at the peripheral chemoreceptor was unaffected by relative hypercapnia at the central chemoreceptor, then the response to hypoxia in A, once the response to hypercapnia in B was subtracted, would be the same as the response to hypoxia in the control protocol C. If the hypoxic response was affected by central hypercapnia then the ventilatory response to hypoxia in protocol A, once B had been subtracted, would initially be greater than the response to hypoxia in the control protocol C, only becoming the same as central eucapnia was restored.

RESULTS

Five non-smoking healthy male subjects (mean age 28 yrs) were studied.

Figure 1 illustrates the experimental results for two of the subjects. Subject 1569 showed evidence for some degree of interaction between peripheral hypoxic drive and central hypercapnic drive. The response to hypoxia was greater in protocol A than in C and remained so for the duration of the hypoxic stimulus. The record on the right (Subject 2402) is typical of the four subjects that showed no evidence for interaction. The quality of end-tidal control was generally the same for the hypercapnic steps in protocols A and B and for the hypoxic steps in protocols A and C for both subjects.

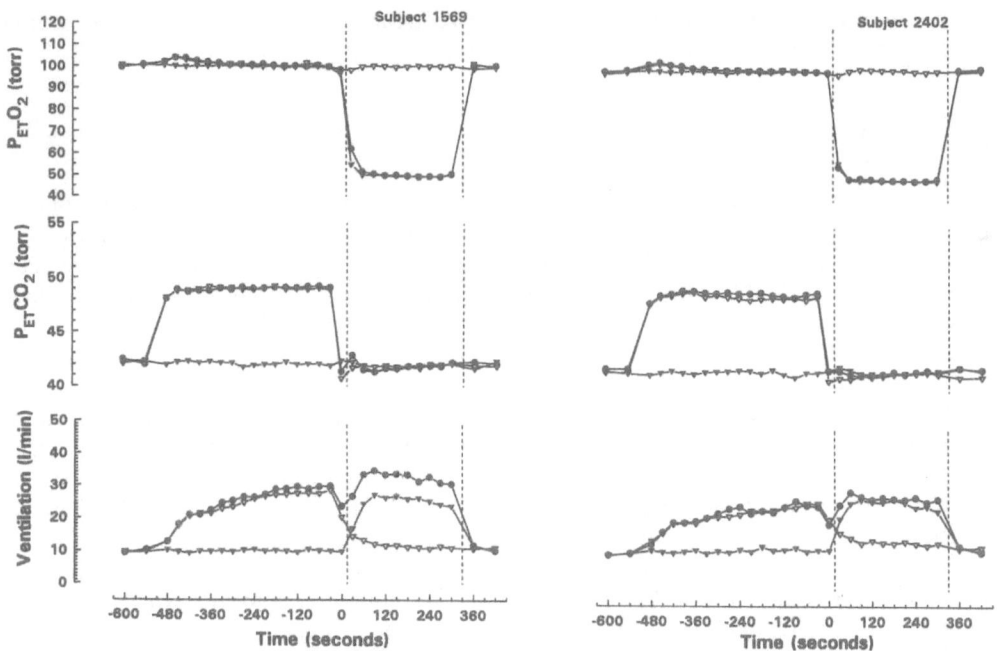

Figure 1. The experimental results of protocols A (closed circle), B (open inverted triangles), and C (closed inverted triangles) for Subjects 1569 (left) and 2402 (right). Top, $P_{ET}O_2$ (torr); middle, $P_{ET}CO_2$ (torr); bottom, $\dot{V}_E$ (l/min). Dashed lines mark the start and end of the hypoxic step.

Table 1. The ventilatory response to hypoxia when the central chemoreceptor is
exposed to diminishing hypercapnia minus the hypoxic response
under control conditions at successive time periods

Time Period (s) (relative to hypoxic step)	Ventilatory Difference (l/min)				
	Subject 2007	Subject 1643	Subject 2374	Subject 2402	Subject 1569
-60	1.87	0.19	1.69	-1.23	1.72
-15	4.22	0.65	1.39	-1.85	2.97
15	1.36	-1.78	-0.08	-1.24	4.95*
45	-1.30	-0.33	0.47	-0.51	5.32*
75	-3.86	1.49	-0.32	-1.42	4.90
105	-3.03	2.09	0.32	-2.60	5.02
135	-1.65	1.64	-0.42	-1.02	4.73
165	-0.97	1.60	0.37	-2.31	5.91*
195	-2.21	1.48	0.40	-2.06	3.63
225	-1.86	0.35	-0.12	0.67	6.36*
255	-0.94	2.11	0.02	-0.51	4.70
285	-1.81	1.58	-0.03	1.21	5.47
+15	0.23	-0.36	-0.72	-1.19	-1.10

*Values are significantly different from zero ($p < 0.05$).

Table 1 lists the results of subtracting the ventilatory response to hypoxia in protocol
C from the difference in ventilation between protocols A and B for each subject. In subject
1569, at the four points indicated by the stars, the ventilatory response to hypoxia was
significantly ($p < 0.05$) greater when the central chemoreceptor was exposed to diminishing
hypercapnia than when both the central and peripheral chemoreceptor environments were
eucapnic. There were no significant differences between the ventilatory responses to hypoxia
under the two conditions in the other four subjects.

DISCUSSION

In four of the five subjects studied, the drives from the central and peripheral
chemoreceptors were independent. Subject 1569 showed some evidence for multiplicative
interaction between the chemoreflexes (4 of 10 points reached significance; Table 1). The
ventilation remained higher throughout the hypoxic period in protocol A, however, rather
than decreasing as central eucapnia was restored as Robbins (9) observed in his two subjects.
Therefore, the possibility that some mechanism other than central-peripheral chemoreflex
interaction was responsible cannot be eliminated.

One potential confounding influence was described by Michel, Lloyd and Cunningham (6). They reported that the in vitro relationship between the pH and bicarbonate in true
plasma does not apply in vivo when the P_ACO_2 is altered by carbon dioxide inhalation. CO_2
breathing gave way to a small metabolic as well as a respiratory acidosis due to distribution
of bicarbonate across the extracellular fluid, and this persisted long after the inspired PCO_2
had been lowered. Due to the time course of the recovery period, Robbins (9) concluded that
this effect could not have influenced his results. If the metabolic acidemia was significant,
the augmentation of hypoxic sensitivity at the peripheral chemoreceptors would last throughout the hypoxic period as seen in subject 1569.

A major assumption of this study is that the 30 s interval was sufficient time for the
discharge from the peripheral chemoreceptor to return to its resting level. If the CO_2 at the
carotid bodies remained high for the duration of the hypoxic step, then the results would be

complicated by the interaction between hypoxia and CO_2-H^+ at the peripheral chemoreceptor. The time constants describing the peripheral chemoreflex response to hypercapnia in awake humans have been reported to be around 8 to 10 s (1,4). Subject 1569 participated in an earlier project in which the time constant of the peripheral chemoreceptor was measured at 23.6 s in euoxia. If we allow an interval of 4 time constants, the peripheral chemoreceptor response to hypercapnia in this subject would not be functionally complete until 94 s into the hypoxic step.

The difference between the results of the current study and Robbins (9) might be explained in part by modifications to the administration of the protocols. These changes were made to ensure that there was sufficient time between each protocol to eliminate the possibility that the stimulus in one protocol might potentiate or diminish the ventilatory response in a succeeding protocol. Robbins (9) administered protocols A and B in one breathing period and two type C protocols in a second breathing period. When the protocols were applied in this manner, it was observed that the hypoxic stimulus administered in the first protocol affected the ventilatory response in the subsequent protocol of the same breathing session. This had the effect of lowering the average response to hypoxia in C and increasing the average ventilatory response to hypercapnia in A or B, thus yielding a larger difference between A and C.

In conclusion, this study demonstrated that in most cases the central and peripheral chemoreflexes are independent of each other.

ACKNOWLEDGMENTS

Supported by Natural Science and Engineering Research Council, Canada

REFERENCES

1. Berkenbosch, A., A. Dahan, J. DeGoede, and I.C.W. Olievier. The ventilatory response to CO_2 of the peripheral and central chemoreflex loop before and after sustained hypoxia in man. *J.Physiol.* 456:71-83, 1992.
2. Clement, I.D., D.A. Bascom, J. Conway, K.L. Dorrington, D.F. O'Connor, R. Painter, D.J. Paterson and P.A. Robbins. An assessment of central-peripheral ventilatory chemoreflex interaction in humans. *Respir.Physiol.* 88:87-100, 1992.
3. Cunningham, D.J.C., P.A. Robbins and C.B. Wolff. Integration of respiratory responses to changes in alveolar partial pressures of CO_2 and O_2 and in arterial pH. In: *Handbook of Physiology, Section 3: Vol II: The Respiratory System*, edited by Cherniack, N.S. and Widdicombe, J.G. Bethesda, MD: American Physiological Society, 1986, p. 475-528.
4. Dahan, A., J. DeGoede, A. Berkenbosch and I.C.W. Olievier. The influence of oxygen on the ventilatory response to carbon dioxide in man. *J.Physiol.* 428:485-499, 1990.
5. Howson, M.G., S. Khamnei, M.E. McIntyre, D.F. O'Connor and P.A. Robbins. A rapid computer-controlled binary gas-mixing system for studies in respiratory control. *J.Physiol.* 394:7P, 1987.
6. Michel, C.C, B.B. Lloyd and D.J.C. Cunningham. The in vivo carbon dioxide dissociation curve of true plasma. *Respir.Physiol.* 1:121-137, 1966.
7. Miller, J.P., D.J.C. Cunningham, B.B. Lloyd and J.M. Young. The transient respiratory effects in man of sudden changes in alveolar CO_2 in hypoxia and in high oxygen. *Respir.Physiol.* 20:17-31, 1974.
8. Nielsen, M. and H. Smith. Studies on the regulation of respiration in acute hypoxia. *Acta Physiol.Scand.* 24:293-313, 1952.
9. Robbins, P.A. Evidence for interaction between the contributions to ventilation from the central and peripheral chemoreceptors in man. *J.Physiol.* 401:503-518, 1988.
10. Robbins, P.A., G.D. Swanson and M.G. Howson. A prediction-correlation scheme for forcing alveolar gases along certain time courses. *J.Appl.Physiol.: Respirat.Environ.Exercise Physiol.* 52(5):1353-1357, 1982.

67

SUBCELLULAR CONTROL OF OXYGEN TRANSPORT

E. Takahashi and K. Doi

Department of Physiology
Yamagata University School of Medicine
Yamagata, 990-23 Japan

INTRODUCTION

According to the Fick's law of diffusion, oxygen pressure of mitochondrial inner membrane (P_{Mt}) is defined by the following equation;

$$P_{Mt} = P_{cap} - \dot{V}_{O_2} \cdot R,$$

where P_{cap}, $\dot{V}_{O_2}$, and R denote Po_2 of capillary blood, flux of oxygen into mitochondria (oxygen consumption rate of the cell), and diffusion resistance of tissue, respectively. R is a lumped parameter and includes diffusion resistance of plasma, capillary wall, extracellular fluid, plasma membrane, cytosol, and mitochondrial inner membrane. Therefore, the oxygen pressure gradient between capillary blood and mitochondria is represented by $\dot{V}_{O_2} \cdot R$. Magnitude of the oxygen pressure gradient *in vivo* appears so large (> 20 - 25 Torr) that intracellular (cytosolic) Po_2 of normal beating heart may be around P_{50} of myoglobin, i.e., ~3 Torr (11). Furthermore, additional oxygen pressure gradients between cytosol and mitochondrial inner membrane result in quite low Po_2 at mitochondrial enzymes.

It is, therefore, presumable that even a decrease of P_{cap} of a small magnitude would considerably interrupt oxygen supply to mitochondria and depress oxidative energy production, if oxygen pressure gradients remain unchanged.

We postulated a mechanism that alters the oxygen pressure gradient ($\dot{V}_{O_2} \cdot R$) in the case of hypoxia (decreased P_{cap}) so that fluctuations of P_{cap} are effectively buffered and oxygen supply to mitochondria is ensured. To estimate the oxygen pressure gradient, we have developed a device which directly quantitates oxygenation of a single cardiomyocyte isolated from the rat. We have examined whether there are adaptive changes in the oxygen pressure gradient in the case of simulated ischaemia in a single cardiomyocyte.

Modeling and Control of Ventilation, Edited by S. J. G. Semple, L. Adams, and B. J. Whipp
Plenum Press, New York, 1995

METHODS

Single cardiomyocytes were isolated from the ventricles of adult male Sprague-Dawley rat weighing 250 - 300 g by the collagenase digestion method. Isolated cells were suspended in HEPES buffer solution (150.0 mM NaCl, 3.8 mM KCl, 1.0 mM KH_2PO_4, 1.2 mM $MgSO_4$, 1.0 mM $CaCl_2$, 10.0 mM glucose, 10.0 mM HEPES, supplemented with 0.1% bovine serum albumin, pH adjusted to 7.4). Ten μl of the cell suspension containing approximately 10^5 cells/ml was placed in the "well" of the measuring cuvette (diameter 10 mm). The cuvette was air-tight and provided gas inlet and outlet ports. Humidified test gas of various oxygen concentrations flowed over the surface of the cell suspension at the rate of 2 ml/min. Po_2 of the effluent gas was monitored at the outlet port by a conventional oxygen electrode. The dead space of the cuvette was < 120 μl.

We estimated oxygen level of intracellular space of an individual single cardiomyocyte by a 3-wavelength microspectrophotometry where myoglobin was used as an endogenous oxygen probe (9). Briefly, transmitted images of a single cardiomyocyte at 406, 420, and 436 nm were captured by a charge coupled device (CCD) camera attached to a microscope. These images were designated as Y_{406}, Y_{420}, and Y_{436}, respectively. Intracellular oxygenation was estimated by a newly defined variable, Z, as follows;

$$Z = (Y_{436} - Y_{420})/(Y_{406} - Y_{420}).$$

Calculation of Z was carried out for each pixel inside the cell image. Using ×40 object lens, one pixel on the computer monitor corresponded to 0.2 μm. Subsequently, histogram of the Z value (Z vs. pixel count) was generated and its mode was assumed to represent the intracellular oxygenation. Theoretical analysis indicated that Z consists of molar extinction coefficients of oxygenated and deoxygenated myoglobin at the 3 wavelengths and fractional oxygen saturation of myoglobin (S). Using reported values for molar extinction coefficients of myoglobin *in vitro*, theoretical relationship between S and Z was almost linear with maximal deviation from the linearity being <3%. Therefore, following appropriate calibrations, intracellular Po_2 could be evaluated by Z.

Firstly, we defined the relationship between extracellular Po_2 and Z histogram. Spectrophotometric images of a single cell were obtained and Z histogram was calculated for different test gases where the fractional oxygen concentration ranged from <0.001% to 20.9%.

Then, we examined whether there were gradients of oxygen inside and/or outside a single cell which were detectable by the present method, and, if detectable, whether the magnitude of oxygen pressure gradients depends on the oxygen flux into the cell. We predicted that changes in the oxygen consumption would alter cytosolic oxygenation for a constant extracellular Po_2, as a result of changes in the oxygen pressure gradient between intracellular and extracellular spaces. To abolish oxygen consumption of the cell (and the oxygen pressure gradient), the measurement was conducted in the presence of 2 mM NaCN. On the other hand, oxygen consumption (and the oxygen pressure gradient) was almost doubled by 1 μM cyanide 3-chlorophenylhydrazone (CCCP, an uncoupler of oxidative phosphorylation) (9). Above experiments were conducted at room temperature (21 - 23°C).

We carried out a simulation of ischaemia in a single cardiomyocyte to demonstrate the physiological role of the oxygen pressure gradient. Isolated cardiomyocytes were placed in the air-tight measuring cuvette and perfused with the HEPES buffer solution at the rate of 3 ml/min at 35°C. Firstly, perfusion was conducted with the HEPES solution equilibrated to 1.0% O_2 (balance N_2, control perfusion). Po_2 of the extracellular fluid may be comparable with Po_2 of the normal cardiac cell *in situ* (3, 11). Then, the perfusion pump was suddenly stopped and the perfusion lines were clamped so as to completely interrupt supply of oxygen

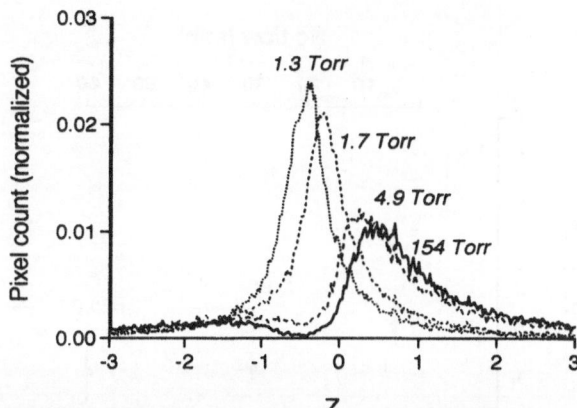

Figure 1. Representative data demonstrating changes in the Z histogram depending on the extracellular Po_2. Mode (position of the peak) shifted to the left when extracellular Po_2 (and intracellular Po_2) was lowered.

to the cell (simulation of ischaemia). This was continued for 60 min, and was followed by perfusion of the cell with the solution equilibrated to 99.999% nitrogen (anoxic perfusion). Measurements of intracellular oxygenation were conducted at 4, 30, and 60 min following the onset of ischaemia, besides during control and anoxic perfusions.

RESULTS

Figure 1 demonstrates dependency of Z histogram on extracellular Po_2.

Figure 2 defines the relationship between extracellular Po_2 and mode of Z histogram. When extracellular Po_2 was decreased, mode Z decreased from 0.79 ± 0.12 (n = 121, mean $\pm$ SD) to -0.20 ± 0.09 (n = 14). Because cytosolic Po_2 and extracellular Po_2 would be identical in the absence of oxygen flux into cell, open circles (NaCN treated cell) represent the relationship between mode Z and cytosolic Po_2. The relationship was fitted to a sigmoid curve to find the Po_2 for half maximal Z (P_{50}). P_{50} for NaCN treated cell was 1.7 Torr.

Presence of oxygen consumption in the control cell slightly shifted the relationship to the right. At extracellular Po_2 of 1.3 Torr, this rightward shift resulted in the mode Z that was significantly lower than that of the NaCN treated cell (Fig. 2). The P_{50} of the control cell was 3.2 Torr. These results may suggest that flux of oxygen established significant Po_2 gradients inside and/or outside the cell. This was further confirmed in the cell where its

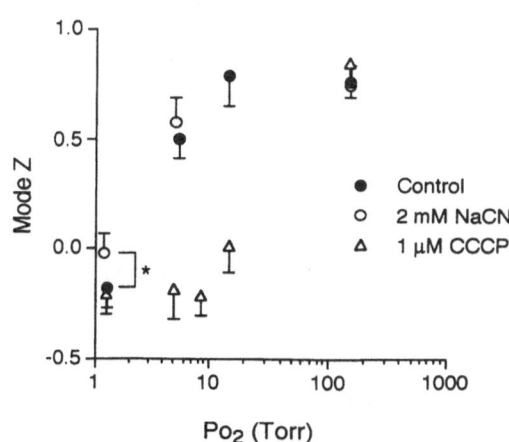

Figure 2. Relationship between extracellular Po_2 and mode of Z histogram. Values are represented by the mean and one standard deviation. *, $p<0.05$, NaCN vs. control.

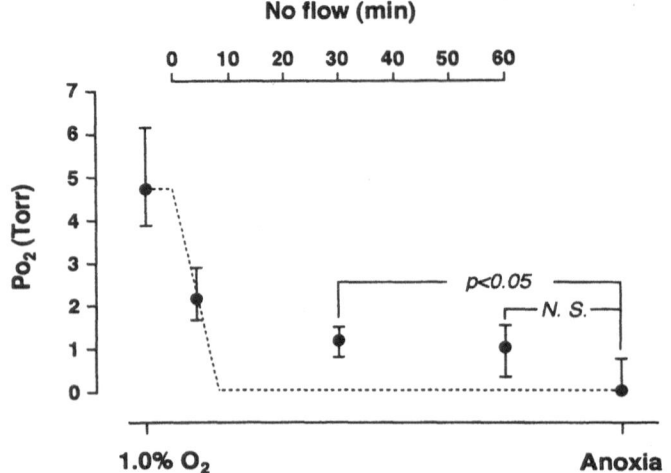

Figure 3. Changes in cytosolic Po_2 during simulated ischaemia (no flow ischaemia). Cytosolic Po_2 was calculated from actually measured Z and Fig. 2.

oxygen consumption was considerably increased by 1 μM CCCP. As seen from Fig. 2, mode Z of the CCCP treated cell was already at the anoxic level for extracellular Po_2 of 8.6 Torr, suggesting considerable oxygen gradients.

Figure 3 represents the time course of intracellular Po_2 during simulated ischaemia. Perfusion of the cell with the solution equilibrated to 1.0% O_2 (control perfusion) slightly deoxygenated the cell. When oxygen supply to the cell was suddenly stopped (simulated ischaemia), intracellular Po_2 rapidly decreased from 4.7 Torr (control perfusion) to 2.2 Torr within 4 min (rate of change, -0.6 Torr/min), due to consumption of oxygen within the cuvette by the myocytes. It had been expected that, if oxygen consumption of the cell remained constant, intracellular Po_2 would have reached the anoxic level within 10 min following the simulated ischaemia. However, intracellular Po_2 was significantly higher than the anoxic level ($1.2^{+0.3}_{-0.4}$ Torr, p<0.05) even 30 min following the stop flow (Fig.3). Intracellular Po_2 was not significantly different from that of the anoxic perfusion 60 min following simulated ischaemia ($1.0^{+0.5}_{-0.7}$ Torr).

DISCUSSION

Fick's low of diffusion predicts gradients of oxygen partial pressure between capillary blood to mitochondrial inner membrane. Measurements of fractional oxygen saturation of myoglobin of the myocardium *in situ* have demonstrated that myoglobin of the normal beating heart is half saturated with oxygen (2, 3). Assuming P_{50} of myoglobin being 2.3 Torr (11), intracellular Po_2 of the normal working heart may be around 3 Torr, implying considerably large (> 20 - 25 Torr) extracellular oxygen pressure gradients *in vivo* (4, 11).

Intracellular oxygen pressure gradients, i.e., those between cytosol and mitochondrial inner membrane, appear to be relatively small in magnitude but their physiological effects are controversial (1). Direct optical measurements of redox state of cytochromes in suspension of isolated cardiomyocytes have demonstrated quite shallow intracellular oxygen pressure gradients which would never exceed 2 Torr even when mitochondrial oxygen consumption was maximally stimulated by uncoupler of mitochondrial oxidative phosphorylation (10). Considering quite high affinity of cytochrome a,a₃ to oxygen ($Km = 0.02 - 0.2$ Torr) (11), these results suggest that, despite presence of oxygen pressure gradients, supply of oxygen to mitochondria would be well maintained even when demand for oxygen is maximal. In contrast, recent optical measurements of mitochondrial redox state in the working canine heart *in situ* have demonstrated that cytochrome a,a_3 of the heart at the resting metabolic level is partially reduced and its redox state is quite susceptible to changes in capillary blood Po_2 (7). It has also been demonstrated that modest desaturation of cytosolic myoglobin results in a reduction of cytochrome a,a_3 in isolated rat cardiomyocytes (5) and in crystalloid perfused heart (8), and a decrease of the level of high energy phosphates in the perfused heart (6). These results seem to suggest critical dependency of mitochondrial oxidative metabolism *in vivo* on supply of oxygen.

In the present study, in a single cardiomyocyte, we have directly demonstrated the presence of significant gradients of oxygen partial pressure. The magnitude of the gradients estimated by the change in P_{50} was approximately 1.5 Torr in quiescent cardiomyocytes, that was considerably increased to >10 Torr by stimulating oxygen consumption. These results have led us to a prediction that, despite quite high affinity of the enzyme to oxygen, oxygen pressure gradients (both extra- and intracellular) would play a significant role in the control of oxidative metabolism in the heart.

To demonstrate possible physiological role of the oxygen pressure gradient, we simulated ischaemia in a single cardiomyocyte. Interruption of oxygen supply to the measuring cuvette resulted in rapid decrease in the intracellular Po_2 estimated by the present method as a result of consumption of oxygen within the measuring cuvette by the cell. The rate of decrease of intracellular Po_2 was so fast (0.6 Torr/min) that oxygen within the cuvette had been expected to be completely depleted within 10 min following stop flow. However, even 30 min following simulated ischaemia, intracellular Po_2 (1.2 Torr) was significantly higher than that of the anoxic perfusion. We interpret these results to mean that, in the case of (simulated) ischaemia, reduction of oxygen flux ($\dot{V}_{O_2}$) resulting from oxygen deficiency at mitochondria decreased the oxygen pressure gradient ($\dot{V}_{O_2} \cdot R$) which then relatively enhanced oxygen supply to mitochondria, and finally retarded the deoxygenation of intracellular space. This mechanism may be regarded as an intrinsic subcellular control of oxygen transport.

REFERENCES

1. Clark, A., J_R., P. A. A. Clark, R. J. Connett, T. E. J. Gayeski, and C. R. Honig. How large is the drop in Po_2 between cytosol and mitochondria? *Am. J. Physiol.* 252:C583-C687, 1987.
2. Coburn, R. F., F. Ploegmakers, P. Gondrie, and R. Abboud. Myocardial myoglobin oxygen tension. *Am. J. Physiol.* 224:870-876, 1973.
3. Gayeski, T. E. J. and C. R. Honig. Intracellular Po_2 in individual cardiac myocytes in dogs, cats, rabbits, ferrets, and rats. *Am. J. Physiol.* 260:H522-H531, 1991.
4. Honig, C. R. and T. E. J. Gayeski. Resistance of O_2 diffusion in anemic red muscle: roles of flux density and cell Po_2. *Am. J. Physiol.* 265:H868-H875, 1993.
5. Kennedy, F. G. and D. P. Jones. Oxygen dependence of mitochondrial function in isolated rat cardiac myocytes. *Am. J. Physiol.* 250:C374-C383, 1986.
6. Kreutzer, U. and T. Jue. ^{1}H-nuclear magnetic resonance deoxymyoglobin signal as indicator of intracellular oxygenation in myocardium. *Am. J. Physiol.* 261:H2091-H2097, 1991.
7. Snow, T. R. and H. L. Stone. A study of factors affecting the oxidation-reduction state of cyt a,a_3 in the in-situ canine heart. *J. Appl. Cardiol.* 3:191-196, 1988.
8. Tamura, M., N. Oshino, B. Chance, and I. A. Silver. Optical measurements of intracellular oxygen concentration of rat heart *in vitro. Arch. Biochem. Biophys.* 191:8-22, 1978.
9. Takahashi, E. and K. Doi. Digital imaging of the oxygenation state within an isolated single rat cardiomyocyte. *Adv. Exp. Med. Biol.* 361: 163-169, 1994.
10. Wittenberg, B. A. and J. B. Wittenberg. Oxygen pressure gradients in isolated cardiac myocytes. *J. Biol. Chem.* 260:6548-6554, 1985.
11. Wittenberg, B. A. and J. B. Wittenberg. Transport of oxygen in muscle. *Annu. Rev. Physiol.*, 51:857-878, 1989.

68

MUSCLE PERFUSION AND CONTROL OF BREATHING

Is There a Neural Link?

P. Haouzi, F. Marchal, and A. Huszczuk

Laboratoire de Physiologie
Faculté de Médecine de Nancy
54500 Vandoeuvre-lès-Nancy, France

INTRODUCTION

When metabolic rate varies, as in exertion, respiratory homeostasis relies on the ventilatory as well as the cardiovascular system in order to match pulmonary and tissue gas exchange rates. It has been consistently demonstrated that the dynamics and magnitude of ventilation can be dramatically reduced in humans and animals when the limb blood flow is impeded during the recovery from (1, 3, 4, 5, 6), as well as at the onset of a muscular exercise (5), with little change in $PaCO_2$. A mechanism has been recently suggested to explain the above observations, that is, changing blood flow in the muscles affects ventilation (5). In other words, a perfusion-linked stimulus to ventilation could originate in the muscular circulation or its immediate vicinity, coupling ventilation to the degree of muscle hyperaemia.

The possibility that the status of peripheral circulation could be a potent factor linking alveolar ventilation to the metabolic demand of the tissues has never really been explored. Anatomical and neurophysiological basis for such a concept exist : 1-Intramuscular receptors with slow conducting afferent fibres (group III and IV), have long been shown to evoke powerful ventilatory responses (9). 2 -Stacey (14) has reported that terminals of many fine afferent units can be found in the interstitial space of the muscle close or within the adventitia of arterioles and venules. 3 -Although muscle type III and IV afferents are not homogeneous with respect to their receptive properties, many can be activated by mechanical stimuli of a low intensity (7, 8, 13). It is therefore plausible that these afferents also encode changes in local pressure related to the distension of the peripheral vascular network.

Here, we attempted to test 2 aspects of this hypothesis. 1) The ventilatory effects of pharmacologically induced vasodilation or vasoconstriction of the isolated hindlimb circulation were studied in sheep. 2) Afferent fibres from the femoral nerve were tested for the presence of a neural activity augmented by intra-aortic injection of vasodilators and inhibited by vasoconstrictor agents in the rabbit.

Modeling and Control of Ventilation, Edited by S. J. G. Semple, L. Adams, and B. J. Whipp
Plenum Press, New York, 1995

METHODS

Ten anaesthetized and tracheotomized sheep (40-55 kg) were prepared for a reversible isolation of the hindlimb circulation. Briefly, two large catheters were inserted in the inferior vena cava using a retroperitoneal approach. They were connected through an extracorporeal circuit (E.C.) to 2 similar catheters placed in the abdominal aorta. The E.C. consisted of a double peristaltic rotor pump and a blood oxygenator (Paediatric hollow fibre). In four sheep, Technetium* injection into the arterial inflow of the isolated preparation did not reveal any contamination of the systemic circulation from the caudal part of the body for at least 5 minutes after the injection. By clamp control, the physiological circulation could be restored after exclusion of the extracorporeal circuit using a short bypass. Breath-by-breath minute ventilation, blood pressure in the carotid artery and in the arterial inflow of the E.C. were measured following a bolus injection of papaverine (2 mg/kg) into the arterial inflow of the E.C. On 10 occasions high dose dopamine (100-200 µg/kg) was injected 2 minutes after papaverine injection to test the effect of a local and sudden vasoconstriction evoked by stimulation of alpha adrenergic receptors.

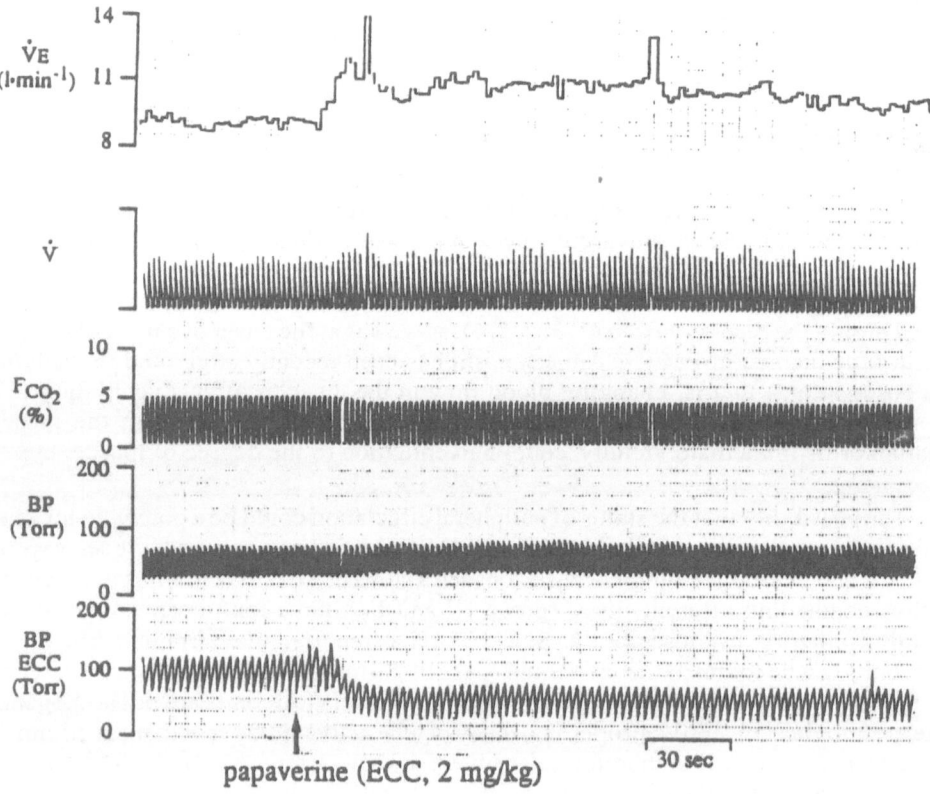

Figure 1. Example of the effect of bolus injection of papaverine (2 mg/kg) in the arterial inflow of the E.C. From top to bottom, the breath-by-breath minute ventilation, expiratory flow, respired CO_2 fraction (F_{CO2}), carotid pressure (BP), and peripheral arterial perfusion pressure in the ECC. The swings in the peripheral pressure are related to the type of pump that was used and the small volume of the reservoir bag. Note that the hypocapnic hyperpnoea coincides with a decrease in peripheral arterial pressure, resulting from the peripheral vasodilation.

Eight rabbits were anaesthetized, tracheotomized, paralysed with pancuronium bromide (0.5 mg/kg i.v.) and artificially ventilated. The left femoral nerve was dissected, severed and placed in warm mineral oil and separated into 3-6 strands. Each strand was tested, one at a time, for the neural signal evoked by an injection of papaverine (1-2 mg/kg) rostraly to the iliac bifurcation. When responsive fibres were identified, the strand was further separated into fine filaments until the activity of a single or a few fibres remained. The effects of hindlimb vasoconstriction was tested by intra-aortic bolus injections of high dose dopamine (100 - 200 µg/kg). The reproducibility of the vasodilator effect and the reversibility of the vasoconstrictor effect were tested by reinjections of papaverine.

RESULTS

Effects of Papaverine and High Dose Dopamine Injections in the Isolated Circulation in Sheep

Bolus injection of Papaverine into the EC induced a sudden rise in ventilation by 1.8 ± 0.21 L.min^{-1} ($25\pm3\%$, $p < 0.01$), as illustrated in figure 1. The increase in ventilation was associated with a drop in mean perfusion pressure in the E.C., which decreased by 24 ± 7 mmHg, reflecting the peripheral vasodilation. Both, the hyperventilation and the peripheral vasodilation occurred ~ 13 seconds after the start of injection. In some tests the hyperventilation was preceded by an apneusis, or an apnoea, that lasted for 5 to 10 seconds. No decrease in systemic blood pressure could be identified during the first minute following the injection. A small rise in systemic BP ($+10\pm4$ mmHg) was observed in half of the tests.

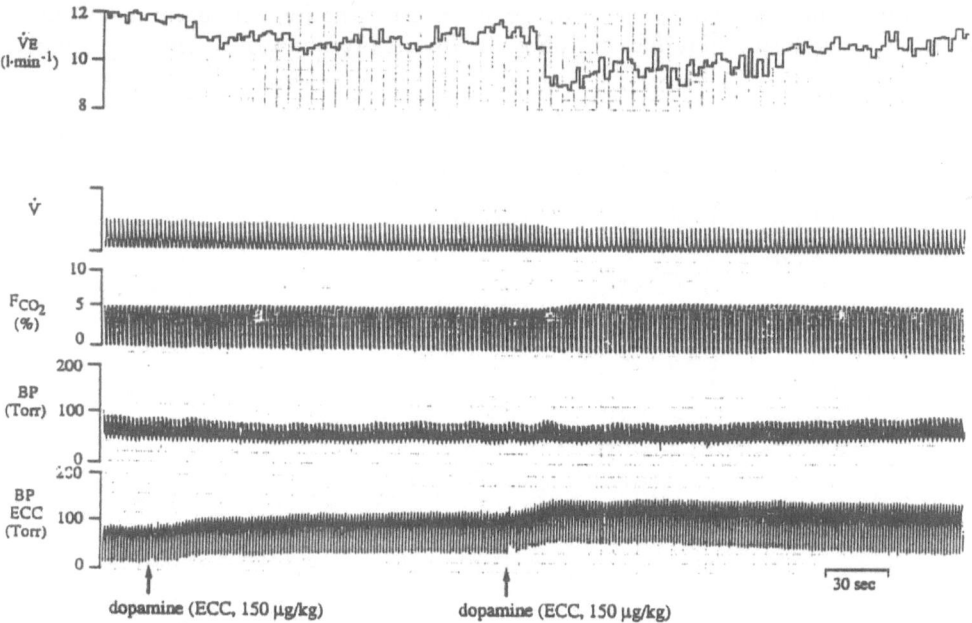

Figure 2. Example of the effect of bolus injection of high dose dopamine (200 µg/kg) in the arterial inflow of the E.C. Traces as in figure 1. Note that the hypopnoea coincides with the rise in peripheral arterial pressure, resulting from the peripheral vasoconstriction.

Injection of high dose dopamine (150-200 µg/kg), 2 minutes after papaverine injection, provoked a rapid reduction in ventilation by 18±4% (p < 0.01), which returned towards control level (figure 2). This ventilatory response was associated with a rise in peripheral pressure by 22 ± 5 Torr, reflecting the hindlimb vasoconstriction. No change in systemic blood pressure was observed.

Femoral Nerve Recording in the Rabbit

After some preliminary experiments, it was clear that the branches of the femoral nerve that contained fibres responding to vasodilation travelled along the femoral vessels. Thirty two strands from these branches were tested. Twenty three contained fibres that presented a clear and prompt increase in the discharge after papaverine injection. The increased activity occurred within seconds following the injection and reached a maximum of 168 ± 76 % of their control rate (figure 3). With single, or few fibre preparations, the activity was often rhythmic, synchronized with the pressure pulse. High dose dopamine, injected on 19 occasions after papaverine, almost completely abolished the activity previously increased during the vasodilation. In some tests, dopamine was found to have an immediate and very transient excitatory effect. Papaverine was administered after dopamine on 18 occasions, and was found to restore the rhythmic pattern of discharge.

DISCUSSION

Injection of papaverine into an isolated hindlimb circulation stimulates ventilation. The site of action of a papaverine-induced hyperpnoea had already been questioned, as this vasoactive agent has long been known to increase ventilation (2, 10). Since papaverine effect on ventilation persisted after complete chemodenervation (2), it was argued that part of the stimulating action of papaverine could be explained by a non-chemoreceptor-related mechanism. This conclusion is further corroborated by the present results, since ventilation rose

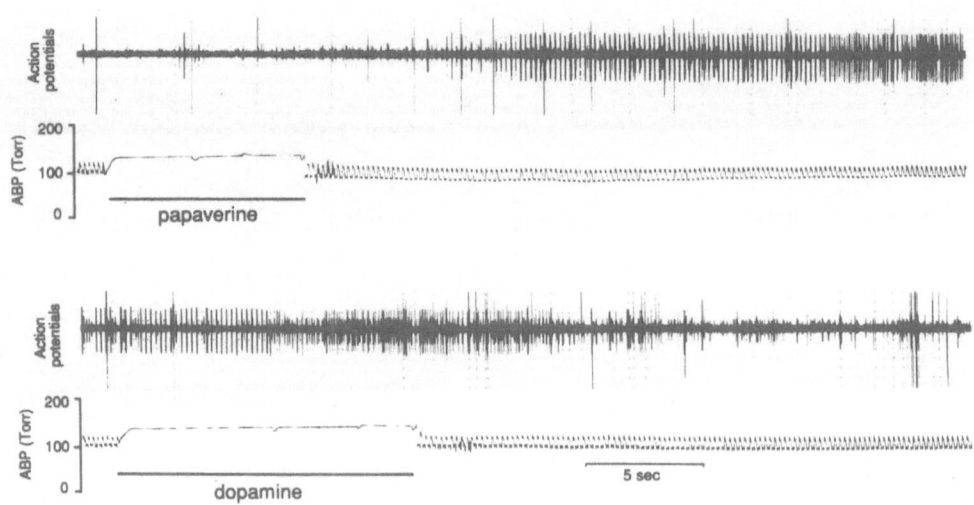

Figure 3. Recording from a few fibre preparation of femoral nerve. Papaverine (1 mg/kg) injected into the abdominal aorta induces a prompt increase in afferent activity. Two minutes later, intra-aortic bolus injection of dopamine (150 mg/kg) inhibits this activity.

even when papaverine was injected directly into the isolated hindlimb circulation. Moreover, because no decrease in systemic blood pressure was observed after injection in the isolated limb circulation, the arterial baroreceptors are unlikely to be involved. The short latency of the ventilatory response, together with the trivial contamination of the systemic circulation by the vasodilator, imply that part of the papaverine induced hyperpnoea originates from the hindlimb tissues. Oren et al. (12) drew similar conclusion using intrafemoral injection of isoproterenol in dogs. They clearly showed that ventilation rises with much shorter delay after the injection directly into the hindlimb arterial inflow, compared to the venous injection. Papaverine dilates the precapillary vessels, transmitting higher pressures to the smaller arterioles and to the venular side of the microvasculature. Since similar ventilatory responses can be obtained during a purely mechanical increase in venous pressure, by impeding the venous return from the limbs in resting sheep (4), these observations suggest that venules or their immediate vicinity could be the site of mechanoreception. Any form of vascular distension within vascularized space could constitute potential stimuli for such a reception. An increased number of venules in the muscles is likely to be recruited due to the smooth muscle relaxation and, consequently, their immediate vicinity exposed to the precapillary pressure, even if this latter decreases following the injection. A similar reasoning can be applied to explain the ventilatory effects following injection of high dose dopamine. Intravenous injection of dopamine has been consistently reported to reduce ventilation in many species. Part of dopamine-induced ventilatory depression is not related to exclusive inhibition of an arterial chemoreception (11, 16), since bilateral carotid and aortic neurotomy did not abolish the ventilatory depression provoked by high dose dopamine (10-20 μg/kg). The reduction in ventilation in these situations was usually associated with an increase in blood pressure, suggesting a predominant stimulation of cardiovascular alpha-adrenergic receptors (16). The inhibitory effects of "high" dose dopamine injected in the isolated hindlimb preparation of our study support these reports and suggest that reducing peripheral blood flow could inhibit ventilation, just like after occlusion of the arterial supply to the hindlimbs (3, 5). The existence of a sensing mechanism responding to muscular contractions and chemical changes has long been demonstrated (9). Stimulation of these endings has been shown to evoke powerful ventilatory and cardiac responses via slowly conducting afferent fibres (15). In the present study, afferent units in the femoral nerve increase their firing rate during vasodilation. The pattern of firing was often in phase with the heart beat during papaverine injections, suggesting a circulatory origin. Conversely, high dose of dopamine abolished the activity of the fibres that had previously been excited by the vasodilator. The conduction velocity of the responding fibres, as well as their origin (skin, muscle) were not examined in this study, but their pattern of response was consistent with those of slow conducting afferent fibres that had been already described. Mense and Stahnke (8) have identified "contraction-sensitive" units, which did not change, or even decreased their activity during arterial occlusion, but were clearly excited after the release. The mechanism underlying this pattern of response, was obviously not compatible with the hypothesis of "metaboreception". On the other hand, these data fit the hypothesis of a mechanical stimulus originating in the peripheral circulation. Interestingly, if the activity of a single unit appears to depend upon the local vascular pressure (or distension), the integrated response from all units in a limb will give a measure of the extent of the overall perfusion and hence the metabolic activity of the tissue.

This study adds further evidence for the vascular origin of the control of ventilation during varying metabolic states and - quite conceivably - during pathological conditions such as dyspnea or sleep apnoea in adults or infants. In the latter, cessation of breathing is associated with a powerful muscular vasoconstriction, and redistribution of a flow to the brain in a pattern resembling diving reflex. The decreased muscular blood flow might therefore shut-off a significant source of ventilatory drive delaying the restoration of eupnoea.

These observations are consistent with the concept of an integrated cardiovascular and ventilatory control relying upon monitoring the perfusion in peripheral tissues and, most likely, acting as a preemptive feed forward mechanism.

REFERENCES

1. Dejours, P., 1964, Control of respiration in muscular exercise, *Handbook of Physiology. Respiration,* sect. 3, vol. I, chapt. 25:631-648.
2. Enders, A., and Schmidt L., 1962. Der Andgriffspunkt der atmungsanregenden wirkung des papaverins. *Arch Int. Pharmacodyn.* 91:157.
3. Haouzi, P., Huszczuk A., Porszasz J., Chalon B., Wasserman K., and Whipp B.J., 1933, Femoral vascular occlusion and ventilation during recovery from heavy exercise, *Respir. Physiol.* 94:137-150.
4. Haouzi, P., Huszczuk A., Gille J.P., Chalon B., Marchal F., Crance J.P., and Whipp B.J.,1995, Vascular distension in muscle contributes to respiratory control in sheep. *Respir. Physiol.* In press.
5. Huszczuk, A., Yeh E., Innes J.A., Solarte I., Wasserman K., and Whipp B.J., 1993, Role of muscular perfusion and baroreception in the hyperpnea following muscle contraction in dog. *Respir. Physiol.* 91:207-226.
6. Innes, J.A., Solarte I., Huszczuk A., Yeh E., Whipp B.J., and Wasserman K., 1989, Respiration during recovery from exercise : effects of trapping and release of femoral blood flow. *J. Appl. Physiol.* 67:2608-2613.
7. Kaufman, M.P., Iwamoto G.A., Longhurst J.C., and Mitchell J., 1982, Effects of capsaicine and Bradykinine on afferent fibers with endings in skeletal muscle.*Cir. Res.* 50:133-139.
8. Mense, S., and Stahnke M., 1983, Responses in muscle afferent fibers of slow conduction velocity to contractions and ischaemia in the cat. *J. Physiol. (London)* 342: 383-397.
9. Mitchell, J.H, and Schmitt R.F.,1983, Cardiovascular reflex control by afferent fibers from skeletal muscle receptors. In: *Handbook of Physiology. The cardiovascular System.*, Bethesda, M.D. : American Physiological Society, Sect. 2, Vol. III, Chapt. 17, p. 623-637.
10. Nims, R.G., Severinghaus J.W., and Comroe J.H.,1953, Reflexe hyperpnea induced by papaverine acting upon the carotid and aortic bodies. *J. Pharmacol. Exptl. Therap.* 109: 58-61.
11. Nishino, T., and Lahiri S.,1981, Effects of dopamine on chemoreflexes in breathing. *J. Appl. Physiol.*, 50(4), 892-897.
12. Oren, A., Huszczuk A., Pokorski M., Ferrer P.H., Whipp B.J., and Wasserman K.,1983, Evidence for a peripheral vascular origin of the isoproterenol induced hyperpnea in the dog. *The Physiologist* 26 : A-95.
13. Paintal, A.S.,1960, Functional analysis of group III afferent fibers of mammalian muscles. *J. Physiol. (London).* 152: 250-270.
14. Stacey, M.J., 1969, Free nerve endings in skeletal muscle of the cat. *J. Anat.* 105: 231-254.
15. Tibes, U., 1977, Reflex inputs to the cardiovascular and respiratory centers from dynamically working canine muscles. *Circ. Res.* 41: 332-341.
16. Zapata, P., and Zuazo A.,1980, Respiratory effects of dopamine-induced inhibition of chemosensory inflow. *Respir. Physiol.*, 40: 79-92.

INDEX